Carbon Capture

Sequestration and Storage

ISSUES IN ENVIRONMENTAL SCIENCE AND TECHNOLOGY

EDITORS:

R.E. Hester, University of York, UK
R.M. Harrison, University of Birmingham, UK

TITLES IN THE SERIES:

How to obtain future titles on publication

A subscription is available for this series. This will bring delivery of each new volume immediately on publication and also provide you with online access to each title via the Internet. For further information visit http://www.rsc.org/Publishing/Books/issues or write to the address below.

For further information please contact:
Sales and Customer Care, Royal Society of Chemistry,
Thomas Graham House, Science Park, Milton Road, Cambridge,
CB4 0WF, UK
Telephone: +44 (0)1223 432360, Fax: +44 (0)1223 426017, Email: sales@rsc.org

ISSUES IN ENVIRONMENTAL SCIENCE AND TECHNOLOGY

EDITORS: R.E. HESTER AND R.M. HARRISON

29
Carbon Capture
Sequestration and Storage

RSC Publishing

ISBN: 978-1-84755-917-3
ISSN: 1350-7583

A catalogue record for this book is available from the British Library

Published by The Royal Society of Chemistry,
Thomas Graham House, Science Park, Milton Road,
Cambridge CB4 0WF, UK

Registered Charity Number 207890

For further information see our web site at www.rsc.org

Preface

It is widely recognised that global warming is occurring due to increasing levels of carbon dioxide and other greenhouse gases in the atmosphere. Methods of capturing and then storing CO_2 from major sources, such as fossil fuel burning power plants, are being developed in order to reduce the levels emitted to the atmosphere by human activities. This book reports on progress in this field and provides a context within the range of natural absorption processes in the oceans and forests and in soil, and of methane emissions from melting permafrost and hydrates. Comparisons of fossil fuels with alternative energy sources such as solar and nuclear are made and policy issues are reviewed.

The opening chapter, by Klaus Lackner of Columbia University, USA, compares the impacts of fossil fuels with alternative energy sources. The ever-growing need for energy to drive economic growth in both developed and developing countries, coupled to an overwhelming dependence on fossil fuels, has led to rising atmospheric levels of CO_2 and to climate change. The various possible strategies to combat this are explored within a wide-ranging discussion of energy which provides a basis for the discussion of carbon capture and storage (CCS). Policy developments related to those CCS technologies considered closest to deployment then are reviewed in Chapter 2 by Jon Gibbins and Hannah Chalmers of Imperial College, London. The need for incentives as well as mandatory requirements for commercial-scale demonstration and deployment of CCS technologies is discussed.

The importance of coal for large-scale power generation has focused attention on CCS as a means of continuing the exploitation of reserves. Australia is a country rich in coal reserves, with a well-developed CCS programme which is reviewed by Allen Lowe, Burt Beasley and Thomas Burly for the Australian Coal Association in Chapter 3. Then in Chapter 4 Dermot Roddy and Gerardo Gonzalez of Newcastle University, UK, describe the potential of underground coal gasification with CCS as a source of clean energy. Recognising the particular problems associated with energy intensive industries, Chapter 5, written by David Pocklington and Richard Leese of the Mineral Products Association,

Issues in Environmental Science and Technology, 29
Carbon Capture: Sequestration and Storage
Edited by R.E. Hester and R.M. Harrison
© Royal Society of Chemistry 2010
Published by the Royal Society of Chemistry, www.rsc.org

addresses the potential for carbon capture and storage in cement manufacture. Geological storage of CO_2, the final stage in most CCS schemes, is reviewed in Chapter 6 by Nicholas Riley of the British Geological Survey.

The next three chapters are concerned with natural processes and examine the potential for enhancement of carbon sequestration as well as the problems associated with rising CO_2 levels. Chapter 7, by Stephen Chapman of the Macaulay Institute in Scotland, explores carbon sequestration in soils and plants; Chapter 8, by Maria Nijnik, also of the Macaulay Institute, reviews CCS in forests; and Chapter 9, by Carol Turley of the Plymouth Marine Laboratory (PML), UK, examines uptake, transport and storage by oceans and the consequences of change. In the final Chapter 10, Vassilis Kitidis of the PML reviews methane biogeochemistry in the Arctic Ocean with particular reference to methane hydrates and permafrost.

The quest for large-scale clean energy is a global concern and this book makes an important contribution to the current debate on how best to utilise the world's huge remaining fossil-fuel resources without adding an unmanageable burden of carbon dioxide to the atmosphere. The book will be of value not only to those scientists, engineers, industrialists and policy makers immediately involved with energy supply and large-scale manufacturing, but also more widely to all concerned with major environmental issues such as climate change, ocean acidification, deforestation and the socio-economic and political choices needing to be made in this fast-moving field.

Ronald E Hester
Roy M Harrison

Contents

Issues in Environmental Science and Technology, 29
Carbon Capture: Sequestration and Storage
Edited by R.E. Hester and R.M. Harrison
© Royal Society of Chemistry 2010
Published by the Royal Society of Chemistry, www.rsc.org

Fossil Power Generation with Carbon Capture and Storage (CCS): Policy Development for Technology Deployment   **41**

Jon Gibbins and Hannah Chalmers

Carbon Capture and Storage (CCS) in Australia   **65**

Allen Lowe, Burt Beasley and Thomas Berly

5.2 Commonwealth Offshore Petroleum and Greenhouse Gas Storage Act 2006 (OPA) 92
5.3 State CCS Legislation 94
6 CCS Challenges in Australia 96

Underground Coal Gasification (UCG) with Carbon Capture and Storage (CCS) **102**
Dermot Roddy and Gerardo González

1 Introduction 102
2 A Brief History of UCG 103
3 The Economic Case for UCG 105
4 An Introduction to UCG Technology 107
4.1 Gasification Configuration and Control 108
4.2 Directional Drilling 108
5 Current Status of UCG Deployment Worldwide 110
5.1 UK and Europe 110
5.2 North America 110
5.3 Asia 111
5.4 Australia 111
5.5 Africa 111
6 Mechanism for Carbon Dioxide Storage in Gasified Coal Seam Voids 111
7 Approaches to Environmental Risk Assessment 115
8 Linking UCG to CCS 118
9 North East England Case Study 121
10 Concluding Remarks on Scale of Opportunity and Challenges 123

Towards Zero Emission Production – Potential of Carbon Capture in Energy Intensive Industry **126**
David Pocklington and Richard Leese

1 Overview 126
1.1 Greenhouse Gas Reduction/Issues for Energy Intensive Industry 126
2 Carbon Dioxide Emissions in Cement Manufacture 129
2.1 Cement Manufacture 129
2.2 Incentives for Carbon Reduction 131
2.3 Costs Associated with Carbon Emissions 135
3 Options for Mitigation 138
3.1 Mitigation in Cement Manufacture 138
3.2 Carbon Capture and Cement Manufacture 140
3.3 Removing Barriers to Development 145
4 Conclusions 150

Carbon Capture and Storage in Forests   **203**
Maria Nijnik

Carbon Uptake, Transport and Storage by Oceans and the Consequences of Change   **240**
C. Turley, J. Blackford, N. Hardman-Mountford, E. Litt, C. Llewellyn, D. Lowe, P. Miller, P. Nightingale, A. Rees, T. Smyth, G. Tilstone and S. Widdicombe

Methane Biogeochemistry and Carbon Stores in the Arctic Ocean: Hydrates and Permafrost   **285**
Vassilis Kitidis

Subject Index   **301**

Editors

Ronald E. Hester, BSc, DSc(London), PhD(Cornell), FRSC, CChem

Ronald E. Hester is now Emeritus Professor of Chemistry in the University of York. He was for short periods a research fellow in Cambridge and an assistant professor at Cornell before being appointed to a lectureship in chemistry in York in 1965. He was a full professor in York from 1983 to 2001. His more than 300 publications are mainly in the area of vibrational spectroscopy, latterly focusing on time-resolved studies of photoreaction intermediates and on biomolecular systems in solution. He is active in environmental chemistry and is a founder member and former chairman of the Environment Group of the Royal Society of Chemistry and editor of 'Industry and the Environment in Perspective' (RSC, 1983) and 'Understanding Our Environment' (RSC, 1986). As a member of the Council of the UK Science and Engineering Research Council and several of its sub-committees, panels and boards, he has been heavily involved in national science policy and administration. He was, from 1991 to 1993, a member of the UK Department of the Environment Advisory Committee on Hazardous Substances and from 1995 to 2000 was a member of the Publications and Information Board of the Royal Society of Chemistry.

Roy M. Harrison, BSc, PhD, DSc(Birmingham), FRSC, CChem, FRMetS, Hon MFPH, Hon FFOM

Roy M. Harrison is Queen Elizabeth II Birmingham Centenary Professor of Environmental Health in the University of Birmingham. He was previously Lecturer in Environmental Sciences at the University of Lancaster and Reader and Director of the Institute of Aerosol Science at the University of Essex. His more than 300 publications are mainly in the field of environmental chemistry, although his current work includes studies of human health impacts of atmospheric pollutants as well as research into the chemistry of pollution phenomena. He is a past Chairman of the Environment Group of the Royal Society of Chemistry for whom he has edited 'Pollution: Causes, Effects and Control' (RSC, 1983; Fourth Edition, 2001) and

'Understanding our Environment: An Introduction to Environmental Chemistry and Pollution' (RSC, Third Edition, 1999). He has a close interest in scientific and policy aspects of air pollution, having been Chairman of the Department of Environment Quality of Urban Air Review Group and the DETR Atmospheric Particles Expert Group. He is currently a member of the DEFRA Air Quality Expert Group, the DEFRA Expert Panel on Air Quality Standards, and the Department of Health Committee on the Medical Effects of Air Pollutants.

Contributors

Burt Beasley *Director Technology and Innovation, Australian Coal Association, PO Box 9115, Deakin, ACT 2600, Australia*

Thomas Berly *Manager Technology, Australian Coal Association, PO Box 9115, Deakin, ACT 2600, Australia*

J. Blackford *Plymouth Marine Laboratory, Prospect Place, The Hoe, Plymouth, PL1 3DH, UK*

Hannah Chalmers *Energy Technology for Sustainable Development Group, Mechanical Engineering Department, Imperial College, Exhibition Road, London, SW7 2AZ, UK*

Stephen J. Chapman *Soils Group, The Macaulay Land Use Research Institute, Craigiebuckler, Aberdeen, AB15 8QH, Scotland, UK*

Jon Gibbins *Energy Technology for Sustainable Development Group, Mechanical Engineering Department, Imperial College, London, SW7 2AZ, UK*

Gerardo González *Sir Joseph Swan Institute for Energy Research, Newcastle University, Devonshire Terrace, Newcastle upon Tyne, NE1 7RU, UK*

N. Hardman-Mountford *Plymouth Marine Laboratory, Prospect Place, The Hoe, Plymouth, PL1 3DH, UK*

Vassilis Kitidis *Plymouth Marine Laboratory, Prospect Place, The Hoe, Plymouth, PL1 3DH, UK*

Klaus Lackner *Department of Earth and Environmental Engineering, Columbia University, 1038D S. W. Mudd Hall, Mail Code: 4711, 500 West 120th Street, New York, NY 10027, USA*

Richard Leese *Mineral Products Association, Riverside House, 4 Meadows Business Park, Station Approach, Blackwater, Camberley, Surrey, GU17 9AB, UK*

E. Litt *Plymouth Marine Laboratory, Prospect Place, The Hoe, Plymouth, PL1 3DH, UK*

C. Llewellyn *Plymouth Marine Laboratory, Prospect Place, The Hoe, Plymouth, PL1 3DH, UK*

Allen Lowe *Consultant to ACA Low Emissions Technologies Ltd., A & SJ Lowe Consulting Services Pty Ltd., 7 Russell Street, Oatley, NSW 2223, Australia*

D. Lowe *Plymouth Marine Laboratory, Prospect Place, The Hoe, Plymouth, PL1 3DH, UK*

P. Miller *Plymouth Marine Laboratory, Prospect Place, The Hoe, Plymouth, PL1 3DH, UK*

P. Nightingale *Plymouth Marine Laboratory, Prospect Place, The Hoe, Plymouth, PL1 3DH, UK*

Maria Nijnik *Socio-Economic Group, The Macaulay Land Use Research Institute, Craigiebuckler, Aberdeen, AB15 8QH, Scotland, UK*

David Pocklington *Director Industry Affairs, Mineral Products Association, Riverside House, 4 Meadows Business Park, Station Approach, Blackwater, Camberley, Surrey, GU17 9AB, UK*

A. Rees *Plymouth Marine Laboratory, Prospect Place, The Hoe, Plymouth, PL1 3DH, UK*

Nick Riley *British Geological Survey, Kingsley Dunham Centre, Keyworth, Nottingham, NG12 5GG, UK*

Dermot Roddy *Science City of Energy, Sir Joseph Swan Institute, Floor 3, Devonshire Building, Newcastle University, Newcastle upon Tyne, NE1 7RU, UK*

T. Smyth *Plymouth Marine Laboratory, Prospect Place, The Hoe, Plymouth, PL1 3DH, UK*

G. Tilstone *Plymouth Marine Laboratory, Prospect Place, The Hoe, Plymouth, PL1 3DH, UK*

C. Turley *Plymouth Marine Laboratory, Prospect Place, The Hoe, Plymouth, PL1 3DH, UK*

S. Widdicombe *Plymouth Marine Laboratory, Prospect Place, The Hoe, Plymouth, PL1 3DH, UK*

Comparative Impacts of Fossil Fuels and Alternative Energy Sources

KLAUS S. LACKNER

1 Introduction

Growing concerns over the consequences of climate change may severely limit future access to fossil fuels. A forced choice between energy and environment could precipitate a major economic crisis, an environmental crisis, or both. Averting such a crisis will be difficult, because fossil energy resources are an essential part of the world's energy supply and climate change is mainly driven by the build-up of carbon dioxide in the atmosphere. Carbon dioxide (CO_2) is the unavoidable product of fossil fuel consumption. Therefore, the use of fossil fuels collides directly with global environmental concerns. Unfortunately, fossil fuels are difficult to replace, but stabilising the atmospheric concentration of carbon dioxide requires a nearly complete transition to a carbon-neutral economy. This implies either the abandonment of fossil fuels or the introduction of carbon capture and storage, whereby for every ton of carbon extracted from the ground another ton of carbon is put back.

This chapter discusses the scope of the required reduction in carbon dioxide emissions and the options available for achieving such reductions. It puts the continued use of fossil fuels, with carbon capture and storage, in context with other approaches toward achieving a carbon-neutral energy infrastructure or otherwise avoiding serious climate change impacts.

The vast scale of energy infrastructures emerges as the central theme. There are very few energy resources that are large enough to cope with modern global energy demand. Any technology that will be able to satisfy these demands will unavoidably interfere with natural dynamic systems. Just like some of the large

Issues in Environmental Science and Technology, 29
Carbon Capture: Sequestration and Storage
Edited by R.E. Hester and R.M. Harrison
© Royal Society of Chemistry 2010
Published by the Royal Society of Chemistry, www.rsc.org

natural cycles, human energy systems are operating on a global scale. It is the vast scale of human energy demand that shapes the available options.

2 Climate Change

The idea that greenhouse gases in the atmosphere control climate is not new. While travelling with Napoleon through Egypt, Fourier was the first to recognise that the composition of a planetary atmosphere regulates a planet's surface temperature.[1–3] Some sixty years later, Tyndall measured the absorption spectrum of CO_2 in the infrared region. His laboratory measurements showed that carbon dioxide is a powerful greenhouse gas, which is largely responsible for the habitable temperature range on Earth.[4] In 1898, Arrhenius was the first to quantify the greenhouse effect and estimate the impact of anthropogenic emissions of CO_2.[5] While extensive research and numerical studies have added much detail to our understanding, his initial ideas remain unchanged.[6] Computer models and observations corroborate the basic insights developed in the nineteenth century.

Fossil fuels provide 81% of the world's commercial energy supply.[7] Consumption of fossil fuels produces nearly 30 Pg (petagram)[i] of carbon dioxide annually. Until now, nearly all of this carbon dioxide has been released to the atmosphere. In the past, the atmospheric sink was considered large enough to accommodate any additional carbon dioxide, but the carbon dioxide content of the atmosphere has now risen by more than a third since the beginning of the industrial revolution, from 280 parts per million by volume (ppm) to 385 ppm today.

Fossil fuel combustion is the single most important contributor to this change. The total carbon dioxide produced in the combustion of fossil fuel since the beginning of the industrial revolution actually exceeds the observed increase in the atmosphere.[8] At present, the carbon dioxide content of the atmosphere is rising by 2 ppm per year,[ii] suggesting that more than a third of the fossil carbon dioxide produced does not stay in the atmosphere.[9]

The rapid increase in the atmospheric concentration of carbon dioxide has raised the spectre of severe climate change, and much effort has gone into understanding the likely scale and the implications of global warming. Today it is generally accepted that doubling of the carbon dioxide in the atmosphere would create serious harm and an often-cited goal for stabilising carbon dioxide in the atmosphere is 450 ppm, which at current rates of increase would be breached in about 30 years.

Carbon dioxide is an important greenhouse gas and the most obvious impact of CO_2 release is global warming. However, CO_2 is also physiologically active in plants and animals, it is of great importance to ecological systems and it is an acid that critically affects the chemistry of ocean water.

---

[i] We chose the petagram (Pg) rather than the Gigaton as a unit of mass, because it is the appropriate SI unit and eliminates all ambiguities over metric *vs.* non-metric tons. One metric Gigaton is equal to 1 Pg.

[ii] 1 ppm of CO_2 in the atmosphere amounts to 2 Pg of carbon or 3.7 Pg of CO_2.

While the focus of the climate scientist is on the impact of CO_2 on global warming, an important focus for the engineer developing a sustainable energy infrastructure is to eliminate the environmental impacts that arise from the release of carbon dioxide to the atmosphere. Even more broadly, the energy engineer has to consider the environmental consequences of generating power. In this context, it is the unintentional mobilisation of large quantities of carbon that needs to be eliminated. With a fossil energy infrastructure, the production of large quantities of oxidised carbon is unavoidable; their release into the atmosphere can and must be avoided.

The climate scientist will lump CO_2 together with other greenhouse gases; the engineer of a sustainable energy infrastructure must find ways of stopping CO_2 emissions. This will eliminate the climate change impact of carbon dioxide, as well as other impacts of excess carbon. The control or elimination of other greenhouse gases may also be necessary for stabilising climate. However, the control of these other greenhouse gases raises rather different issues and may occur outside of the energy sector. Thus, their management should be considered separately.

Unlike other emissions, carbon dioxide is not a problem at the point of emission. Carbon dioxide rarely reaches concentrations that constitute a local hazard. The ambient background level of CO_2 is so high that mixing of CO_2-rich plumes with the atmosphere reduces excess concentrations to a small fraction of the background already in the vicinity of the source.[iii] Carbon dioxide differs from other power-plant emissions like sulfur dioxide (SO_2), because it is not the local impact of CO_2 emissions, but the impacts arising from the accumulation of CO_2 in the environment that need to be controlled. In the past, when the local impact of other sour gases was recognised as a serious hazard, dilution of CO_2 still provided an adequate solution. Today, the CO_2 emissions from power plants have become so large that their impact on the entire mobile carbon pool can no longer be ignored.

Conceptually it is useful to consider the various carbon pools on earth and separate them into stable pools that are isolated from other pools, and mobile pools that interact rapidly. Carbon is either tied up in permanent and stable carbon pools, like carbonate rocks or coal seams deep underground, or it is part of the mobile carbon pools on the surface of the Earth. The stable pools are much larger than the mobile pools. The mobile carbon pools consist of the atmosphere, the biosphere carbon and the ocean. These three reservoirs are in rapid exchange with each other, but are essentially decoupled from the other carbon pools.

Before the industrial revolution, the atmosphere contributed less than 600 Pg (*i.e.* 600 x 10^{15} g) to this pool, today it is 800 Pg. The biomass contribution is

---

[iii] As an example, consider traffic on a freeway releasing CO_2 in the wind blowing across the road. Let us assume a high traffic density of 10 cars per second passing a stationary observer, an emission rate of 2×10^{-4} kg m^{-1} of CO_2 per car (roughly 0.0871 km^{-1} of gasoline or 27 miles to the gallon), a slow breeze at a wind speed of 2 m s^{-1} and mixing depth of about 50 m which, given the turbulence created by the freeway, should be achieved within a few hundred meters of the freeway. This leads to a total emission rate of 2×10^{-3} kg m^{-1} s^{-1}, which is diluted into 100 m^3 (a column 50 m tall and 1 m×2 m wide) of air, raising the CO_2 content by 2×10^{-5} kg m^{-3} as compared to a background concentration of 7×10^{-4} kg m^{-3}.

also around 600 Pg. Soil carbon provides another 1500 Pg. The ocean contains about 39 000 Pg of dissolved inorganic carbon, which is part of the mobile pool, but cannot easily be changed. The ocean carbon pool may be mobile in the sense that any carbon atom can enter or leave, but it is persistent in the sense that it cannot be increased or decreased by large amounts. The amounts that could be added to the ocean by, for example, doubling the partial pressure of CO_2 over the ocean are between 1000 and 1400 Pg.[10] Thus, the total flexibility in the mobile surface carbon pool is several thousand petagrams (Pg).

Fossil fuel consumption adds to the mobile carbon pool. Fossil carbon which is taken from stable carbon pools is oxidised and released to mobile pools, particularly to the atmosphere. Past fossil fuel consumption has already added 350 Pg. This is a substantial amount.[8] The exchange between the mobile carbon pool and the naturally sequestered permanent pool is very small, involving a small fraction of a petagram of carbon per year. As a result, human influences completely dominate the change in size of the mobile pool, even if the transfer rates between the various parts (*e.g.* between the biomass pool and the atmospheric pool) are far larger than the annual human input to the pool. It will take several tens of thousands of years before the total mobile carbon pool will re-establish its equilibrium with the permanent carbon pools.[11]

This deviation from equilibrium matters beyond just climate change. For example, excess carbon leads to the acidification of the ocean.[12,13] It has been shown that such a modification of the ocean chemistry stunts coral growth.[14,15] Excess CO_2 in the atmosphere also leads to the eutrophication of terrestrial and oceanic ecosystems. While environmental concern over climate change may be the leading reason for managing anthropogenic carbon, climate change is only one concern of many.

At the heart of the problem is the introduction of excess carbon into the mobile carbon pool. Any human infrastructure which ignores the continued build-up of excess carbon in the mobile carbon pool cannot be sustained. Technologies which purport to stop global warming, while allowing the rise in the mobile carbon pool to continue, are at best emergency measures to bridge a gap, but they are guaranteed to fail over time. Albedo engineering, for example by adding sulfates to the stratosphere, can fix one symptom but it does not address the underlying problem.

Practical solutions will need to stop or even reverse the build-up of CO_2 in the environment. The build-up of carbon must be stopped not just in the atmosphere, but also in the surface ocean and throughout the entire mobile carbon pool. This means stopping the mobilisation of additional carbon, or compensating for the mobilisation of carbon by demobilising an equal amount.

3 The Urgent Need for Energy

Energy is central to economic growth. Without access to adequate energy supplies, a world population of six to ten billion people would not be possible. Empirically, economic growth and energy consumption are closely linked, even allowing for the fact that the energy efficiency of most industrial and commercial processes can be improved, and indeed is improving. The dependence of a modern society on metals and synthetic materials, on transportation and information processing, makes access to energy paramount. Every sector of the economy requires energy and even the most basic needs of humanity could not be supplied without access to plentiful energy. Energy is necessary in the production of food and in the provision of clean water.[16]

If the environmental constraints on fossil energy resources cannot be overcome, the resulting serious shortfall in energy would very likely precipitate a crisis of unprecedented proportion. Even without the added concerns of climate change, the world's energy systems are in a precarious state. Rapid economic growth is constantly pushing the existing infrastructure to its limits. It would be extremely difficult to provide sufficient energy for rapid world economic growth while at the same time phasing out fossil energy for environmental reasons.

Energy demand, which had been outstripping supplies in the last few years, led to enormous price increases, even though the bottleneck was only a few percent of the total supply. This shows how little flexibility there is in the energy supply sector and how difficult it is to increase the world's energy supply. Even the recent sudden drop in demand only makes the point how inelastic the world's energy supplies are, but this time with the opposite sign. If these relatively small variations in energy supply and demand can have such a dramatic impact, consider what might happen if over a few decades 80% of the entire energy base became off-limits, and if the most cost-effective source of energy could no longer be deployed in the construction of a new energy infrastructure.

While it may be necessary to learn how to manage with much less energy, if the development of a sustainable energy supply fails, the highest priority should be given to developing energy solutions that can provide plentiful energy for everyone in the decades to come. Supporting a growing world population and their demand is paramount to political stability and the eventual stabilisation of the world population. Unfortunately, it is very difficult to achieve this goal while phasing out fossil fuels.

Higher living standards and increased energy consumption are intrinsically related. Fighting poverty worldwide will require a means of raising the world's living standards to levels the developed nations take for granted. This will involve the introduction of a basic energy infrastructure and consumption patterns that are not very different from those found today in developed countries, where these infrastructures have been built over the last hundred years.

It has been suggested that developing nations might stop at a level of about 2 kW of primary energy per person.[17,18] At this level, basic human needs are satisfied and consumption would still be only a fraction of that in Europe or in the United States.[19] However, it seems unlikely that countries would voluntarily give up their potential for growth, particularly as long as there are other countries that enjoy a much higher standard of living.

Even though one can expect significant improvements in efficiency and a generally reduced energy intensity of the world economy, it is unlikely that developing countries could find a way to leap-frog developed nations and arrive at a far less energy-intensive economy that nevertheless delivers a high standard of living. Developing basic transportation infrastructures, a decent housing stock with the attendant need for heating and cooling, the development of basic food supplies and basic manufactured goods will likely require an energy infrastructure of similar size to that built up in the developed countries.

It may well be possible to reduce some energy consumption by applying more advanced technology, but this is likely to remain the exception rather than the rule. For example, the need to build a wire-based telephone network may well be avoided, but most of the infrastructure will be similar in energy intensity to those known from the developed nations. Furthermore, rapidly growing economies tend to be less efficient in their implementations of technology, because there is a large opportunity cost in squeezing out the last bit of efficiency. Indeed it is generally the case that the energy intensity (*i.e.* the ratio of energy consumption to GDP) is lower in developed countries than in developing nations.

Much of what will be implemented initially in a developing country are low cost, and hence often less efficient versions of technology than those already deployed elsewhere. Over time, both developing and developed countries will adopt similar technologies and the two types of economies will converge. Catching up with today's developed countries would increase world energy demand by a factor of five to ten.

It is difficult to see why rich countries would refrain from raising their living standards. Economic slowdowns are typically fought at great cost and policy makers have every incentive to keep their local economics on a growth trajectory. Economic growth will bring with it additional energy demands, which are difficult to predict. In the developing countries, one can assume that, at least to some approximation, the development is likely to retrace the steps already taken by the developed nations. However, much of future growth in the developed countries will arise around new and innovative technologies that either do not yet exist, or that are still in their infancy. It is not clear what will be the next technological wave and how much energy it will demand. It is, however, worthwhile to point out that the last technological wave, which was focused on computation and information processing, was exceptional in its low energy intensity. Thus, the past trend of a continued reduction in energy intensity[20] may not be maintained over the next few decades. This would put in disarray all predictions of future energy consumption, as it has been taken for granted that the energy intensity of the world economy should drop at a rate of at least 1 percent per year.[21]

Energy is so central in supporting economic well being, that it seems highly unrealistic to expect that humans will forgo the use of energy, unless circumstances make any other course impossible. Just for the world to catch up to the standard of living taken for granted in the developed world implies an economic growth that could be ten-fold. Annual growth of 2.3% would result in a ten-folding over a century as well. Past growth in the developed nations resulted in an increase in energy consumption by a factor of greater than ten during the 20th century.

4 The Environmental Impact of Energy

A modern society's energy consumption is so large that it cannot help but have an impact on large natural cycles and the environment. For example, based on US averages, the human energy consumption in the State of New Jersey, with a population density of $438 \, km^{-2}$, exceeds the photosynthetic productivity of a similar, equally sized, but pristine area. In other words, human activities have created energy flows which match those of entire ecosystems. Environmental impacts from human power generation systems are therefore bound to be profound. In very real ways, engineered systems begin to shape the dynamics of large-scale natural systems. In engineering energy infrastructures it is important to recognise these impacts and to develop designs which minimise them.

The understanding that energy industries can do harm to the environment is not new, and environmental concerns over the use of fossil fuels in power generation have already shaped the way these industries operate. Carbon dioxide concerns simply broaden the agenda. Impacts of energy consumption, ranging from thermal pollution of rivers to smog and acid rain, have shaped a large part of the environmental agenda of the twentieth century. Now the release of CO_2 has been added to the long list of environmental impacts that are caused by the use of fossil energy resources. Fossil fuels are not just a large source of greenhouse gases, but they are also a source of many different streams of pollutants, including heavy metals and fine particulates. Their extraction from the ground also adds to their environmental impact.

Attempts to address the environmental consequences of fossil fuel use have resulted in technological advances. Technology changes have led to a drastic reduction in the pollution from fossil fuel-based power. While there is still a gap between what can be done and what is done, it is clear that current technology can eliminate many of the major concerns over fossil fuel-based energy sources. However, the large-scale production of CO_2 has not yet been addressed.

Carbon dioxide may have less of an acute impact at the point of emission, but it is produced in exceedingly large quantities. Unlike sulfur, which is a trace constituent of fossil fuels, the oxidation of carbon to carbon dioxide is central to energy extraction and thus cannot be avoided. Today, the primary obstacle to the use of fossil fuels is the carbon dioxide emission associated with its use.

Given the huge scale of energy generation in a modern society, it is not surprising that all modern energy systems have some impact on the local and global environment. The specifics differ for each case, but a few general themes emerge. Most energy systems cause changes in the system from which they harvest the energy. Mining impacts are large, but there are also concerns about the sheer size of the windmill parks necessary to replace fossil fuels. Many of the alternative energy systems also release pollutants or toxins into the environment.

In order to create a sustainable, global energy infrastructure, the world must solve several problems simultaneously. First among them is the need to provide a strong energy foundation for a path out of poverty and toward rapid economic development. This will require abundant and cheap energy. Another problem is to minimise the environmental impacts of energy extraction, energy conversion and energy consumption. First and foremost, this requires a solution to the problem of climate change and of excess carbon in the environment. The need for economic growth makes it unlikely that the world will give up on readily available fossil fuels.

To find a solution to these seemingly contradictory demands requires a revolution in the energy infrastructure. 'Business as usual' cannot resolve this conundrum.[16] Today 81% of all energy comes from fossil carbon and the world emits approximately 1.34 mol of carbon dioxide for every megajoule (MJ) of primary energy consumed, or about 60 g of CO_2 for each MJ of primary energy.[iv] The amount of CO_2 produced must be drastically reduced, while energy consumption will continue to grow. The difference between maximum allowed world-wide emissions and *per capita* emissions in the developed countries suggest that the long-term goal has to be carbon neutrality. As a result, it is necessary either to avoid the use of fossil fuels, which are one of the cheapest and one of the most abundant resources of energy, or to move to carbon capture and storage, which would make it possible to continue the use of fossil fuels as it removes the most immediate environmental threat associated with their use.

5 Carbon Capture and Storage

The concept of carbon capture and storage is quite simple: for every ton of carbon taken from the ground another ton of carbon has to be stored permanently and safely, and away from the mobile carbon pool. While the conventional term for this concept is "carbon storage," in reality we are referring to a large-scale disposal problem. The only way the term storage rather than disposal could be justified is that storage needs to be maintained and the responsibility for keeping the carbon out of the mobile carbon pool will indeed remain for a long time. Assuring continued storage may demand action from now into the distant future. On the other hand, the most desirable storage options are those that do not require long-term maintenance.

The creation of a waste stream is unavoidable. The world is consuming energy-rich forms of carbon for the primary purpose of generating power. This results in large quantities of energy-poor forms of carbon, *i.e.* carbon dioxide or various carbonates. The oxidation of carbon is an unavoidable outcome of extracting the energy available in carbon and thus it cannot be undone. The consumption of hydrocarbons produces carbon dioxide and water.

The production of water in fossil fuel consumption does not pose a problem because the amount is insignificant compared to the amounts already on the earth's surface. By contrast, the atmosphere is not a reservoir of sufficient size to accept all the CO_2 that is produced. It is therefore necessary to find safe and

---

[iv] Calculated from a world primary energy consumption of 468.671 quadrillion BTU, and world CO_2 consumption of 29 195.42 Mt, as reported by the US Department of Energy's Energy Information Administration, http://tonto.eia.doe.gov (last visited July 4, 2009).

permanent ways of storing the CO_2, or possibly the carbonate that can be formed, without investing large amounts of energy.

In a perfect world, industrial processes would not have waste streams, but instead all outputs of a process would be used either directly as commercial products, or as inputs for some other industrial processes that could take advantage of these materials. Indeed industrial ecology strives toward this goal, and often there is the possibility of finding a good use for a by-product.[22] Unfortunately, the scale of energy consumption is so large that by-product utilisation becomes severely strained or outright impossible. The *per capita* production of CO_2 in the US is 20 Mg per year. This far exceeds all commercial demand for CO_2 or for carbonates, which would have more than twice the mass of the equivalent amount of CO_2.

For example, if waste carbonate from a coal-fired power plant proved viable as a substitute material for producing wallboard, then a single Gigawatt power plant would create enough carbonate to saturate the entire wallboard market in the United States.[23]

Thus, fossil fuel consumption is one example of an industrial process that unavoidably creates a waste stream and it raises the question as to the best possible disposal strategy.

This problem is not unique to carbon. Similar issues also arise for sulfur. Sulfur is found in excess worldwide, because it is embedded in most hydrocarbons. As an indirect consequence of increased energy consumption, the last sulfur mine in the US closed in 2001. Other examples include arsenic, which is readily supplied through copper production. In this case, one can observe a gradual reduction in demand for arsenic that is driven by environmental concerns. Over the years, demand for arsenic has shifted from agricultural use to wood preservatives, which are now also being phased out.[24]

Waste streams that cannot be used as practical products must be disposed of in a safe and permanent manner. For some streams this is easily accomplished, because the materials involved are relatively inert and small in mass or volume. The problem becomes challenging if the waste streams are highly toxic or very large in volume. The latter is the case for carbon dioxide disposal from energy production. In all cases, one must render these waste streams harmless. An initial step in this direction is to transform these materials chemically, so that they are changed into their thermodynamic ground state. This avoids their gradual and uncontrolled conversion in an open environment. Often this transformation process has the added benefit of releasing useful energy, since thermodynamically stable states are often the energetically lowest states.

For the actual disposal there remain two options: either dilution into a large stream or reservoir, or permanent and safe storage in a closed site. Sometimes it is possible to create thermodynamically stable forms of a material that are also common in nature. In this case, it may be possible to dilute the waste material to the point that it is of no further concern. This is an ideal outcome in that the waste stream is simply hidden behind a large natural background. In some cases this might be possible. For example, Rappold[25] has suggested adding anthropogenic sulfur to the oceans in the form of dissolved sulfates. In this case, the entire world production of sulfur cannot change the total sulfur in the ocean by more than one part in a million. In the case of fossil fuel consumption, the water by-product is simply absorbed into the natural water cycle; it is such a small part of the water cycle that it can be safely ignored. Generally, dilution is only possible if the waste material is already naturally present and the amounts discarded are small compared to the amounts already present.

Unfortunately, in many cases, the human production of such waste streams exceeds the capacity of natural environments to absorb the volumes produced. Even large reservoirs with large natural occurrences of a waste product may not be able to absorb more. For example, changing the CO_2-content of ocean water by more than a few percent would result in a dramatic change in pH.

There are only two practical reservoirs to add CO_2 or carbonate to and store in diluted form. One is the atmosphere, the other is the ocean. So far, dilution in the atmosphere has been the method of choice, but the capacity of the atmosphere has proven to be insufficient as CO_2 concentrations are increasing rapidly worldwide. The second large-scale reservoir for dilution is the ocean. Roughly a third of all anthropogenic emissions have moved into the surface ocean. Here too, the capacity to store CO_2, which dissolves as carbonic acid, seems more limited than that which is required. A possible alternative could be the storage of dissolved carbonates or bicarbonates. This could be accomplished by transforming gaseous CO_2 into a water-soluble mineral carbonate or bicarbonate that could then be added to the ocean. The storage capacity of the ocean for dilute magnesium bicarbonate or (as has been suggested) calcium carbonate, completely dwarfs the ocean's capacity for storing carbonic acid.

The total volume of liquid CO_2 that could plausibly be produced during the course of the 21st century is on the order of Lake Michigan. Lake Michigan contains 5000 Pg of water; the world's CO_2 emission has reached 30 Pg per year, or 3000 Pg per century. While large, such numbers do not create an insurmountable obstacle in finding long-term storage sites. Indeed humans produce similar amounts of groundwater.[26] As a result, storage options for CO_2 have emerged below the ground, either below the continents or below the ocean floor.

Another alternative is to store concentrated carbon-rich materials as carbonate. Soluble carbonate, such as sodium and potassium carbonates, will end up in the ocean. However, solid materials can be piled up in a large site. This is the concept of *ex situ* mineral sequestration.[27]

6 Stabilising Atmospheric Carbon Dioxide Concentrations

Rather than starting with climate change and phrasing the problem as one of greenhouse gas emissions, in the energy sector it may be more fruitful to think about the problem as one of mobilising excess carbon. Conceptually, thinking in terms of excess mobile carbon in the environment simplifies finding a solution. By itself, excess carbon will redistribute itself between the mobile pools, but its removal rate, or rate of natural sequestration, is very small. Essentially all the carbon that has been mobilised since the beginning of the industrial revolution is still part of the mobile carbon pool. Thus, setting a limit on the excess in the carbon pool effectively sets a limit on the total amount of carbon that can be released by human enterprises.

If one restricts one's accounting to the atmosphere, this argument does not quite apply. Since atmospheric CO_2 interacts rapidly with the ocean reservoir, much CO_2 deposited in the atmosphere will eventually be withdrawn; however, at the price of ocean acidification. Nevertheless, even for the atmosphere it is not a bad approximation to consider the excess in the atmosphere a more-or-less linear function of the amount of CO_2 emitted. However, for this approximation to be useful, one needs to assume that half of the CO_2 that enters the atmosphere is withdrawn on a relatively short timescale. At present, the annual rise in CO_2 is approximately 60% of the rate of emissions. Even when equilibrium is eventually reached about a quarter of the CO_2 will remain.

Thus, setting a stabilisation level for the atmosphere is nearly equivalent to suggesting a finite total budget for carbon dioxide emissions.[28] If CO_2 is to be stabilised at 450 ppm, this budget is very small. The world would have less than 30 years to achieve the turnaround necessary.[29]

What is an acceptable level of CO_2 in the atmosphere is likely to be debated for some time. There is no doubt that the impact of excess CO_2 in the atmosphere already has been detected and that its presence has environmental consequences. The severity of these impacts will grow as the CO_2 concentration in the air rises. Quantification of these impacts is complex. It is not sufficient to understand the physical consequences of excess carbon in the environment; the evaluation of these impacts also depends on social constructs which tell which impacts are tolerable and which are not. In effect, one needs to understand people's pain threshold and their willingness to absorb damage.

It is also far from settled as to what extent a modern society will leave it to future generations to deal with damage that is caused by them today, but will only become manifest much later.

In addition, one will have to consider the public's risk aversion, when it comes to small, but not insignificant, risks of sudden and severe changes which result in grave consequences. In this category, one could consider positive feedback based on large-scale methane releases from the permafrost regions in the Northern Hemisphere, or the collapse of the annual Monsoon in large parts of Southern Asia, which rely on predictable rainfall for agricultural production.

What is considered a "safe" level of carbon dioxide emissions varies widely, from numbers as low as 350 ppm (as suggested by James Hansen)[30] to numbers well above 550 ppm. It appears that the current consensus is around 450 ppm. If 450 ppm is the right level, this represents an exceedingly ambitious target. The often-quoted target of 450 ppm, which is the CO_2 equivalent greenhouse gas forcing of all greenhouse gases combined, is virtually unattainable. The current equivalent greenhouse gas level is already at 435 ppm[31] and it is rising by more than 2 ppm$_e$ per year.

If we assume that roughly half of the CO_2 remains in the atmosphere, 4 Pg of carbon represents an increase of 1 ppm in the atmosphere. A stabilisation target of 450 ppm of CO_2 would thus leave a remaining budget of 260 Pg of carbon. The world is consuming 8 Pg of carbon per year. With an exponential decay in consumption and a decay time of less than 33 years, the total emissions would indefinitely stay below 260 Pg of carbon. This is a very ambitious target, requiring an annual world-wide reduction in emissions of 3%. In the context of electric power generation, this target implies not only that all new plants must not emit CO_2, but that existing plants have to be rebuilt or phased out, because typical turnover times in the power-plant stock are substantially longer than the required decay rate in emissions.

If the stabilisation level were to be set much higher, *e.g.* 800 ppm (which is generally considered to be a harmful level), it would still be necessary to start reducing emissions today. Even with such a relaxed target, the above analysis suggests that annual carbon emissions would have to be reduced by approximately 1% per year, which stands in stark contrast to the current annual growth.

A 3% annual reduction in emissions needs to be compared to the aspiration of a 3% annual growth in economic activity, which suggests that the improvements in the world's carbon intensity have to be in the order of 6% per annum. Even fixing the target at 800 ppm would require steady improvements in carbon intensity in the order of 4% per year.

Thus, at this time, a debate over the correct level of stabilisation is misguided. The major focus of any effort has to be an immediate emission reduction. Even if the level of stabilisation were as high as 800 ppm, it would still be necessary to cease creating energy infrastructures which lock into additional emissions. For example, a Gigawatt coal-fired power plant basically commits to a seventy-year long emission, which adds up to half a petagram of CO_2.

We conclude that it is impossible to solve the climate change problem without curbing carbon dioxide emissions. This requires not only capture at the large sources, but also capture at the small sources, which ultimately is best done by capture from air. In order to solve the climate change problem, the world will have to achieve a large-scale transition from the current energy infrastructure to one that is essentially carbon neutral.

There are three policy and technology options with which to approach this goal. The first option is to use less energy. Energy savings, energy conservation and improved energy efficiency can help in reducing the carbon problem, but they cannot by themselves solve the problem considering the huge growth in demand. The second option is to eliminate current fossil sources and replace them with non-fossil sources of energy that can fill the gap. This option is, in principle, feasible, but it would eliminate the foundation of the current energy infrastructure. The third option is to prevent the carbon dioxide that is associated with the consumption of fossil fuels from accumulating in the atmosphere.

It is likely that all three options will play a part in the transition from today's energy infrastructure to that of a future world economy. It is exceedingly unlikely that any one of these three options will completely dominate the transition. For that, the constraints, even in the absence of climate change, would just be too large.

There are, in principle, two more options. The first one is to adapt to the temperature and climate change. The second is to look for engineering solutions that can stop climate change in the presence of increased greenhouse gases in the atmosphere: the so-called geo-engineering options.

7   Geo-Engineering as a Means of Stabilising Climate

There are several geo-engineering approaches to managing climate or to stop climate change.[32–34] In the context of climate change, geo-engineering attempts to prevent or counteract the warming associated with the greenhouse effect. This could be done by various means. For example, one could inject sulfate aerosols into the higher atmosphere so as to reflect more incoming sunlight back to space.[35,36] It would even be possible to intercept some of the sunlight aimed for Earth in outer space and reflect it away from Earth.[37] The resulting reduction in solar flux reaching the ground would cause a general cooling that counteracts the warming due to greenhouse gases.

Another approach to managing climate is to modify the existing, natural carbon cycles on earth so as to remove more carbon from the atmosphere. It is, for example, possible to fertilise the oceans in order to increase their CO_2 uptake. Other options include changing the alkalinity of the ocean, or to rapidly grow biomass on land and store the resulting carbon. For these methods to be successful, it is important that they remove carbon from the atmosphere and store it in other carbon reservoirs. For example, the production of charcoal or Terra Preta[38] relies on the fact that certain forms of graphitic materials are quite stable in the environment and can thus be used to store enormous amounts of carbon.

Climate stabilisation or carbon-cycle engineering are only some aspects of geo-engineering. In general, any large-scale engineering effort that aims to modify natural dynamical systems on Earth should be considered geo-engineering. Some authors consider it necessary that these changes are deliberate and not coincidental.[33] Otherwise, the use of fossil fuels, with climate change as an unintentional byproduct, would be considered geo-engineering. In practice, even intentional changes on a global scale are usually not considered geo-engineering unless the purpose has been a global change. Otherwise, agriculture, which has changed entire continents but with a distributed and localised decision-making process, should be considered geo-engineering.

Nevertheless, an argument could be made that geo-engineering is already on its way. There are large-scale human operations which impact global dynamical systems, sometimes purposefully, sometimes by accident. Of course, the boundary between geo-engineering and large local operations is vague. For example, restructuring ecological systems and water flow in nearly continental scale basins, should qualify as geo-engineering, particularly if one considers the goal to be the provision of agricultural goods at a truly global scale.

It is likely that geo-engineering will quietly be introduced for a variety of purposes. Often these will be uncontroversial purposes. Consider the reduction of impacts from hurricanes, the steering away of hurricanes from land, the irrigation of large desert lands, the re-routing of rivers in order to improve climate and weather. These are all examples of incipient geo-engineering.

However, the currently popular discussions of global engineering for avoiding climate change are applying geo-engineering concepts to a problem to which geo-engineering is exceedingly ill-suited. If the earth were simply too hot or too cold, then geo-engineering might be able to create a world in which this imbalance has been addressed. This, however, is fundamentally different from a geo-engineering approach, which must compensate for an ever-increasing imbalance in greenhouse gases. It is a little like solving a municipal waste and sewage problem not by managing effluents and waste streams, but by putting houses on stilts to make room for the garbage and propose an annual raising of the stilts.

While countering the greenhouse effect may work for a while, the larger the effect, the less likely the two countervailing drivers will actually cancel out. At first sight, global warming and global cooling cancel each other out. However, second-order effects are likely to be quite different. For example, the greenhouse effect will lead to the largest amount of warming in Polar Regions, while a simple change in albedo will lower temperatures most dramatically where the sun shines.

Geo-engineering methods that create a change and do not need to be maintained are more easily managed than geo-engineering methods that require continuous maintenance. In the case of managing the climate, not only does one require continuous maintenance, but also a constantly increasing level of albedo modification. As long as greenhouse gas concentrations rise, the albedo modification must increase in order to match the increasing greenhouse effect.

Continuous maintenance is costly and introduces large risks. It should always be considered a second-best option and one which puts larger and larger demands on future generations. This raises issues of inter-generational equity. Geo-engineering for climate change demands that future generations will work to counter this generation's environmental impact. It is perfectly acceptable to suggest that future generations may want to continue geo-engineering efforts that provide for a better climate for their era and for all future generations. It may even be necessary for future generations to maintain such a climate because their population size may not be sustainable otherwise. In this case, these future generations contribute to an effort which benefits them directly. However, in canceling out the greenhouse effect, the burden on future generations is solely to clean up after the present generation. Does the present generation have the right to force future generations into large-scale geo-engineering efforts solely to cancel out climate forcing which the present generation neglected to eliminate?

The argument has been made that the requirement for a maintained geo-engineering effort also creates dangers to a modern society. Consider, for example, that in times of war or economic turmoil, the world may not be able to maintain sulfates in the stratosphere for albedo management. Whatever crisis befalls the world, it would be exacerbated by its inability to manage the climate. Temperatures may suddenly jump to much higher levels.[39]

While this argument has some merit, it does not introduce a fundamentally new exposure to danger for humankind. Modern societies depend completely on the availability of advanced infrastructures; without them, the ability to sustain large populations would collapse. For example, modern societies critically depend on uninterrupted food supplies based on high-tech agriculture, on the delivery of clean water and the functioning of modern information-processing systems, without which, for example, the banking system would collapse.

Thus, maintaining an albedo shield would add yet another risk factor, but it would not qualitatively change humanity's dependence on advanced technology. A more serious argument is that if all generations simply pass their responsibilities down to future generations the burden on those generations may simply become too large. After all, the assumption that future generations will be wealthier and more advanced than our present generation is questionable.

Geo-engineering approaches do not solve the problem of excess carbon but instead mask the symptoms for a time. However, as a bridging strategy they should be seen in a different light. For example, the world may decide that the current level of CO_2 in the atmosphere is already higher than is affordable. Even with advanced technologies for removing carbon dioxide from the atmosphere, it would take decades to reduce the greenhouse gas concentrations to an acceptable level. If, during this time, one could reduce global warming effects, or ocean rise, by actively steering the planet to a cooler climate, then geo-engineering could serve a useful purpose.

Even then, one must carefully consider the risks and the unintended side-effects of large-scale geo-engineering. Climate change concerns may tempt the world into jumping into geo-engineering with both feet. It is an untested technology, however, and may not work as well as predicted, or it may introduce unintended side effects which could be very difficult to manage. It is quite possible that geo-engineering becomes commonplace over time. It should be introduced in a gradual manner though, whilst always retaining the option of retreating to a prior state as necessary.

Jumping into such a program in an emergency is likely to create unanticipated hazards that are difficult to manage. Geo-engineering provides an attractive approach to tackling climate change because it is perceived as a low cost option. However, geo-engineering is a low-cost solution precisely because it tries to leverage small effects into large impacts. The consequence of such an approach is that if things do go wrong there is no way to control the system. Furthermore, since geo-engineering does not solve the root cause of the problem, it will still remain necessary to rebuild the world's energy systems so that they become carbon neutral. Thus, geo-engineering will create large risks without the concurrent benefit of large cost savings.

8　Energy Sources, Energy Carriers and Energy Uses

The entire energy value chain is about converting energy from one form to another. One typically starts with an energy source that is naturally available but needs to be collected or harvested. This raw form of energy is then converted into a form of energy that can be transported to the point of use and is made available as needed. It is typically far cleaner than the original form of energy. It therefore seems useful to logically separate the energy chain into *energy sources*, *energy carriers* and *energy applications*. To give a specific example, the coal coming out of the ground is an *energy source*, the electricity produced is an *energy carrier*, and the motive power generated at the user's location feeds directly into the *energy application*. This chapter mainly deals with energy sources, but (as for example in the transportation sector) it is important that the energy carrier is well suited to the application, and the choice of the energy carrier will often predetermine the choice of energy source.

Energy sources can be categorised as:

- Chemical sources.
- Thermal sources.
- Electromagnetic radiation sources.
- Mechanical sources.
- Nuclear sources.

There are many different examples for each of these categories. Most natural chemical energy sources are in the form of reduced carbon found in nature. These include: fossil fuels, like coal, gas, oil, tar or shales. Chemical energy sources also include biomass products, ranging from wood to algae matter and to municipal waste which is often rich in biomass. It is, of course, possible to broaden the category of chemical energy sources to the food people eat, and to the salt used in the colder regions of the world to melt substantial quantities of snow and ice.

Thermal sources vary widely in terms of temperature, size and quality. In effect, thermal sources of energy always involve at least two thermal reservoirs; one of them is often the ambient surroundings.[40] Therefore, one is often only concerned with the second thermal reservoir. Usually, but not always, it is at a higher temperature. These reservoirs include geothermal sources, which often involve very high temperatures, and ocean thermal resources, which take advantage of the temperature difference between the ocean surface and the deep ocean.

Direct sunshine is the only significant source of electromagnetic radiation energy, but it overwhelms all other sources of energy by a considerable margin. The Earth is exposed to 170 000 TW of solar radiation, which completely dwarfs human primary energy consumption at about 15 TW.[40]

Mechanical sources of energy are often derived from sunshine. These include: hydro-energy and wind energy. Mechanical energy can also be derived from other sources. Tidal energy, for example, is ultimately derived from the gravitational energy stored in the Earth–Moon system.

While an argument could be made that sunshine is fundamentally a nuclear energy source, we distinguish it from energy sources that are directly based on nuclear energy harnessed here on earth. These are virtually limited to the use of isotopes of uranium and thorium, and certain isotopes of hydrogen and lithium that could provide the basis for future fusion energy.

Energy carriers come in two distinct forms: networks that carry energy, like pipelines and electric wires, and the physical implementations of chemical energy that may flow through pipelines or are carried in individual containers, like the gasoline tank on board of a vehicle or the battery pack in a computer or a hybrid car.

Batteries, flywheels and ultra-capacitors are all means of carrying energy. So are liquid hydrocarbon fuels. The advantage of the latter is that hydrocarbons can carry up to 50 MJ per kilogram of fluid, or close to that amount if one accounts for the weight of the storage container. By contrast, flywheels, batteries and even hydrogen storage lead to storage capacities which are far less. Electric storage and mechanical storage is usually substantially less than $1\,MJ\,kg^{-1}$.

Finally, energy applications involve a last transformation of the energy into the form the consumer requires, as, for example, motive power, heat, light or chemical energy.

9   A Matter of Scales

For an energy resource to be important and successful in the 21st century, it must be able to provide substantially more energy than any of the large energy sources provide today. This is a formidable challenge, considering that all energy sources are already considered hard pressed to satisfy current demand. Difficulties arise not only because the resource base may simply be too small, but also because it is difficult to maintain the high rate of growth necessary to keep up with demand. Concerns of this nature are not new. Already a hundred years ago, there was a substantial concern that oil shortages would develop because supply could not keep up with rapidly growing demand.[41]

Estimates of future demand vary, in part because of different assumptions about how much of a reduction in energy intensity can be achieved and how much economic growth the world can hope to see. However, a growth in energy demand by a factor of four over the course of the century would be a conservative estimate. Even such a slow demand growth would drive the primary energy consumption rate to approximately 60 TW.

Even though it is unlikely that a single source of energy will end up satisfying nearly all of the demand, it is equally difficult to see how the world could stitch together a global energy infrastructure that is a quilt of small sources of many different types that are loosely held together. Today, 81% of all energy comes from fossil fuels. Large slices of energy seem necessary to support a stable global energy system. Such slices would be on the order of 10 TW of primary power.

Very large slices that could, in principle, provide 100 TW of primary power would be able to support the entire world's energy demand and should thus receive particular consideration in future energy research and development.

Most energy resources would simply be unable to support such large energy consumption. In some cases, if such a source were harnessed at the maximum possible rate, it would still fail to satisfy the energy demand. In other cases it could rise to a large output rate, but would be consumed in a very short time. Even if the scale of the available resource were sufficiently large, environmental concerns could directly limit its use. Fossil carbon energy sources without CCS technology would fall into this category. Even though there is plenty of carbon in the ground, the associated emissions would not be tolerated.

It is difficult to predict the constraints that will affect large-scale use of a potential energy resource that is currently used at only a fraction of its potential scale. While it may be possible to identify some issues, others will only become obvious during scale-up.

Some looming obstacles may prove to be much less of a problem than initially thought. Technological changes can easily make simple extrapolation to much larger scales obsolete. Indeed, one should expect that a straightforward extrapolation from today's systems to those which are larger by one or more orders of magnitude would result in systems that are impractical. A current design will aim at current scales and may not incorporate features necessary to operate on a significantly larger scale. Problems which one may have identified, and which initially seem insurmountable, may have easy solutions which may be addressed long before the new larger scale has been achieved.

On the other hand, a critical constraint on a technology may simply not be recognised until the system has reached a very large scale. For example, at a small scale of operation an effluent stream, that proves hazardous on a very large scale, may have been too insignificant to matter.

Climate change again provides a good example. Climate change is a direct consequence of having reached a limit where the dilution of CO_2 into the atmosphere is no longer sufficient. Before one reaches this scale, the problem is virtually invisible. In hindsight, we celebrate those researchers who recognised the problem early, but there is a tendency to forget that many others argued that the ability of the atmosphere to absorb CO_2 was much greater than it ultimately proved to be. It is worth noting that Arrhenius, who did understand the greenhouse effect, welcomed it as a positive attribute rather than a downside of fossil fuels.[42]

As operations grow from kilowatt test stands to gigawatt utilities, and move on to terawatt energy infrastructures, new issues constantly arise. Issues that seemed trivial or too small to notice can suddenly become significant. For example, it took a large number of cars to see the impact of cars on air pollution, and it took even longer to understand the subtle connections between sunshine and ozone production. These problems were not anticipated, but were resolved once automobile traffic had reached a large scale.

The use of lead in gasoline provides another good example, where an initially reasonable idea implemented at full scale causes serious problems. In the end, lead was removed from gasoline because it interfered with catalytic converters, but the most important advantage may have been to public health.[43] The realisation that lead solder in discarded electronics could prove to be problematic required large quantities of discarded computers.[44]

In harnessing energy sources it is, of course, important to consider the total size of the resource and, if this proves to be adequate, one should consider the density of the resource, as it becomes difficult to harness a resource for which the size of the collection facility dwarfs the size of other human infrastructures. Finally, one should consider all the environmental impacts of new energy infrastructure as much as one possibly can, and consider fluxes of effluent or releases to the atmosphere which are literally orders of magnitude larger than those of today.

If energy systems harness energy flows in large-scale dynamical systems, the question of feedback needs to be carefully considered. For example, it has been shown that a large tidal power plant in the Bay of Fundy would have noticeable effects on tides in the Boston area.[45] In short, it is important to think ahead, identify problems before they occur, but be willing to be flexible if problems find solutions, and also be prepared to encounter problems that were initially not even considered.

10 Small Carbon-Neutral Energy

10.1 Ocean Tides, Waves and Currents

Ocean waves, ocean currents and ocean tides would be unable to provide the power the world is consuming today. Individually they will always fall far short of a 10 TW slice. Even together they could not add up to 10 TW without severely changing ocean dynamics.

The actual dissipation of tidal energy can be measured because the slow-down in the Earth's rotational speed is detectable. Estimates for tidal dissipation are around 3–4 TW,[46] which is less than the world's primary energy consumption of around 15 TW.

Wave energy systems aim to intercept the energy that otherwise dissipates along the shores. Typically the power delivered by ocean swells approaching a coastline ranges from 20 to $70 \, \mathrm{kW \, m^{-1}}$ of coast. A high wave power is in the order of $50 \, \mathrm{kW \, m^{-1}}$.[47] Wind transfers energy into ocean waves, and waves can travel the width of the ocean. Indeed, it requires a long fetch for wave energy to build up. For example, the wave energy arriving at the US East coast is far smaller than that arriving at the US West coast, reflecting the prevailing wind patterns.[48] The energy content of waves in small ocean basins is significantly smaller than along open coastlines facing prevailing wind. Smaller bodies of water cannot support large waves.

To obtain a rough estimate, we assume that one can intercept the waves along five different longitudes, *viz.* on the North and South American West coast, along Europe and Africa, across the mid Indian Ocean and once more before getting back to the West side of the Pacific Rim. This would create a linear collector size of 100 000 km, which at $50 \, \mathrm{kW \, m^{-1}}$ of input power would pass 5 TW of wave energy through the collectors.[49]

Of course, only a tiny fraction of this energy could be practically harvested. Otherwise one would effectively convert ocean shore into lakefront. Stopping waves before they reach the shore or shallow water on a larger scale will impact sedimentation and the ecology of beaches.[50]

Ocean currents carry kinetic energy, but the total amount is too small to qualify it as a large energy source. Kinetic energy levels average between 50 and 100 J m^{-3} (ref. 51). Much higher levels are limited to small regions in the ocean.

If we generously assume that this level of kinetic energy can be found in the top 10% of the ocean, then we can estimate the total energy available.[v] The total volume of the ocean is 1.35×10^{18} m^3, resulting in a total kinetic energy 1.35×10^{19} J. If this were extracted in the course of 1 year, the total available power would be only 0.45 TW. It is worth noting that most of this energy is not in the large currents in the ocean, like the Gulf Stream, but that most of it is in the smaller eddies. Larger rings have an estimated spin-down time in the order of a year. This suggests that at 0.45 TW, one would be extracting a large fraction of the energy that is injected into the ocean by external forces.[52]

Even large-scale ocean current collectors situated in places where the ocean moves quickly, at $2 \, \text{m s}^{-1}$, would have to be very large. The flow provides only $4 \, \text{kW m}^{-2}$. Hence a collector that goes to 100 m depth would still need 2500 m width to encounter 1 GW. Accounting for the limited efficiency of a turbine, the width probably would exceed 10 km. In effect, in order to collect 1 GW of power, one would need to stop the water in a 100 m-thick layer over a square kilometer every three minutes.

Ocean currents, while locally useful, certainly do not add up to a large slice of the world's energy demand. Even here one has to carefully consider the footprint of such an operation, as it is quite easy to reach scales of operation at which the feedback onto the ocean currents cannot be neglected.

10.2   Hydroenergy

Hydroelectricity is a well-established, cost-effective form of renewable energy. Wherever it has been implemented, it has helped support energy-intensive industries like aluminum smelters that rely on particularly cheap forms of electricity. Hydroelectricity can also play a role in energy storage. However, there is not enough hydro-energy potential to run the world economy on hydroelectricity.

Simple 'back of the envelope' calculations show that hydroelectricity supply is fundamentally limited. For example, the entire potential energy of the rain that falls on the ground in the US during a year is less than the energy people consume in the US.[53] Indeed, it is remarkable how close the US has come to harnessing the entire potential of this enormous energy source, and it may well be that hydroelectricity is already overused.

As a result, hydroelectric energy is likely to be deployed wherever it is readily available. Whether it is harnessed in large installations or in small 'run of the

---

[v] Estimates similar to this "back of the envelope" approach can be found in ref. 44.

river' operations, the total energy available is simply not enough to solve the world's energy problems.

10.3   Wind

There are plenty of analyses that show that the world's wind potential far exceeds current consumption and even likely future consumption.[54] Yet, harnessing the wind pushes against various environmental limits. Even without direct environmental consequences to consider, there is a noticeable backlash that is driven by the large size of wind installations. Wind energy is very dilute, and hence its collection requires large installations.

A more serious concern is the feedback that different wind installations will have on the wind field. Taking wind energy out of the atmosphere will change the dynamics of the atmosphere. While it is true that the total taken out is still small compared to the total wind power added by solar input, the change is large enough to start feeding back on the dynamics of the system.

We estimated the total kinetic energy in the ocean currents at about 10^{19} J. The total kinetic energy in the air is about $1.3\,\mathrm{MJ\,m^{-2}}$ or roughly 7×10^{20} J (ref. 55), *i.e.* two orders of magnitude larger than for the ocean. Thus at a 10 TW extraction rate, it would still take two years to extract all the kinetic energy present in the wind field. This is substantially longer than the spin-down time. The power dissipation from the wind field to the ground is approximately 300 TW (ref. 40). Hence, a 10 to 60 TW wind power system could noticeably affect the wind field.

The interactions between energy extraction and energy injection into the wind field are quite complex. In effect, large scale wind installations increase the roughness of the ground and thereby increase the rate of energy dissipation from higher altitude to the ground. Rather than having to transfer energy through very thin boundary layers, large installations are efficient in extracting wind energy at larger vertical scales, increasing the efficiency of energy removal. This, in turn, will affect the wind field, and model calculations have shown that the indirect impacts on the climate of these changes are also significant. Indeed it has been suggested by Keith *et al.*[56] that gigawatt for gigawatt, the climate effects of wind energy and fossil energy are quite comparable. This suggests that wind, too, is unlikely to provide a 10 TW slice of power.

10.4   Biomass

Photosynthesis has proven itself as a large-scale carbon cycle in which approximately 100 Pg of carbon move back and forth between the atmosphere and biological matter on an annual basis.[57] Much of this carbon, like leaf detritus, is only maintained for a very short time; some part of the biological cycle is stored for many decades, and a tiny proportion is tied up on geological timescales. Coal seams appear to be ancient biomass. On the other hand, the entire reserve of fossil fuels represents the biomass production of a few decades.

Accelerating biomass growth, or inducing enlarged storage of biomass, provides another avenue for eliminating the human carbon footprint. There are several issues one needs to consider carefully. Firstly, will the added storage lead to the production of different and potentially more potent greenhouse gases? Crutzen et al.[58] have suggested that enhanced plant growth would raise the emissions of N_2O, which is a far stronger greenhouse gas than CO_2 and would thus cancel out the total CO_2 reductions.

Another obvious concern is the production of methane that could be released to the atmosphere. As biomass decays, a fraction will be released as methane rather than CO_2. Methane is a much more powerful greenhouse gas than CO_2 and even if only a few percent of this additional biomass find their way into the atmosphere as methane, the net advantage of storing organic carbon may be cancelled out.

A third concern is long-term stability. The carbon stored in this way is not thermodynamically stable and it needs to be protected from decay. It may be possible to achieve long-term storage by carefully treating the biomass. The production of charcoal has been suggested.[59]

A special implementation of this approach is the creation of Terra Preta, a particularly inert charcoal-rich form of soil that has been introduced in the Amazon region. Whether the long-term stability of this method is sufficient, or whether bacterial action or, for example, termite action, could remobilise the carbon so stored, is far from clear.

Lastly, there is the issue of how much land one would need to store CO_2. For example, all standing biomass contains around 600 Pg of carbon and soil carbon adds around 1400 Pg of carbon. Hence, these reservoirs do not appear large enough to absorb all carbon.

Finally, there is the option of reusing the carbon from biomass. In effect, biomass can displace fossil fuels and be used to create more fuel. This is a combination of CO_2 capture from air and fuel production. The question remains whether one can do it on the necessary scale. Biomass conversion at best occurs at $3\,Wm^{-2}$, which means that a 10 TW effort would require land comparable in size to all the agricultural land. It is difficult to see how the world could commit that much agricultural land at a time when food production itself is considered under pressure.

10.5   Geothermal

The geothermal potential is very large.[60] At present, the technology is limited to special locations with very steep thermal gradients, resulting in very large temperature differentials between the surface and the geological reservoir. Iceland, Kenya and some locations in California are good examples. In such locations of unusually high heat flux it may even be possible to treat the reservoir as a renewable source of energy.

It is also possible to think of geothermal heat as a non-renewable resource. One is in effect mining heat energy from rock. In principle this can be done anywhere, and the scale is large but not completely unlimited. A cubic kilometer of rock contains roughly 3×10^{15} J of energy for every degree Kelvin of temperature change. In a densely populated area ($> 300 \, \text{km}^{-2}$) and with 10 kW of primary consumption, which amounts to an energy need of $3 \, \text{MW km}^{-2}$, this could be satisfied with a 1 km-thick layer for approximately 300 years. The unknown is the rate at which one could extract energy and the consequences of the wholesale hydrofracture of entire regions. Note too, that this approach does not lend itself to large-scale operations. A gigawatt power plant would have to draw on a 25 km by 25 km area, if it were to reduce temperatures by 10 K through a thickness of 1 km over the course of 60 years. Use of low-grade heat, or heat at extreme depth, is at present not quite feasible. What impacts such large-scale cooling of underground reservoirs would have is largely unknown.

Ocean thermal energy has similar scale limitations. Assuming one can utilise a 25K thermal gradient at 5 percent efficiency while causing a temperature drop of 10K, one would require roughly $500 \, \text{m}^3$ of water per second, or 40 million tons a day. Nevertheless, OTEC has been estimated as a large potential source of energy, in the order of 10 TW.[61]

In summary, many of the possible energy resources are simply not large enough to operate at the scale necessary to satisfy a substantial fraction of the world's energy demand. In some cases, like geothermal energy and wind energy, they are large enough, but raise serious questions about the environmental impact of large-scale deployment.

11 The Three Truly Big Energy Resources

There are three sources of energy that seem large enough to satisfy the world's energy demand without being stressed and without having obvious large-scale environmental impacts that unavoidably would make their use at the global scale impossible. These three options are nuclear energy, solar energy and fossil energy combined with carbon capture and storage. All three are plausible candidates for being the dominant energy source of the future, but none of them is actually ready to provide energy at the necessary large scale without substantial advances over the current state of the art.

11.1 Nuclear Energy

The closest to being ready for large-scale deployment is nuclear energy. Nuclear energy already has been demonstrated on a large scale. In 2000, France covered most (77%) of its electricity demand with nuclear energy, Japan generated 29% and the United States 20% of its electricity from nuclear power generation.[62]

Even though operation at large scale has been demonstrated, there still remain serious questions about the nuclear waste material that is produced. No country has reached a fully institutionalised and accepted solution of this problem. While it may be technically feasible to dispose of the waste, in practice this has remained an obstacle in most countries using nuclear energy.

Furthermore, natural uranium resources may not be large enough to support the much larger demand, which would arise if nuclear fission were to take on an important role in the world's energy supply. At ten-to-fifty times current consumption, the proven reserves would be depleted in a few decades.[63] On the other hand, reserve numbers in the absence of demand should always be considered uncertain. A large expansion of nuclear energy may well go hand in hand with a transition to fusion or, at a very least, with a transition to breeding fuel from natural uranium and natural thorium. Fuel reprocessing combined with breeding would greatly reduce the size of the waste disposal issue and it would completely change the resource calculus. Utilisation of these two additional isotopes would raise resource estimates by two orders of magnitude, mainly because of much higher fuel utilisation and the addition of thorium, which is more abundant than uranium.[63]

Breeding and fuel reprocessing raise serious questions about the security of the fuel supply, and proliferation resistance of the technology becomes the highest priority. There are technological and political issues that would need to be sorted out, since it is, in principle, possible to create weapon-grade fissile material in the reprocessing cycle and the dangers of this process need to be addressed.

Security will likely require the internationalisation of the fuel cycle. This is foremost a political problem. Thus, the future of nuclear energy, to a large extent, will depend on the ability to create the institutional framework in which it can thrive.

11.2 Solar Energy

The introduction of solar energy is mainly limited by cost and distribution issues. It is likely that solar energy will begin to play a larger and larger role, but its dilute nature makes it intrinsically more expensive than other options. For example, at 20% efficiency, a solar panel in a desert climate would still need $20\,m^2$ to produce on average power output of $1\,kW$.[vi] A car engine, which comfortably fits under the hood of a car, is easily capable of producing in excess of $100\,kW$ of power. This suggests that solar power could always feel strong competition from fossil fuels.

The intermittent nature of solar energy also requires that solar electricity be cheaper than electricity that can be delivered on demand. How much cheaper it needs to be will depend on the cost of electricity storage which, at present, is not available at the necessary scale. On the other hand, chemical conversions become feasible if the primary electricity that is produced at a solar panel is roughly three times cheaper than the electricity that is generated to satisfy an immediate demand. Thus the long-term goal of solar energy development should not just be to reach parity with coal electricity, but to undercut coal electricity by about a factor of three.

[vi] Average insolation in a desert climate is between 200 and 250 W m^{-2}.

The dilute nature of the solar flux brings with it a potential for environmental limitations. Large installations will occupy large amounts of land which certainly will impact local environments. On the other hand, while the land-use numbers are large, they are much smaller than for biomass production. Surprisingly, they are not much larger than for coal mining operations, specifically for surface mining.

This land-use argument can be seen by comparing the energy collected by a solar panel in 30 years with the energy content of a 1 m-thick seam of coal below the land. In effect, the panel may produce about 30 GJ of electric power, while the coal seam may contain 30 GJ of chemical energy. Therefore, the land use in the two situations is not as different as it may appear at first sight. In mining a coal seam, it may take decades for the land to be reclaimed and in some cases it will never be reclaimed. In a matter of decades a solar panel which occupies a similar amount of land will produce a similar amount of energy.

At present, solar electric power is in nearly all situations still too expensive to compete against other options. The only way for it to compete is with price support as, for example, in Germany.[64] It is likely, however, that prices will come down dramatically, particularly for the production of photovoltaic cells. A more difficult question is whether these dramatic reductions will give solar energy a competitive edge or whether other energy sources will take their intrinsic advantages and also reduce cost.

Solar energy conversion technologies range from photovoltaic to low-grade solar heat. Apart from the large land-use concern, it is difficult to anticipate specific environmental issues as each approach to solar energy will raise its own and idiosyncratic environmental concerns. Photovoltaic cells made of silicon may create upstream environmental issues during production, as they involve substances that are hazardous,[65] but silicon itself is a common element and it poses no serious environmental concerns if it is left exposed to the elements. On the other hand, gallium arsenide, cadmium telluride and similar heavy metals will pose serious environmental concerns that will need to be addressed. As a consequence, First Solar, a photovoltaic company, has promised that it will take back the solar panels after they reach the end of their useful life.[vii]

11.3 Fossil Fuels with Carbon Dioxide Capture and Storage

The third option that is not limited by resource constraints is fossil energy. While there may be a real worry about limitations in the oil and gas supply there are no real limitations in the coal supply. Using the Hubbert curve[66] and applying it to coal seems foolhardy, as there is plenty of evidence that the cost of coal mining has been driven mainly by other concerns than resource limitations. The fact that Britain, France and Germany have severely depleted

---

[vii] First Solar Inc. has announced on http://www.firstsolar.com/recycle_modules.php that anyone in possession of a First Solar Panel can return it for recycling free of charge. First Solar is setting aside funds for this transaction. See also Reuters, First Solar to Supply Modules for Australia's Largest Solar PV Installation http://www.reuters.com/article/pressRelease/idUS11909 + 29-Apr-2009 + BW20090429.

their original coal supplies is not necessarily proof that coal is running out. In part, the exploitation of coal in these countries with a small land area has been so intensive that an extrapolation from their scale to the full world scale would actually result in unrealistically high estimates of mineable coal.

If land areas all over the world could yield as much coal as land in Britain, which has extracted 28 Pg of coal,[67,68,viii] the total world coal resources would be 50 000 Pg of coal, roughly ten times larger than typical coal estimates. Germany may have more or less mined out the coal seams in the Ruhr Valley, but it still has much larger resources in terms of lower quality brown coal. While higher grade fossil fuel resources may be more desirable, there are no fundamental obstacles in introducing lower grade hydrocarbon fuels as the basic source of energy.

Estimates of fossil fuel reserves vary and they are naturally highly uncertain.[69] However, estimates uniformly suggest that the resource base is measured in thousands (if not tens of thousands) of petagrams ($1 \text{ Pg} = 10^{15}$ g) of carbon. This should be compared to a past consumption of around 350 Pg of carbon since the beginning of the industrial revolution, which marks the start of the fossil fuel era. Fossil energy resources are not running out. In spite of some critics' points of view,[70,71] it is important to realise that there is still a considerable resource base even if the proven reserve is a very uncertain number.

This view of a nearly unlimited supply of fossil fuels is fully compatible with a finite resource that eventually will run out. There is good reason to believe that there is at least 5000 Pg of carbon that could be extracted from the ground.[72] Improved extraction technology, additional discoveries, or a combination of these two, could easily drive these estimates even higher. Nevertheless, the total consumption of fossil fuels would be limited at about 10 000 Pg of carbon, even if more carbon were available. At this point oxygen depletion of the air would become serious. A kilogram of oxygen supports approximately 14 MJ of heat of combustion. (The variation between coal, oil and gas is small as the difference in their heat content scales directly with the amount of oxygen required). Hence at 100 TW it would take five thousand years to burn through the entire atmosphere (10^6 Pg of oxygen). Thus, if we somewhat arbitrarily set the limit at a 5% reduction in oxygen, the oxygen reserve would last less than 250 years. If the oxygen were used for the combustion of natural gas (*e.g.* from methane hydrates), this would have led to the consumption of 10 000 Pg of carbon.

Clearly, the carbon dioxide limit hits much earlier than the oxygen limit. Therefore, access to fossil fuels is predicated on the availability of carbon capture and storage technology, which in turn depends on a capacity to store CO_2. Rather than being limited by fossil carbon resources, access to fossil fuels may well be limited by CO_2 storage capacity.

Within the community of researchers who investigate underground geological storage of CO_2, there seems to be a general consensus that CO_2 storage volumes are sufficient to store all the CO_2 that could possibly be produced.[73,74]

[viii] Total production of coal in Great Britain is obtained by adding up all entries in references 67 and 68 and interpolating log-linearly on missing data in the early years.

It is necessary, however, to consider the sheer size of the storage volumes. For example, the IPCC report on Carbon Capture and Storage[74] states explicitly that Alberta alone could store 4 000 Pg of CO_2 in the sedimentary basin. This would inject a volume into the underground formation that spread over the entire province of Alberta would be a layer 6 m thick. In a recent paper, we suggested that an equivalent amount of water was removed and desalinated so that it could be used as part of the conventional water cycle. The additional cost would be small compared to the total cost of CO_2 disposal, but it would add another cost item to the overall cost of fossil energy.[75]

Ultimately, fossil fuels will have to compete with other energy resources in the market. Today, fossil fuels are cheaper than most other alternatives, but the price of fossil energy does not include the cost of dealing with climate change. If fossil fuel consumption is combined with carbon dioxide capture and storage, the environmental concern over carbon dioxide emissions is removed, but at a financial cost. At present this cost is considered quite high. Some suggest that CCS will not be introduced until prices reach $100 per ton of CO_2. Certainly in the long term one can expect lower prices, somewhere between $30 and $50 per ton of CO_2.[74] If this cost goal is indeed reached, then fossil fuel resources could compete with other energy sources.

The use of fossil energy resources raises other environmental issues, apart from climate change, that also will need to be addressed. For example, mining impacts, particularly for coal and tar extraction, tend to be large. The management of ash, sulfur, fine particulates and heavy metals will also need to be considered. Combining CO_2 capture with zero emission power plants may provide a particularly attractive way of solving these problems.[76,77] Underground coal gasification may reduce the mining impact by avoiding the extraction of unwanted material, but it also mobilises gaseous species that need to be contained.

11.4   Summary

In summary, it is far from clear which of these three big energy resources is likely to dominate future energy supplies. All three have the resource depth to supply the necessary energy and, in this regard, they stand out among all possible options.

All three require future development before they are ready to operate at the full scale. Advocates of nuclear energy will have to solve its acceptance problem, find ways of dealing with a full-scale breeding programme and manage the security and safety of nuclear material cycle, without significantly raising cost.

Solar energy, particularly in the electricity sector, will have to become much more cost efficient before it can compete with other energy resources. While there is no obvious floor to the price of solar energy, progress in the last few decades has been slow.

Finally, fossil fuel technology would have to rely on a largely untested carbon capture and storage technology that has not yet been proven at scale. Even though CCS appears feasible, it needs to be proven that CO_2 capture and storage can be performed at affordable prices.

12   Capture of Carbon Dioxide Directly from Ambient Air

The continued use of fossil fuels could be made more acceptable, if carbon dioxide capture from the atmosphere proves to be a practical solution. Such air capture would make it possible to leave the current infrastructure intact and introduce carbon dioxide reductions without having to modify the existing energy infrastructure.

It is technically feasible to capture carbon dioxide directly from the air. The technology has been used for decades to remove carbon dioxide from the air inside submarines or spacecraft. Technologies for removing carbon dioxide from ambient air have been used in the past to create carbon dioxide-free air prior to air liquefaction. Processes using strong alkaline solutions have been investigated by a number of authors.[78–82]

We have recently shown that it is possible to develop very low cost processes for the capture of carbon dioxide from the air,[83] which could make carbon management by direct CO_2 capture feasible.

The direct capture of CO_2 from air (or 'air capture') is in many ways analogous to collecting energy from the wind. Windmills reduce CO_2 accumulation in the air by avoiding emissions. Air capture devices extract CO_2 from the air and thus also reduce the CO_2 load of the atmosphere. By removing all the CO_2 from a cubic meter of air, one enables the carbon-neutral combustion of a small amount of gasoline, enough to produce 10 000 J of heat. By capturing all the kinetic energy from a cubic meter of air, one avoids the CO_2 which is made in the generation of 20 J of energy. Hence, removing CO_2 from the air is far more powerful than extracting kinetic energy from the air. By using the heat of combustion of gasoline as a conversion factor, we find that CO_2 collection requires about 500 times less air than producing an equivalent amount of energy with a windmill. This suggests that contacting the air for CO_2 is far cheaper than contacting the air for wind energy.[84] The cost of the air capture process is not in contacting the air, but in the cost of managing the sorbent cycle.

The basic process is conceptually very similar to the scrubbing of flue gas in a chimney stack. In both cases, one uses a chemical sorbent to bind carbon dioxide as the gas flows over sorbent surfaces. Early attempts used sodium hydroxide, but the binding energy of CO_2 to NaOH is unnecessarily high. The challenge to the designer is to find a sorbent that binds carbon dioxide strongly enough to remove it from ambient air, yet does not bind so strongly as to make the regeneration of the sorbent expensive. Since the minimum required binding energy scales only logarithmically with the concentration of CO_2 in the effluent of the collector, the difference in binding energy between what is required for CO_2 capture in the flue stack and what is required at air capture is small. Indeed most sorbents that work in a flue stack are also capable of collecting CO_2 from the air. One takes advantage of the fact that air capture does not need to deplete the air of CO_2, whereas the flue gas scrubber needs to remove essentially all of the CO_2 that is present in the flue gas.

One of the most interesting sorbents used in flue gas scrubbing is ammonia, which is cycled between ammonium carbonate and ammonium bicarbonate. This approach to flue gas scrubbing takes advantage of the fact that it is far easier to recover carbon dioxide from a bicarbonate, than it is from a carbonate that would have to be converted into a hydroxide.

The air-capture sorbent we developed also takes advantage of a swing that operates between carbonate and bicarbonate.[83] Here the base is a solid, an anionic exchange resin, which is typically used for water preparation purposes. The material is a strong base, with a quaternary amine attached to a poly-styrene backbone. One can think of the quaternary amine ion as the analog to an NH_4^+, where every hydrogen atom has been replaced with an organic chain that is anchored in the polymer matrix. Since the material cannot donate a proton, it is a strong base that mainly differs from a strong sodium hydroxide or sodium carbonate solution in that the positive ions are firmly attached to the polymer structure. The charge density of the material is approximately $1.7 \, mol \, kg^{-1}$. In its capacity to hold carbon dioxide, the resin thus resembles a 1.7 molar solution of sodium hydroxide.

The cycle is run by changing the resin from a dry state to a wet state. We found experimentally that the resin when it is dry wants to hold CO_2, and when it is wet it gives it back. Therefore, we refer to the cycle as a humidity or moisture swing. The moisture swing is performed by moving the sorbent filter from an open air stream into a closed chamber. After pumping the residual air out of the chamber, the resin is exposed to moisture and subsequently it releases CO_2 into a low pressure gas stream, which is then pressurised, cleaned and made ready for re-use or sequestration.[83]

Air capture becomes naturally a part of a more broadly applied CCS strategy. Air capture is a form of capture and it is the capture of last resort. If capture at the point of emission is not possible or excessively expensive, air capture at any other place in the world offers a real alternative. Since air mixes rapidly, it is possible to cancel out an emission in North America in the Australian desert. Air capture would typically occur at the disposal site. The additional cost of air capture can thus be partially offset by much smaller transport costs.

The price of air capture would put an upper limit on the price of managing carbon. If all else fails, air capture is always a viable capture option and thus it can be applied to all types of emissions. It can even deal with emissions that occurred in the past.

Once the price of air capture becomes affordable, the cost of dealing with all CO_2 emissions becomes manageable. Rather than ending up with a cost curve that moves steeply up as the world economy approaches carbon neutrality, the cost curve with air capture will rise from the low cost of other forms of capture in small niches to eventually the cost of air capture. This cost should be more or less independent of the total amount captured and the total amount stored. Presumably the price of air capture will gradually drop if more is done. In any event, air capture which provides a method of last resort can be used to mop up after cars, air planes, or any other emitters with a difficult local capture situation. On occasion that could include a power plant which cannot be affordably retrofitted, because the space to do so is not available, or because it is located in an urban setting, which makes running a new pipeline prohibitively expensive.

Generally speaking, the lower the cost of air capture, the larger its potential market in sequestration. If it comes close to the cost of retrofitting, then it could develop into a viable alternative to retrofitting. Otherwise it may be limited to special applications where no other options are available. Air capture is particularly well suited to recapture the carbon dioxide released from the transportation sector.

Air capture has been categorised as a form of geo-engineering in the public press and by a number of authors.[85] However, air capture in its simple initial applications should not be considered geo-engineering. It is far more akin to capture at the tailpipe than geo-engineering. It is about balancing out specific emissions rather than engineering the composition of the atmosphere. The purpose of air capture is to prevent the accumulation of excess carbon in the environment. The only unusual feature of the approach is that one can take the CO_2 back, even after it has been released into the atmosphere.

However, air capture becomes a tool of geo-engineering when it is used as part of a control system that attempts to set the carbon dioxide in the air to a particular level, especially if this level is different from the pre-industrial level. For example, the world could decide that going back to 280 ppm is not only too expensive, but also not desirable. If that were the case, air capture which allows one to hold a constant CO_2 level regardless of changes in the natural carbon cycle would become a tool for geo-engineering.

This, however, is futuristic. Such control can only be achieved, once all the carbon in the modern human infrastructure is managed in some form or another. Only after all emissions have been compensated for would it become possible to hold the CO_2 concentration in the atmosphere constant or to drive it toward a particular level.

Any emission to the atmosphere will stay in the air for a very long time. Hence, it is not possible to close the carbon cycle in the presence of air emissions of CO_2, unless one deploys air capture technologies. In effect, air capture is a necessary part of the anthropogenic carbon cycle as long as fuels are combusted and the CO_2 is released to the air. The cycle may be closed by taking carbon out of the ground and sequestering it after capture, or the cycle may be closed, just as in the case of biomass, by recreating the fuel which produced the CO_2 with the input of additional non-fossil energy.

The basic concept of fuel re-use is that there are well established means of creating synthetic fuels from carbon monoxide (CO) and hydrogen (H_2). A mixture of these two gases is commonly known as synthesis gas (or 'syngas') and it provides the starting point for many different synthesis routes to various energy-rich compounds which can be used as fuels. Fischer–Tropsch reactors can produce synthetic diesel or gasoline from synthesis gas. Synthesis gas, or even a mixture of CO_2 and H_2, can be used as the starting point for methanol synthesis. Pure carbon can be produced from carbon monoxide through the Boudouard reaction. ($2CO \rightleftharpoons C + CO_2$). Plenty of other substances can be produced from these starting points.

There are a number of ways of creating H_2 and CO from water and CO_2. We believe that the thermal decomposition of H_2O or CO_2 is expensive and thus should be avoided. However, technologies have been developed for the thermal decomposition of CO_2.[86] Electrochemical separations are usually more efficient. One requires at least one electrochemical step in the cycle, where one produces H_2 from H_2O, or CO from CO_2, or possibly a combination of the two from a mixture of CO_2 and H_2O.

In effect, one uses an electrochemical means of freeing the oxygen that is tied up with carbon and hydrogen. It is only necessary to perform electrolysis on one of the two compounds because it is quite easy to shift the remaining oxygen from carbon to hydrogen or *vice versa*. Moving oxygen from water to CO, to produce CO_2, is known as the water–gas shift reaction. The transfer of oxygen between hydrogen and carbon is feasible in either direction. As a result, one can produce H_2 and CO, by electrolyzing either H_2O or CO_2.

Creating synthetic fuels from carbon dioxide that has been collected by technical air capture would parallel the concept of a hydrogen economy, resulting in a non-fossil, but carbon-based energy economy. In this case, air capture would move away from carbon capture and storage and instead support carbon capture and re-use. Since CO_2 is a gas that, at least in the transportation sector, is best released into the air, the only practical way to close the carbon cycle is the use of air capture. Fuel production is the only viable re-use option, as otherwise the accumulation of oxidised carbon will unavoidably result in a large waste stream.

The advantage of making carbon-based fuels is that they have more desirable properties than any other form of energy storage. They are typically liquids and, even counting the weight of the container tank, they can reach remarkable energy densities when compared to hydrogen tanks or batteries. The ease of handling and the high energy density of liquid fuels make them highly desirable. If the world were running on hydrogen or batteries, liquid carbon-based fuels would be considered a major advance.

Air capture thus becomes an important building block in the development of sustainable energy options. Air capture combined with carbon sequestration makes it possible to let the current infrastructure live out its natural life. Combined with renewable energy, air capture can build a new carbon energy infrastructure. In a world where primary energy comes in the form of electric power, liquid chemical fuels will be at a premium. Air capture makes it possible to create these fuels without recourse to fossil fuel resources.

13 A Revolution in the Energy Sector

To move from a carbon-emitting energy infrastructure to a carbon-neutral energy infrastructure requires a revolution in the energy sector. If we follow

Pacala and Socolow's suggestion, then it is necessary at the very least to hold CO_2 emissions constant for the next fifty years.[87] This is not a very ambitious goal, which would leave us with about 480 ppm of CO_2 at the end of fifty years, and with an additional increase to about 550 ppm over the following fifty years being almost unavoidable. Thus, it would lead at minimum to a level of 550 ppm for the stabilisation point.

Even with this rather unambitious scenario, however, one still will need to build a new energy infrastructure over the next fifty years which is substantially larger than the existing energy infrastructure, and which has to be, for all practical purposes, carbon neutral. The new energy infrastructure may emit some CO_2 if the existing infrastructure is upgraded to reduce its emissions by an equivalent amount. More likely than not, the total energy demand in fifty years from now will have more than doubled. A simple doubling would imply an anaemic annual growth rate of 1.4%. In the last 20 years this rate was closer to 2% per year,[ix] which would result in a total growth in demand by a factor of 2.7.

In laying out a scenario for the future one can discern several stages.[21] First, the electricity sector will have to move toward carbon neutrality. Then one will have to address emissions in the commercial sectors. Cement plants and steel plants must become carbon neutral. Emissions from homes and office buildings have to be reduced gradually to zero. Finally the transportation sector must be addressed, either by introducing different fuels or by introducing air capture technology.

For the electricity sector to move toward carbon neutrality, one is in effect proposing a transition to power plants which have no carbon dioxide emissions to the atmosphere. This could be because they are nuclear power plants, or are based on renewable energy, or because they capture the carbon dioxide that they produce. Such a transition, in itself, is a major challenge, because it is counter to the current, worldwide trend of building preferentially the lowest-cost, pulverised coal-fired power plants.

While developed countries are wringing their hands over China's ambitious electrification project, in proportion to their population size they are not doing much better. Consider, for example, Germany, a stalwart of environmental protection, which apparently has 25 large coal-fired power plants on the drawing board or in the early stages of construction.[88] Most of these plants will be built within the next three years, giving the country 7–8 new power plants per year. Prorated by its size of population, Germany has a similar rate of construction of power plants to China.

Thus, the world has not yet seriously embarked on the transition of the electricity sector, which is just the first step toward holding CO_2 levels at about 550 ppm. If stabilisation goals were to become more ambitious, it would clearly be insufficient to simply make new plants carbon neutral, a complete transition

---

[ix] During the 26 years in the EIA data set, the growth has been 1.9%, leaving out the first six years eliminates the economic slowdown of the early 1980s and results in an annual growth of 2.0%. http://website www.eia.doe.gov.

to carbon neutrality for all plants would be required. However, even with carbon capture and storage installed, today's power plant designs are not carbon neutral. Currently, scrubbing in existing power plants aims to get 90% of the CO_2 back. If a power plant lost 30% efficiency, the avoided CO_2 is in the order of 85% of current emissions. Hence the residual emissions are still 15% of the original emissions. Assuming that the entire electricity sector reduces its CO_2 emissions to 15% of the current level, overall emission reductions will reach about 35%.

Further reductions are necessary, and this requires the gradual introduction of other options. The concept of capture at the source can be easily extended to cement plants, steel plants and to some refinery operations. However, for small sources, the introduction of carbon capture and storage becomes quite difficult.

For the commercial and the home sectors it is possible to achieve a substantial amount of decarbonisation by switching to electricity, or carbon-free fuels. One might expect that the introduction of the all-electric home is more likely than the hydrogen-heated home. Either approach would greatly reduce CO_2 emissions, provided that the central generation station eliminates or greatly reduces its CO_2 emissions. This transition does not necessarily result in lower energy efficiency. For example, heat generation could be covered by heat pumps, which for small temperature differences tend to have coefficients of performance well in excess of one, and often the heat transfer exceeds the chemical energy content of the fuel consumed at the power plant.

High temperature applications may be an exception to this rule. Except for very large installations, the use of hydrogen or electricity may be the only alternatives to decarbonising the process directly on site. Another option to consider in such cases is air capture technology.

Decarbonising the transportation sector is more difficult than decarbonising other sectors, because here the advantage of liquid hydrocarbon fuels are evident. Gasoline has a substantial advantage over batteries and hydrogen both in the way it is stored and can be used. Air planes depend on jet fuel and it would be extremely difficult to take its advantages away. For cars, gasoline is clearly the preferred choice. The advantages of liquid hydrocarbon fuels not only make it difficult to decarbonise the transportation sector, it is also the root cause of the world's excessive dependence on oil.

If it proved too difficult in the past to move away from petroleum to avoid the political costs of an over-dependence on oil, one must wonder whether the concerns over climate change will be sufficient to succeed where oil cartels failed. It may, however, be argued that both concerns together will be more successful in reducing the use of oil in the transportation sector.

Hydrocarbon consumption by cars could be significantly reduced even without abandoning fossil fuels. Fuel mileage has improved over the years, and the introduction of high efficiency diesel engines in Europe and hybrid cars in Japan and the US has created a noticeable improvement in efficiency. Better battery technology and the introduction of plug-in hybrids may herald a time when liquid fuel consumption on board of cars will be limited to long trips.

Short trips, which represent the bulk of all travel, could well be covered by electric power supplies.

On the other hand, the convenience of liquid hydrocarbon fuels should not be underestimated. To make their use carbon-neutral requires air capture, either in the guise of biofuels or in a system that is based on synthetic fuels. However, realistically, synthetic fuels will not be introduced until the price of non-fossil electricity has come down significantly or, alternatively, until the price of carbon has become much higher than it is today.

In short, the energy infrastructures supporting electricity generation, large industrial users, commercial and residential energy uses, and energy use in the transportation sector will have to be completely revamped in order to stabilise CO_2 levels in the atmosphere. The need for a large-scale transition will force change. Change, in turn, will make it possible to rethink energy systems in many different ways. It is rare that the entrenched competitor is unable to compete; as a result, new ideas can be introduced and old ideas that could not compete in the past may be given a second chance.

14   Conclusions

There is no serious shortage of energy because of resource limitations. Shortages could be precipitated by reckless policies or by creating supply bottlenecks in an economy that is driven by long business cycles. Large energy sources that could support the world energy demand for centuries include solar energy, nuclear energy and fossil energy. Even if other energy sources may gain importance, there likely will remain a role for fossil fuels for quite some time, because the rate of growth is limited for all options.

The limits to sustainability will come not from resource depletion, but will ultimately derive from the environmental impact of the use of these resources. Technologies can help overcome these limits.

Fossil fuels are a serious competitor for other energy sources. At present, cost considerations and the availability of a large resource base drive energy infrastructures towards fossil fuels. The major problem for fossil energy sources is their large environmental footprint. This footprint is large not only in terms of the amount of carbon mobilised and emitted as carbon dioxide, but also in terms of other environmental impacts.

After carbon dioxide emissions, mining is probably the second biggest impact which will need to be addressed if carbon dioxide capture and storage is to be developed. It is important to think ahead, because it would not be politically wise to spend large efforts on making fossil fuels climate-neutral, if one had no possibility to overcome the next levels of impacts.

However, it is worth realising that the environmental impact of any energy technology that operates at the tens of terawatts level must be seriously considered. The reasons that wind and biomass energy have little environmental impact is mainly attributable to the fact that they do not yet produce significant amounts of energy. In a world where all other energy competitors are an order of magnitude smaller than fossil fuel sources, it is easy to overlook their potential for environmental harm.

It is critical to develop options for the future. Neither the energy industry nor society as a whole has fully realised the urgency of the situation, which is far more precarious than is usually acknowledged. In the realm of energy development, technological advances are necessary. In contrast to water and food supply, in the energy sector the worldwide implementation of the state of the art would not be a solution to the problem but would exacerbate it. Energy development is on a collision course between energy demand and environmental constraints.

A brand new infrastructure needs to be built within the next fifty years. The technological and institutional obstacles are formidable. Yet the urgency of this situation has not yet sunk in.

In developing solutions one must keep in mind the sheer scale of the problem. There are very few energy options that are big enough to cope with world demand. A scenario, in which solar energy resources, nuclear energy resources and fossil energy resources fail to provide a basis for the energy infrastructure, will be a scenario with drastic reductions in economic output and one in which living standards will fall back to much lower levels. This, in turn, has the potential for creating economic and political strife on a global scale.

It would be oversimplified to argue that the big three options for energy boil down to three simple well-defined paths. Indeed, under the large umbrellas of fossil fuels, nuclear energy and solar energy are many distinct sub-options that all could be pursued independently. For example, nuclear energy includes conventional fission plants, inherently safe Generation-IV plants, which encompass a large family of options, various breeder designs and, last but not least, fusion. On the other side, solar energy can be pursued as photovoltaic energy with a plethora of photoelectric materials, or as solar thermal energy, where again one can pursue a path of relatively cheap low-temperature heat, or high-grade high-quality heat.

However, even outside this limited set, there are many more options that need to be considered, and should be considered even though their priority for governmental support should be lower as they, by themselves, cannot solve the energy conundrum. Nevertheless, adding these energy sources to the palette of options offers diversity in approach, competition and cost reductions. The presence of these other options will ultimately help in pushing the entire energy system toward an optimum. Furthermore, one cannot entirely rule out that a large number of small solutions ends up carrying the day.

Finding a global sustainable energy solution is a complex task. It involves solving technical problems, creating institutions and changing people's attitude toward the need of energy. Overly simplistic solutions are unlikely to work. It is very unlikely that a solution to the energy conundrum will develop solely from a change in attitude about energy. Indeed, if anything, the prevailing attitude that energy consumption is the problem hinders an approach which sees energy as an important ingredient in solving the sustainability problem. To support a stable population near ten billion people will require large amounts of energy; probably much more than the world consumes right now.

The issue is not the use of energy, but the environmental consequences of the current approach to providing this energy. It's the environmental footprint of energy consumption that is unsustainable, not the energy use *per se*. As a consequence, technological fixes ought to be welcome, as they can remove obstacles without eliminating access to energy.

The opposite approach, which starts out by eliminating energy consumption, will close doors that may need to be kept open. For example, access to water and food may be made easier by providing energy. In a world in which energy is made expensive, the adequate supply of food and water may simply prove impossible. A break-down of the transportation infrastructure due to lack of fuel would be a disaster of unprecedented proportions.

The often-disparaged end-of-pipe solutions may well be the ones that can provide the largest improvements and at the lowest cost. If the goal is to eliminate the environmental impact of energy use, they may provide the most efficient and most direct path toward solving the problem.

References

1. J.-B. J. Fourier, *Ann. Chim. (Paris)*, 1824, **XXVII**, 136–167.
2. J.-B. J. Fourier, *Mémoires Acad. Royale Sci. Inst. France*, 1827, **VII**, 570–604.
3. R. T. Pierrehumbert, *Nature*, 2004, **432**(7018), 677–677. (An annotated translation of Fourier's paper, Ref. 2 above, is included in the supplementary online information to this essay.).
4. J. Tyndall, On radiation: *The 'Rede' Lecture*, delivered in the Senate-house before the University of Cambridge on Tuesday, May 16, 1865, D. Appleton and Co., New York, 1865.
5. S. Arrhenius, *Philos. Mag.*, 1896, **41**(251), 237–276.
6. *Climate Change 2007: The Physical Science Basis. Contribution of Working Group I to the Fourth Assessment Report of the Intergovernmental Panel on Climate Change*, ed. S. Solomon, D. Qin, M. Manning, Z. Chen, M. Marquis, K. B. Averyt, M. Tignor and H. L. Miller, Cambridge University Press, Cambridge, United Kingdom and New York, 2007.
7. *OECD Factbook 2009: Economic, Environmental and Social Statistics*, OECD, 2009.
8. T. A. Boden, G. Marland and R. J. Andres, 2009. *Global, Regional, and National Fossil-Fuel CO₂ Emissions*, Carbon Dioxide Information Analysis Center, Oak Ridge National Laboratory, U.S. Department of Energy, Oak Ridge, TN, U.S.A.
9. P. Forster, V. Ramaswamy, P. Artaxo, T. Berntsen, R. Betts, D. W. Fahey, J. Haywood, J. Lean, D. C. Lowe, G. Myhre, J. Nganga, R. Prinn, G. Raga, M. Schulz and R. Van Dorland, in *Climate Change 2007: The Physical Science Basis. Contribution of Working Group I to the Fourth Assessment Report of the Intergovernmental Panel on Climate Change*, ed. S. Solomon, D. Qin, M. Manning, Z. Chen, M. Marquis, K. B. Averyt,

M.Tignor and H. L. Miller, Cambridge University Press, Cambridge, United Kingdom and New York, 2007, ch. 2, pp 129–234.

10. K. S. Lackner, *Annu. Rev. Energy Environ.*, 2002, **27**(1), 193–232.
11. D. Archer, *J. Geophys. Res.*, 2005, **110**, C09S05, doi:10.1029/2004JC002625.
12. R. A. Feely, C. L. Sabine, K. Lee, W. Berelson, J. Kleypas, V. J. Fabry and F. J. Millero, *Science*, 2004, **305**(5682), 362–366.
13. C. Turley, *Mineral Mag*, 2008, **72**(1), 359–362.
14. C. Langdon, T. Takahashi, C. Sweeny, D. Chipman, J. Goddard, F. Marubini, H. Aceves; H. Barnett and M. J. Atkinson, *Global Biochem. Cycles*, 2000, **14**(2), 639–654.
15. J. A. Kleypas, R. A. Feely, V. J. Fabry, C. Langdon, C. L. Sabine and L. L. Robbins, *Impacts of Ocean Acidification on Coral Reefs and Other Marine Calcifiers*, Report of a workshop sponsored by NOAA, NSF and USGS, June 2006.
16. K. S. Lackner, *Asian Economic Papers*, 2006, **4**(3), 30–58.
17. V. Smil, in *OECD Global Science Forum: Conference on Scientific Challenges for Energy Research*, Paris, 2006.
18. B. McKibben, *New York Times Magazine*, July 23, 1995, 24.
19. V. Smil, *Annu. Rev. Energy Environ.*, 2000, **25**(1), 21–51.
20. M. R. Raupach, G. Marland, P. Ciais, C. Le Quéré, J. G Canadell, G. Klepper and C. B. Field, *Proc. Natl. Acad. Sci. U. S. A.*, 2007, **104**(24), 10288–10293.
21. K. S. Lackner and J. D. Sachs, *Brookings Papers Econom. Activity*, 2005, **2**(2), 215–284.
22. B. M. Trost, *Science*, 1991, **254**(5037), 1471–1477.
23. R. Sharpe and G. Cork, in *Industrial Minerals and Rocks, Commodities, Markets and Users*, ed. J. E. Kogel, N. C. Trivedi, J. M. Barker and S. T. Krukowski, 7th Edition, Society for Mining, Metallurgy and Exploration Inc. (SME), Littleton, Colorado, 2006.
24. W. E. Brooks, in *Mineral Commodity Summaries*, United States Geological Survey, US Government Printing Press, Washington, 2007. See also *Minerals Handbook Volume I, Metals and Minerals*, United States Geological Survey, US Government Printing Press, Washington DC, 2007, and T. D. Kelly, G. R. Matos, D. A. Buckingham, C. A. DiFrancesco, K. E. Porter, in US Geological Survey Data Series 140, Version 3.0 Online Only USGS Information Service, Denver, Colorado, 2009.
25. T. Rappold and K. S. Lackner, *Disposal of Waste Sulfur at Large Scales: Sequestration in Sulfates and the Fuel Utilization of Hydrogen Sulfide*, submitted.
26. T. E. Reilly, K. F. Dennehy, W. M. Alley and W. L. Cunningham, *Ground-Water Availability in the United States: U.S. Geological Survey Circular 1323*, 2008.
27. K. S. Lackner, C. H. Wendt, D. P. Butt, E. L. Joyce and D. H. Sharp, *Energy*, 1995, **20**, 1153–1170.
28. W. S. Broecker, *Science*, 2007, **315**(5817), 1371.

29. 60 ppm would be equivalent to 240 Pg of C, or 30 years of current consumption.

30. J. Hansen, M. Sato, P. Kharecha, D. Beerling, R. Berner, V. Masson-Delmotte, M. Pagani, M. Raymo, D. L. Royer and J. C. Zachos, *Open Atmos. Sci. J.*, **2**, 217–231.

31. T. J. Blasing, Current Greenhouse Gas Concentrations, CDIAC Carbon Dioxide Information and Analysis Center Oak Ridge National Laboratory, HYPERLINK "http://cdiac.ornl.gov/"http://cdiac.ornl.gov/ (last visited July 8, 2009).

32. See the entire issue of *Trans. Royal Soc.*, which is prefaced by B. Launder and J. M. T. Thomson, *Philos. Trans. A: Math., Phys. Eng. Sci.*, 2008, **366**(1882), 3841–3842.

33. D. W. Keith, *Annu. Rev. Energy Environ.*, 2000, **25**(1), 245.

34. P. J. Rasch, S. Tilmes, R. P. Turco, A. Robock, L. Oman, C.-C. Chen, G. L. Stenchikov and R. R. Garcia, *Philos. Trans. A: Math., Phys. Eng. Sci.*, 2008, **366**(1882), 4007–4037.

35. M. I. Budyko, *Izmeniya Klimata, Gidrometeoizdat*, Leningrad, U. S. S. R., 1974. (Translated: *Climatic Changes*, Am. Geophys. Union, Washington, D. C., 1977).

36. P. J. Crutzen, *Climatic Change*, 2006, **77**(3), 211–220.

37. E. Teller, L. Wood and R. Hyde, *Global Warming and Ice Ages: I. Prospects for Physics-Based Modulation of Global Change*, Lawrence Livermore National Laboratory, Livermore, CA, August 15, 1997.

38. J. Lehmann, J. Gaunt and M. Rondon, *Mitigat. Adaptat. Strategies Global Change*, 2006, **11**(2), 395–419.

39. V. Brovkin, V. Petoukhov, M. Claussen, E. Bauer, D. Archer and C. Jaeger, *Climatic Change*, 2009, **92**(3–4), 243–259.

40. W. A. Hermann, *Energy*, 2006, **31**(12), 1685–1702.

41. *Nature*, 2007, **448**(7154), 653–653, quoting from Nature, 1907, **76**(1971), 373.

42. G. E. Christianson, *Greenhouse: The 200-Year Story of Global Warming*, Walker and Company, New York, 1999.

43. H. von Storch, M. Costa-Cabral, C. Hagner, F. Feser, J. Pacyna, E. Pacyna and S. Kolb, *Sci. Total Environ.*, 2003, **311**(1–3), 151–176.

44. M. Macauley, K. Palmer and J. -S. Shih, *J. Environ. Manag.*, 2003, **68**(1), 13–22.

45. C. Garrett, L. R. M. Maas, *Oceanus (United States)*, 1993, **36**(1), 27–36.

46. W. Munk and C. Wunsch, *Deep Sea Res. Part I: Oceanogr. Res. Papers*, 1998, **45**(12), 1977–2010.

47. A. Clément, P. McCullen, A. Falcão, A. Fiorentino, F. Gardner, K. Hammarlund, G. Lemonis, T. Lewis, K. Nielsen, S. Petroncini, M. T. Pontes, P. Schild, B. -O. Sjöström, H. C. Sørensen and T. Thorpe, *Renewable Sustainable Energy Rev.*, 2002, **6**(5), 405–431.

48. H. E. Krogstad and S. F. Barstow, *Coastal Eng.*, 1999, **37**(3–4), 283–307.

49. Other references, e.g. Falnes Lovseth, suggest a total on the order of 1 TW. J. Falnes and J. Lovseth, *Energy Policy*, 1991, **19**(8), 768–775. See also Hermann (ref. 40 above).

50. E. G. Leigh, R. T. Paine, J. F. Quinn and T. H. Suchanek, *Proc. Natl. Acad. Sci., U. S. A.*, 1987, **84**, 1314–1318.

51. K. Wyrtki, L. Magaard and J. Hager, *J. Geophys. Res.*, 1976, **81**(15), 2641–2646.

52. D. B. Olson, *Annu. Rev. Earth Planetary Sci.*, 1991, **19**(1), 283–311.

53. *The Energy Sourcebook*, ed. R. Howes and A. Fainberg, American Institute of Physics, New York, 1991.

54. C. L. Archer and M. Z. Jacobson, *J. Geophys. Res.*, 2005, **110**, D12110.

55. J. Wallace, in *Encyclopedia of Atmospheric Sciences*, ed. J. R. Holton, J. A. Curry and J. A. Pyle, Elsevier, 2003, **vol. 1–6**.

56. D. W. Keith, J. F. DeCarolis, D. C. Denkenberger, D. H. Lenschow, S. L. Malyshev, S. Pacala and P. J. Rasch, *Proc. Natl. Acad. Sci., U. S. A.*, 2004, **101**(46), 16115–16120.

57. J. Grace, *J. Ecol.*, 2004, **92**(2), 189–202.

58. P. J. Crutzen, A. R. Mosier and K. A. Smith, *Atmos. Chem. Phys.*, 2008, **8**(2), 389–395.

59. J. Lehmann, J. Gaunt and M. Rondon, *Mitigat. Adaptat. Strategies Global Change*, 2006, **11**(2), 395–419.

60. J. W. Tester, B. J. Anderson, A. S. Batchelor, D. D. Blackwell, R. DiPippo, E. M. Drake, J. Garnish, B. Livesay, M. C. Moore, K. Nichols, S. Petty, M. N. Toksöz and R. W. Veatch, *The Future of Geothermal Energy*, Massachusetts Institute of Technology, Cambridge, MA, 2006.

61. R. Pelc and R. M. Fujita, *Marine Policy*, 2002, **26**(6), 471–479.

62. S. Ansolabehere, J. Deutch, M. Driscoll, P. E. Gray, J. P. Holdren, P. L. Joskow, R. K. Lester, E. J. Moniz and N. E. Todreas, *The Future of Nuclear Power*, Massachusetts Institute of Technology, Cambridge, MA, 2003.

63. *Forty Years of Uranium Resources, Production and Demand in Perspective, "The Red Book Retrospective"*, Nuclear Energy Agency, Organisation for Economic Co-operation and Development, 2006.

64. S. Jacobsson and V. Lauber, *Energy Policy*, 2006, **34**(3), 256–276.

65. V. M. Fthenakis, H. C. Kim and E. Alsema, *Environ. Sci. Technol.*, 2008, **42**(6), 2168–2174.

66. M. K. Hubbert, *Nuclear Energy and the Fossil Fuels*, Publication No. 95, Shell Development Company: Houston, TX, June, 1956.

67. S. Pollard, *Econom. History Rev.*, 1980, **33**(2), 212–235.

68. UK Department for Business Enterprise & Regulatory Reform, Historical coal data: coal production, availability and consumption 1853 to 2007, http://www.berr.gov.uk/files/file40592.xls (July 6, 2009),.

69. C. L. Brierley, F. P. Burke, J. C. Cobb, R. B. Finkelman, W. Fulkerson, H. J. Gluskoter, M. E. Karmis, K. S. Lackner, R. E. Mitchell, R. V. Ramani, J. -M. M. Rendu, E. S. Rubin and S. A. Wolfe, *Coal: Research*

and Development to Support National Energy Policy, National Research Council of the National Academies, Washington DC, 2007.

70. K. S. Deffeyes, *Hubbert's Peak: The Impending World Oil Shortage,* Princeton University Press, Princeton, NJ, 2001.
71. D. Rutledge, *Eos Trans. AGU,* **89**(53), Fall Meet. Suppl. Abstract U42A-04, 2008.
72. H.-H. Rogner, *Ann. Rev. Energy Environ.,* 1997, **22**, 217–262.
73. *Carbon Sequestration Atlas of the United States and Canada,* National Energy Technology Laboratory, US Department of Energy, Albany, OR, Fairbanks, AK, Morgantown, WV, Pittsburgh, PA, Tulsa, OK, 2007.
74. B. Metz, O. Davidson, H. D. Coninck, M. Loos and L. Meye, *Mineral Carbonation and Industrial Uses of Carbon Dioxide, IPCC Special Report on Carbon Dioxide Capture and Storage,* Cambridge University Press, New York, 2005.
75. K. S. Lackner and S. Brennan, *Climatic Change,* 2009, **96**(3), 357–378.
76. K. S. Lackner, *Science,* 2003, **300**(5626), 1677–1678.
77. T. M. Yegulalp, K. S. Lackner and H.-J. Ziock, *Int. J. Surface Mining, Reclamat. Environ.,* 2001, **15**(1), 52–68.
78. J. B. Tepe and B. F. Dodge, *Trans. Am. Inst. Chem. Eng.,* 1943, **39**, 255.
79. G. Astarita, *Mass Transfer with Chemical Reactions,* Elsevier, Amsterdam, London, New York, 1967.
80. K. S. Lackner, P. Grimes and H. J. Ziock, in *Proceedings of the 24th Annual Technical Conference on Coal Utilization and Fuel Systems,* 1999, 885–896.
81. F. S. Zeman and K. S. Lackner, *World Resource Rev.,* 2004, **16**(2), 157–172.
82. J. K. Stolaroff, D. W. Keith and G. V. Lowry, *Environ. Sci. Technol.,* 2008, **42**(8), 2728–2735.
83. K. S. Lackner, in *Energy Supply and Climate Change,* ed. W. Blum, M. Keilhacker, U. Platt and W. Roether, *Eur. Phys. J. Special Topics,* **176**, 2009, 93–106.
84. K. S. Lackner, P. Grimes and H.-J. Ziock, *Carbon Dioxide Extraction from Air?* Los Alamos National Laboratory, LAUR-99-5113, Los Alamos, NM, 1999.
85. Geo-engineering Research, in *Postnote,* Parliamentary Office of Science and Technology, London, 2009.
86. A. J. Traynor and R. J. Jensen, *Ind. Eng. Chem. Res.,* 2002, **41**(8), 1935–1939.
87. S. Pacala and R. Socolow, *Science,* 2004, **305**(5686), 968–972.
88. Spiegel & Wikipedia, Spiegel Wissen, das Lexikon der nächsten Generation: References on German Power Plant Construction. http://wissen. spiegel.de/wissen/dokument/dokumentdruck.html?id = Liste_geplanter_ Kohlekraftwerke&top = Wikipedia (last visited July 10, 2009).

Fossil Power Generation with Carbon Capture and Storage (CCS): Policy Development for Technology Deployment

JON GIBBINS* AND HANNAH CHALMERS

1 Introduction

In recent years, there has been growing concern that carbon dioxide (and other greenhouse gas) emissions from fossil fuel use could cause dangerous climate change, with serious negative impacts on human activities.[1,2] In this context, it is expected that the net CO_2 emissions from fossil fuels must be significantly reduced within the next 10–20 years and ultimately remain at or near zero, probably indefinitely.[3] Three options can be identified to achieve this goal:

- Leave fossil fuels in the ground; and/or
- Use the fossil fuels, but capture and safely store the CO_2 produced; and/or
- Use the fossil fuels, but 'offset' the CO_2 released by removing CO_2 from the atmosphere elsewhere.

Although a number of approaches to use alternative energy sources are being developed to replace fossil fuels, many energy system studies, including the 2008 International Energy Agency World Energy Outlook,[4] suggest that fossil fuels will retain at least some of their, currently significant, role in providing energy for human activities for several decades or more. The CO_2 emissions associated with this use of fossil fuels would be unacceptable in the context of the current scientific consensus for limiting further CO_2 additions to the atmosphere, in order to have a reasonable chance of avoiding some of the worst

*Corresponding author

Issues in Environmental Science and Technology, 29
Carbon Capture: Sequestration and Storage
Edited by R.E. Hester and R.M. Harrison
© Royal Society of Chemistry 2010
Published by the Royal Society of Chemistry, www.rsc.org

potential impacts of dangerous climate change. It is, therefore, likely to be necessary to develop and successfully deploy technologies that allow the CO_2 emissions associated with continued use of fossil fuels to be significantly reduced. The alternative is an unacceptably risky increase in the stock of CO_2 in the atmosphere.

Carbon capture and storage (CCS) involves the removal of CO_2 from other streams produced at a power plant (or other large source of CO_2, *e.g.* steel or cement works), followed by transport of the captured CO_2 to safe long-term storage, *e.g.* in a deep geological formation. A number of detailed reviews of CCS technology are available in the literature, including a 2005 special report by the Intergovernmental Panel on Climate Change.[5] Current cost estimates do not give a clear indication of a 'winning' technology for CO_2 capture for use in the most likely applications, since the technology options that are sufficiently close to deployment to assess appear to have similar costs, within likely uncertainty ranges.[6] It is therefore possible that site-specific factors (*e.g.* coal type, water requirements, electricity market features) may determine the capture technology that would be used in many cases. A brief introduction to the technologies for CO_2 capture closest to commercial deployment at the time of writing is provided in Appendix A, and CO_2 storage is discussed in more detail in Chapter 6 of this book.[7]

This chapter will focus on a range of policy options that could be used to support deployment of power generation technology options that would allow CO_2 to be successfully captured and stored. The power sector is chosen as a case study since CO_2 emissions from power plants made up 41% of the global total in 2006[4] and many studies, including the first report of the UK Committee on Climate Change,[3] suggest that deep decarbonisation of overall energy supplies could be achieved by widespread use of electricity following a rapid reduction in CO_2 emissions from electricity generation.

Although much of the literature and public debate on CCS is currently dominated by coal, CO_2 capture can be (and probably, eventually, will have to be) applied to any plant producing CO_2. The scope of this chapter, therefore, also includes one 'offset' option: the combined use of biomass and CCS. As biomass grows it removes CO_2 from the atmosphere. If this CO_2 is not re-released to the atmosphere when the biomass is used, then it is possible for net negative emissions to be obtained, although the overall net benefit depends partly on the greenhouse gas emissions associated with the whole biomass lifecycle, including production and transport to the point of use.

Since it is expected that costs for CCS will normally be dominated by capture costs, this chapter focuses on improving the economic case for installing and operating CO_2 capture at power plants. It will, of course, also be necessary for all aspects of a CCS project to be economically viable, if operated by separate entities, and to be regulated effectively. Useful reviews of the regulations required for complete CCS projects can be found in the literature.[8,9] Although further significant work is required to complete the implementation of appropriate regulatory frameworks, some general principles are beginning to emerge and enter into law in some jurisdictions. For example, the use of an approach to

CO_2 storage site operation, closure and transfer of liability to the state that is based on an ongoing assessment of site-specific risks, is likely to be crucial for developing a reasonable operating environment that is acceptable to both regulators and investors. A detailed review of regulatory issues is beyond the scope of this chapter, however, partly due to rapid ongoing developments in this area at the time of writing.

2 Reasons for Incentivising CCS Capture Projects

Implementing CCS will generally increase the costs of power generation, since additional equipment is required, in addition to energy for operation. Some regulators expect that, in the medium to longer term, CCS projects will be supported by carbon price alone, so driven entirely, for example, by the EU Emissions Trading Scheme.[10] It is likely, however, that most CCS options will require additional support for initial integrated, commercial-scale demonstration and deployment, as illustrated in Figure 1. It is also worth noting that while other mechanisms, currently typically to support renewables, are in place they can be expected to significantly distort a carbon market and make it more difficult for prices to rise to a level where a range of technologies can compete effectively.

It should, of course, be noted that the schematic illustration of Figure 1 omits many likely additional features in the actual future development of both CCS costs and CO_2 prices. For example, some volatility in CO_2 prices can be expected, although a general upward trend is likely to occur if serious global (or local) action to reduce greenhouse gas emissions has been agreed and is successfully implemented. Also CCS costs are, in practice, likely to vary across a

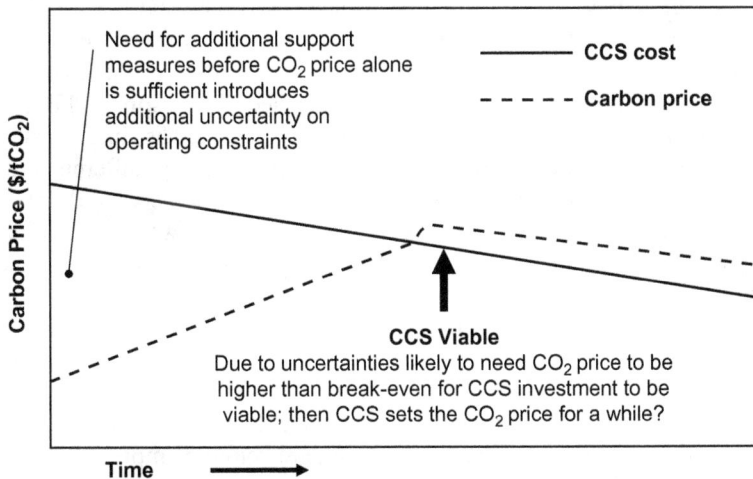

Figure 1 Schematic diagram of CCS cost and general CO_2 price development over time.

potentially wide range due to general price fluctuations (in, for example, steel or labour), site-specific variation and differences between technologies. It is also possible that estimated costs may increase before they reduce, due to the 'appraisal optimum' that can occur in theoretical studies undertaken (as now) before any commercial-scale plants are deployed.[11]

As already noted, one aspect of Figure 1 that is likely to be reflected in any real development of CCS is an initial period where support that is additional to any general carbon price is required to allow projects to be economically viable. Before discussing the potential form of such incentives, it is important to first understand the context in which these measures may be applied. This section will, therefore, outline an ideal model for commercial-scale CCS development, consisting of two tranches of integrated CCS projects before large scale roll-out. Some differences between power plant CCS projects and other CCS applications will also be discussed, using a classification system based on their respective expected CO_2 mitigation potential.

2.1   Tranches Model for Commercial-Scale Development and Deployment

At the time of writing, there is very limited international experience of integrated commercial-scale (*ca.* 1 Mt CO_2 yr^{-1} and above) operation of CCS, and no projects at this scale that use CO_2 generated by a power plant. The very early commercial-scale CCS projects that are enabled by suitable support will therefore have a different role for technology innovation and learning in general than later projects. Figure 2, based on previous work by the authors,[12] illustrates this point by separating initial integrated CCS projects undertaken prior to general rollout into two tranches.

Although it is expected that the first tranche of power plants with CCS will work, it is also very likely that a number of technology-related 'teething' problems with designs that are yet to be deployed at this scale in this environment will be identified. In addition, it can also be expected that lessons will be learned in project management and design/operating approaches as, in at least some cases, new groupings of original equipment manufacturers, different industry sectors (particularly utilities and oil and gas), project financers and regulators work together to deliver complete CCS projects. There would, therefore, be relatively high levels of risk associated with moving directly from this first tranche of plants to full scale global rollout, and almost certainly sub-optimal results from a rollout based only on such a limited build programme.

Instead, it is likely that a second tranche of plants that also receive additional support for initial deployment of commercial-scale CO_2 capture at power plants (and other large point sources of CO_2) would be a valuable, and possibly essential, interim step. As well as trialling some technology improvements based on learning from the first tranche, this second tranche of plants could be expected to resolve most, if not all, of the initial problems identified in the first tranche of plants. They should, therefore, provide a suitable basis for proving commercial

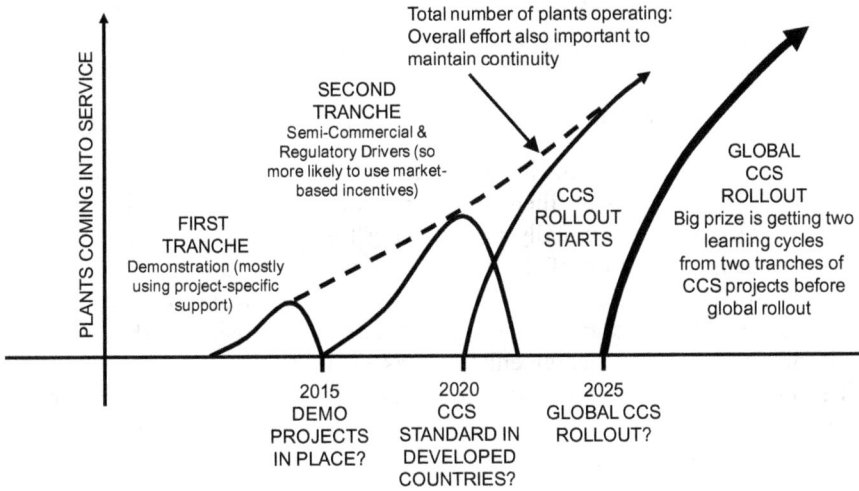

Figure 2 Schematic diagram of tranches for commercial-scale integrated CCS deployment (after Gibbins and Chalmers, 2008; ref. 12).

guarantees and developing the first reference plant designs, as examples of plants that went into service and performed as expected after 'normal' commissioning periods. It seems very likely that politicians and investors would be unwilling to rely on CCS as a significant potential contributor to global CO_2 emissions cuts until such successful second tranche projects are in evidence.

The time required for two tranches of initial commercial-scale deployment to be undertaken can be expected to vary between different CO_2 capture options. For example, it is possible that post-combustion capture could be developed rapidly using retrofits to existing plants.[13] This could be valuable since, provided that at least this widely-applicable CCS option has been proved sufficiently well to justify policy-maker reliance on CCS, then a more ambitious framework for developing and deploying all CCS options is justified as a viable route to pursue.

Retrofitting of CO_2 capture to the existing fleet could also be important during the CCS roll-out phase, as well as for earlier demonstrations, since overall electricity sector emissions (in tonnes $CO_2 \, yr^{-1}$) cannot be reduced unless existing fossil-fired plants are either retrofitted with CCS or are closed down, possibly prematurely. Recent work suggests that, not surprisingly, retrofit of CO_2 capture could be more economically effective than shutting down existing plants with reasonable life expectancy and replacing them with entirely new plants.[14,15]

Appropriate use of retrofits is also likely to significantly reduce the construction activity required during rapid decarbonisation of the electricity network, since far fewer new base power plants have to be built in addition to capture plants and storage system infrastructure. For plants built without capture in the interim period before full CCS rollout, it should be possible to design power stations so that they are well suited for retrofit, *i.e.* 'carbon capture ready' (CCR), with the essential requirements expected to increase the total capital cost of the plant by no more than 1% (ref. 16,17). The potentially important role of retrofits and CCR is discussed further in Section 5 of this chapter.

2.2   Classes of Climate Change Mitigation Benefit with CCS

If CCS is being supported for its expected contribution to mitigating the risk of dangerous climate change, then it is important to consider whether all CCS projects will contribute equally to achieving this goal. In particular, if near-complete decarbonisation of the energy sector is required to limit cumulative emissions of CO_2 to the atmosphere and hence the risk of dangerous climate change,[3] it will be necessary to develop methods to use fossil fuels that emit very little or no CO_2 (or other greenhouse gases). In this context it is useful to differentiate between CCS projects, by the extent to which they add (or remove) CO_2 produced by fossil fuel use to the atmosphere, rather than making a tacit assumption that all CCS will achieve the same climate benefit:

- Class 1 (carbon positive) – Projects producing carbon-containing fuels which are likely to be used without CCS and hence result in the release of a significant amount of the original fossil carbon to the atmosphere as CO_2. These residual emissions are too high to be acceptable if large CO_2 emission reductions are expected to be required.
- Class 2 (near-carbon neutral) – Projects producing carbon-free energy vectors (*i.e.* electricity, heat, hydrogen) and hence able to achieve low residual emissions.
- Class 3 (carbon negative) – Class 3B CCS projects using suitable biomass fuels and with a suitable product mix to give a net removal of CO_2 (absorbed by the biomass) from the atmosphere. Some options for direct removal of CO_2 from the air are also under development (*e.g.* Mahmoudkhani *et al.*, 2009)[18] and can be classified as class 3A projects.

The ultimate fate of CO_2 produced by a CO_2 capture unit should also be considered in determining the classification of a project. For example, some projects may use CO_2 for enhanced oil recovery (EOR). In these cases, the net CO_2 captured will depend on the operating approach of the EOR project. The counterfactual option will also need to be carefully considered. For example, if EOR oil was going to be produced anyway (or will be used with CCS), then the change in climate benefit for the CCS project can be argued to be different to a project where the availability of CO_2 for EOR leads to oil production that would not otherwise have occurred. The key criterion is whether the sum of all cumulative CO_2 emissions to atmosphere is reduced or not.

Since different classes of CCS project are able to make different contributions to mitigating the risk of climate change, regulators should consider whether these different classes should have different rules and incentives applied. For example, it may be decided that Class 1 projects should be required to use CCS with no additional incentive support provided. In contrast, it may be appropriate to provide special incentives for Class 3 projects that are able effectively to remove

CO_2 from the atmosphere, since this offset may be very valuable if the stock of CO_2 in the atmosphere becomes too high and also as a potential offset for high value activities that are difficult to decarbonise (*e.g.* air travel).

It should also be noted that many Class 1 projects (*e.g.* coal-to-liquids and similar plants) may have relatively low $ per tonne CO_2-abatement costs, since CO_2 is often separated from other streams within the plant as part of the normal processes required to produce desired products. Although it is sometimes argued that supporting these low cost projects is a good way to kick-start the CCS industry, there are likely to be some significant limits in what can be learned from Class 1 projects for plants that are producing electricity (and some other carbon-free energy vectors) when significant differences in the CO_2 capture technologies are involved. The main benefit of Class 1 projects could be in supporting the establishment of a CO_2 transport network, if plants are in suitable locations, and proving CO_2 storage sites that could then be used for other CCS projects too.

3 Features of Effective Incentives for Power Plants with CCS

The cost penalty for CCS, compared to plants without CCS, can typically be broken down into the following elements, which effective incentives must seek to cover:

- Reduction in the electricity output of the a plant (for the same fuel input) due to the additional energy required to operate CO_2 capture and compression equipment.
- Extra capital and operating expenditure.
- Increased risk due to novel technology not yet proven in this application and additional regulatory risks.

As discussed above, CCS could be commercially viable in the longer term if legal and regulatory systems are put in place that make it possible for CCS project developers to recover their costs as part of global (or local) action to mitigate the risk of dangerous climate change. Such measures could include a 'cap and trade' scheme, or a carbon tax which introduces a financial penalty for emitting CO_2, or a performance standard limiting CO_2 emissions associated with fossil fuel use. It is likely that there will be a funding gap between such measures and costs for early commercial-scale CCS projects, requiring a supplementary payment be made if CCS is to be made available for widespread rollout from around 2020. Equation (1) presents an ideal model for such a payment as a supplement to the EU Emission Trading Scheme, based on CO_2 captured and stored as a direct measure of plant performance that is also consistent with the EU Emissions Trading Scheme. Figure 3 contains an illustrative example of how support levels might vary under such a supplementary payment scheme over the first 15 years of operation of a coal-fired power plant fitted with CCS.

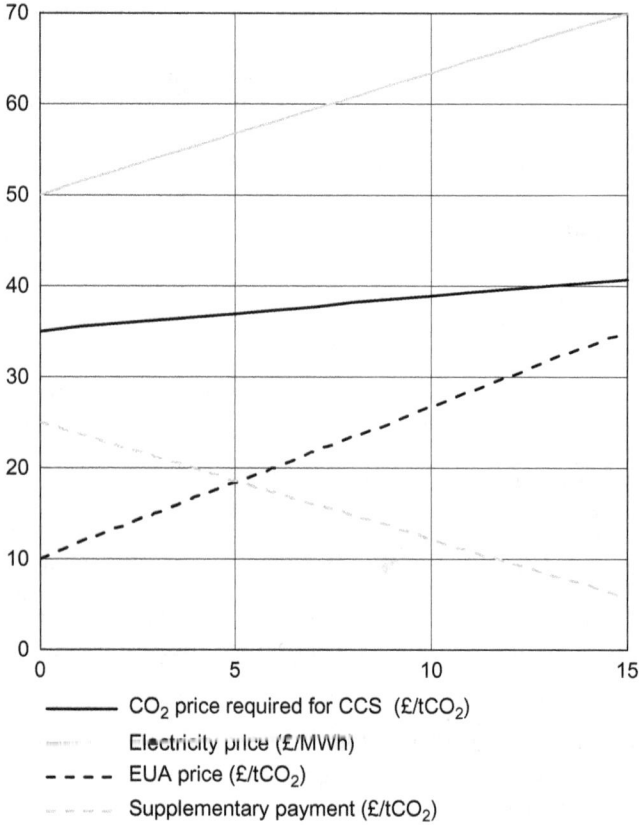

Figure 3 Illustrative example for supplementary payment required to support CCS on a coal-fired power plant as electricity and carbon prices vary (fuel specific emissions 350 kg CO_2 per MWh, capture level 90%, capture energy penalty 9 percentage points – results are independent of power plant efficiency without capture).

$$\text{Supplementary payment} = \left[\frac{\delta.POE}{f.c} + \frac{fcc}{N} + vcc \right] - \text{EUA price} \qquad (1)$$

POE	£/MWh	Time averaged price for electricity
N	hr	Annual operating hours
f	tCO_2/MWh	Specific emissions factor for heat from the fuel
δ	% age points LHV	Capture penalty
c	–	Fraction of CO_2 captured
fcc	£/(tCO_2/hr captured)	Specific fixed capital charges for CCS
vcc	£/(tCO_2 captured)	Specific variable costs for CCS (inc. storage charges)

A more detailed discussion of this type of approach to determining required support levels for early CCS projects, and another illustrative example, can be found in previous work by the authors.[19] One key feature of this type of supplementary payment formula is that additional support reduces as the cost of carbon emissions increases, due to other broader measures to reduce CO_2 emissions, such as the EU Emissions Trading Scheme. This should allow a smooth transition at the end of the initial support period. Supplementary payments for CCS projects including power plants should also increase or decrease in line with electricity prices, to cover the varying cost of the lost electricity output. Taken together these measures would go a long way to cover the additional non-technical risks which early CCS projects would face, while at the same time avoiding the prospect of government support levels being higher than necessary over the life of a project.

A consideration of fair distribution of the costs (and benefits) of demonstration and deployment of CCS, also suggests that the involvement of a range of different fossil fuel users is justified and this should be encouraged by any regulations or incentives. Some of the expected outcomes of the first two tranches of deployment of CCS are likely to make a contribution to CCS from a much broader range of projects, and 'first movers' are unlikely to be able to be the only beneficiaries of the costs and risks they would incur if they financed their projects independently. For example, infrastructure development, regulatory regime and technology learning are all likely to be transferable to some extent. As outlined in the next section, it seems likely that tradable compliance using a sectoral CCS standard could be well suited to this requirement and, importantly, should also allow for flexible operation of CCS schemes.

By spreading costs and risks (and placing risks with those best placed to manage them) during commercial-scale integrated demonstration of CCS in the initial tranches of deployment, it is possible that the total costs to society could be decreased, since overall support requirements could be reduced for the same level of experience gained and CO_2 captured and stored. Another important feature of a mechanism that facilitates a sector-wide approach to CCS demonstration and deployment, is that schemes which are based on optimum capture levels (probably around 90% capture capacity from treated flue gas for initial commercial applications and then higher later) may be more likely to be implemented. Without multiple funding sources it is more likely that capital constraints for individual project development may lead to smaller projects, possibly with suboptimal capture levels or at smaller scales, and hence higher costs, than is necessary from a technical perspective. In the worst case it is possible that plant operators could fit a partial capture process which made it difficult or impossible to increase capture levels later, although this should be avoided if plants are also designed according to 'carbon capture ready' principles so as to facilitate subsequent retrofit of additional CO_2 capture.[16]

Another important feature of an incentive mechanism to support rapid development of CCS is likely to be its ability to complement other measures, so that it is able to provide a bridge between initial demonstration projects and 'business as usual' commercial deployment in the context of global action to achieve significant reductions in CO_2 emissions. Thus, clarity in how an incentive mechanism will encourage developments between these starting and ending points will be required. It is also likely to be useful to avoid being overly specific about which projects are chosen for support.

There may, however, be a case for 'banding' support levels to ensure that a variety of technology options are able to proceed, since different CCS technologies are at different stages of development, and there could be benefits associated with giving support across this range using the same basic mechanism. This approach could follow principles established in the UK Renewables Obligation to allow a single mechanism to provide differentiated support for different approaches for generating electricity from renewable sources.[20]

Finally, as already noted, some clear differentiation in support level is likely to be important between different classes of CCS, not just because they have different climate benefits, but because Class 1 CCS, capturing CO_2 in conjunction with hydrocarbon production, is likely to have much lower unit costs than carbon-neutral Class 2 CCS (and carbon-negative Class 3 projects). In addition, Class 1 projects are unlikely to demonstrate all of the necessary features that will be required for Class 2 and Class 3 projects. Without differentiation, market forces would probably tend to produce a much greater preponderance of Class 1 CCS projects for early large-scale demonstration projects, and these projects might also receive a higher level of support than is necessary.

4   Example CCS Incentives for the Electricity Sector

As discussed previously, the examples used in this section are focussing on incentives that would supplement generally-applicable carbon pricing established by an emission trading scheme or carbon tax, in order to accelerate the deployment of Class 2 and Class 3B CO_2 capture at power plants. In particular, this section considers site and project-specific funding options that may be used for first tranche plants, likely implications of relying on electricity emissions performance standards to incentivise CCS and the potential for a sectoral CCS standard to complement other measures.

4.1   Site and Project-Specific Funding Options for First Tranche Plants

Since a price for carbon (either through a carbon tax or 'cap and trade' scheme) is not likely to be sufficient to get electricity CCS projects started before perhaps 2020, additional incentives are needed to overcome the inherent inability of current carbon markets to compensate early movers. Providing additional funding to initial projects through a technology-specific, and possibly project-specific, mechanism seems to be becoming accepted in the context of broader electricity/energy policy in many jurisdictions. For example, current evidence suggests that both the additional costs and the climate change mitigation benefits of electricity CCS projects could be roughly comparable to those for generation from renewable sources,[21] although of course this depends to a great extent on assumptions made about fossil fuel price levels.

In the shorter term, as a measure to overcome inherent market failures associated with the early introduction of new technologies, in many jurisdictions special support measures (*e.g.* portfolio standards, feed in tariffs, *etc.*) have been introduced for renewables, and it, therefore, seems reasonable to assume that analogous support could and should be made available for some CCS initial projects too. In the longer term it seems likely that progress towards a near-zero emission energy system could be accompanied by a reduction in fossil fuel prices. Since costs of fossil fuel production are generally significantly lower than typical recent selling prices, it can be expected that fossil fuel prices will reduce towards production costs as far is necessary to allow them to compete with non-fossil energy, including with the costs of CCS and/or carbon pricing taken into account.[22]

Support for a first tranche of CCS projects and confirmation of a rollout date for CCS was included in the Joint Statement by G8 Energy Ministers following the Aomori Summit in Japan on 8 June 2008,[23] as shown below:

"We strongly support the recommendation that 20 large-scale CCS demonstration projects need to be launched globally by 2010, taking into account varying national circumstances with a view to supporting technology development and cost reduction for the beginning of broad deployment of CCS by 2020."

More recently, Australia has established the Global Carbon Capture and Storage Initiative (GCCSI)[24] with funding of up to $100 million annually *"to act as a catalyst for accelerating projects to deliver the G8's goal through facilitating demonstration projects, and identifying and supporting necessary research."* (ref. 25). At the time of writing, a limited number of large scale electricity CCS projects have been announced in the public domain and reached an advanced enough stage for them to be potential contributors to the possible first tranche of plants supported by the G8 Energy Ministers. A number of different mechanisms have driven the development of these projects, generally including:

- Direct government support; and/or
- In suitable markets, payment for CO_2 for enhanced oil recovery (EOR), possibly plus additional incentives for increased oil production; and/or
- Funding by groups of stakeholders (usually for relatively small projects); and/or
- Funding by regulated or publicly owned utilities for projects that would be allowed to pass the costs of CCS on to their ratebase.

One key requirement for funding mechanisms for first tranche plants is that they are able to deliver finance rapidly, so that projects are able to progress as quickly as possible. In this context, the inclusion of CCS in stimulus packages that require rapid spending of available funds in response to global economic concerns could make a valuable contribution to timely delivery of first tranche plants. Of course, the economic downturn that has led to stimulus packages has also endangered many of the other routes identified above. For example, reductions in oil price in response to reduced demand make EOR-based investments less attractive, and concerns over energy prices may reduce the willingness of regulators to allow any 'unnecessary' increase in rates, however modest, through the inclusion of CCS costs. Carbon prices in the EU Emissions Trading Scheme have also fallen sharply. On the up side, however, steel and construction prices have fallen (although at the time of writing these have not stabilised) and lower fossil fuel costs also lead to reduced running costs for fossil-fired power plants, making CCS power much more cost-competitive with non-fossil sources.

4.2   Electricity Emissions Performance Standards (EPSs)

It is relevant to discuss electricity emissions performance standards since some commentators have suggested that they could be used as an incentive for CCS.[26] Electricity emissions performance standards (EPS) set a limit on the CO_2 emission allowed per unit of electricity produced. At the time of writing, they are topical because such a standard was introduced in California[27] in order to prevent utilities entering into long-term contracts for coal-generated electricity imported from neighbouring states. The so called 'Schwarzenegger Clause' was set at $1100 \, lb \, CO_2 \, (MWh)^{-1}$ ($500 \, kg \, CO_2 \, (MWh)^{-1}$), a value selected so as not to preclude any reasonably-efficient unabated natural gas plant.[28] Subsequently a similar standard was introduced in Maine[29] and Washington[30] and proposed in Europe for new power plants.[31] European policy-makers have considered an EPS reduced to $350 \, kg \, CO_2 \, (MWh)^{-1}$, while a new plant standard was also proposed in the March 2009 draft of the 'American Clean Energy and Security Act of 2009',[32] initially also at $1100 \, lb \, CO_2 \, (MWh)^{-1}$, and falling to $800 \, lb \, CO_2 \, MWh^{-1}$ ($\approx 360 \, kg \, CO_2 \, (MWh)^{-1}$) in 2020.

It is worth reflecting, however, that while such standards are likely to prevent developers building and operating new coal plants without CCS, particularly if it is assumed that compliance must be over relatively short time periods (*e.g.* every hour) and cannot be traded with other plants, they can do little to make CCS happen. If CCS were considered economic (*e.g.* in response to a general CO_2 tax or to market-wide CO_2 prices from a 'cap and trade' scheme) it would happen without such a performance standard. If CCS is not considered economic, then it seems likely that developers will choose not to build any fossil-fired plants at all, if the only change in the regulatory/incentive regime is that CCS is required by a performance standard if a new plant is built. The case for any investment in a fossil-fired power plant has been made more difficult, but without any financial support to facilitate the required addition of CCS.

Two important considerations in determining whether an EPS could be an effective mechanism to support deployment of CCS are the scope of coverage and whether new plants must be built at all. On the latter point, much of the current debate seems to assume implicitly that the main uncertainty is what type of power plants will be built, rather than whether plants will be built at all.

Especially in competitive markets is it important to note, however, that there is no requirement for electric utilities to build new plants. It seems likely, therefore, that in many jurisdictions the only way that such a performance standard alone might make CCS economic, is if the EPS was extremely widely applied and effectively led to operation with CCS becoming the only way in which coal (and, eventually, all fossil fuels) could be used for power generation, including on existing plants.

As discussed above, if an EPS was applied in a jurisdiction where power plants with CCS were constructed and operated, then it might be possible that coal prices would drop low enough for coal plus (possibly partial) CCS to compete successfully with natural gas-fired power plants (probably without CCS, at least for the first few years of operation of coal-fired power plants with CCS). Following the introduction of the EU Emissions Trading Scheme, some arbitrage between coal, gas and carbon prices already seems to be the norm in Europe. It should be noted, however, that in many locations, coal prices are now set by imports and, hence, prices are less elastic than in markets relying on domestic coal alone, since a number of different prospective users compete to buy the same coal. It therefore seems likely that all countries that compete with the UK for coal in the internationally-traded coal market, probably including China, would need to have a similar EPS in place before coal prices would be expected to be sufficiently low, for coal with CCS to be competitive with other power generation options available in the UK.

In markets such as the USA, using almost entirely indigenous coal and with limited natural gas to use as an alternative, it might be considered that a performance standard covering the whole electricity sector would be more likely to be successful in delivering CCS through incorporation of CCS costs in coal pricing. The same wide reliance on coal would, however, make the introduction of such a standard impractical in the short-term. The paradox of relying entirely on an EPS to deliver CCS would be that widespread mandating of CCS is not possible until it is proven. But CCS will probably not be regarded as proven until two smaller tranches of large-scale plant development have been undertaken, as discussed above and in previous work by the authors.[12]

This limitation of an EPS to implement CCS has been recognised in the March 2009 draft of the 'American Clean Energy and Security Act of 2009',[32] which proposes "a payment per ton of carbon dioxide captured and sequestered" over a limited number of years for a series of tranches of projects, with each tranche comprising a certain MW of generating capacity. Projects will be supported on "a first-come, first-served basis". It is specified that the payment per ton of CO_2 will fall with succeeding tranches, a process that would be consistent with learning and also with rising carbon prices. As noted above, such a measure could eventually be expected to taper into support for CCS through market-wide carbon pricing. The level of additional support required would also be lower if the parallel EPS allowed the same flexibility as a carbon market, *i.e.* averaging of emissions over a period of a year or longer, as discussed below. The total proposed amount of such support is, however, not stated, and little is said about the source for the funds required. A logical funding hypothecation would be to say it arises from the auctioning of emission allowances (or a carbon tax). It is also not clear what would happen if the payment per ton of CO_2 offered was insufficient to induce enough CCS to meet the desired rate of development.

Finally, it should be noted that if an EPS is to be used as an option for encouraging CCS, then it is likely to be most effective at encouraging CCS (rather than the choice of a non-CCS alternative) if the emission level of the plant can be averaged over an extended period. For example, the EU Emissions Trading Scheme requires compliance over a period of a year (sufficient allowances must be surrendered only once a year). Also typical allocations of 'caps' to limit emissions have been for a number of years and, in some cases, banking of allowances between phases has also been allowed so that emissions are effectively being averaged over 10 years or longer. From the perspective of climate change, flexibility such as this is sensible since limits to cumulative CO_2 emissions over a number of decades are the relevant approach for mitigating the risk of dangerous climate change.[3] At least during the initial phases of CCS deployment, allowing flexibility in when CO_2 is captured should maximise the volume of CO_2 stored given the inevitable constraints on funding available to cover the additional costs of CCS and other mitigation activities.[33,34]

4.3   A Sectoral CCS Standard

As noted previously, one approach to providing additional support to power plants with CCS until general CO_2 prices (from taxes or markets, for example) are high enough, is to make supplementary payments for CO_2 captured and stored. As well as the proposals made in the March 2009 draft of the 'American Clean Energy and Security Act of 2009',[32] Europeans are considering this approach. The third phase of the EU Emissions Trading Scheme has set aside 300 million allowances to be used to provide additional support to CCS projects (and some innovative renewable energy projects). At the time of writing, the comitology process to determine how allowances should be allocated and distributed is under discussion. One commentary on a potential approach for the European support mechanism from the European Zero Emissions Technology Platform[35] is that:

"The process by which the EUAs are disbursed is not yet clear. We assume that the projects will be selected and a number of EUAs allocated to each. The EUAs of the New Entrant Reserve will then be sold in part or as a whole in the market via the auctioning process by an intermediary to make cash or a series of cash amounts over time. The cash would then be offered to the winning projects. If that money was sufficient for the project to proceed then the investment decision would be made by the organisation leading the project and it would go ahead."

As with the proposed US legislation, some potential problems for direct supplementary payments using this approach are that the total value of support is still uncertain, although the number of allowances available for this phase of support has been agreed, and also there is a lack of clarity on what happens if the payment offered is not sufficient to induce the desired level of CCS deployment.

Both of these uncertainties can, to a large extent, be addressed by specifying a sectoral standard related to CCS. For it to be significantly different from a plant-based emissions performance standard, a sectoral standard has to allow compliance to be shared across the sector and so, by implication, requires that compliance units can be traded. This makes a sectoral standard for CCS more flexible. In principle any CCS standard should be incremented gradually over time to increase its stringency without undue shocks to the market. In practice, however, for an individual plant it is likely to be best in most cases to either have no capture or, once the technology is proven, always to fit a high level of capture. For example, once any level of partial capture is implemented (*e.g.* $500 \, kg \, CO_2 \, (MWh)^{-1}$ is typically about 50% capture on a coal power plant, against a likely maximum of 90% or more) then any significant increase in capture level beyond the operating envelope of the equipment fitted initially is likely to require a major retrofit activity (although technically feasible if plants are designed appropriately).

With a sectoral performance standard, any desired level of stringency can be met by adjusting the number of plants implementing 'full' capture – of course these will also include any new plants that may also be subject to a plant emissions performance standard, although it seems likely these new plants will then operate with capture levels that are higher than those required by the standard alone. In effect, the EPS retains a role as a mechanism to constrain the construction of new fossil-fired plants without CCS, with the addition of a sectoral performance standard (applied to existing and new plants) providing a greater incentive to build new plants with CCS (or retrofit plants in the existing fleet), rather than to continue to operate existing plants that would not have their CO_2 emissions constrained by a new plant EPS. The sectoral performance standard overcomes the perverse incentive to continue operating existing plants unabated rather than building new plants with CCS, that is associated with a new plant EPS alone.

One metric used to set a sectoral CCS standard could be based on an average emissions level per MWh of electricity generated. Then plants achieving lower levels of emissions per MWh would generate a certain tonnage of compliance units to trade with plants that had higher levels of emissions per MWh. Another metric could be an aggregate amount of CO_2 captured and stored from power plant CCS projects, possibly with the required amount being assessed on the basis of the carbon in the fuel used by each power plant – effectively a required average capture level. A uniform emission level standard would fall more heavily on existing, lower efficiency power plants than would a capture level standard, but the former standard might well be adjusted by plant type in practice to ease the transition.

In either case, the type of standard and the way it is shared during some transition period is inevitably expected to be a largely political decision. The overall effect is, however, more rigidly constrained if overall emission objectives are to be met. For example, in the UK, the Committee on Climate Change has reported that average emission levels across the whole electricity sector should be well below $100\,kg\,CO_2\,(MWh)^{-1}$ by 2030 and below $50\,kg\,CO_2$ $(MWh)^{-1}$ by 2050 (ref. 3). Although the level of CCS on fossil fuel generation will be affected by the fraction of non-fossil generation at these times and also by the extent to which emissions can be offset by trading – again largely a political decision – it is apparent that in the longer term (and possibly as soon as 2030) a very low level of emissions and a correspondingly high level of capture will be required on all fossil power plants that operate for any length of time, at least in the UK.

With either sectoral performance standard metric, once tonnes of CO_2 compliance units can be produced and traded this also offers a means of covering the costs of CCS. Such a market would be expected to have many similarities to that for Renewable Obligation Certificates (ROCs) in the UK.[36] Unlike the ROC market, it seems less likely that there will be oversupply to collapse the price during the period when additional support for CCS is required to supplement general mechanisms for reducing CO_2 emissions, such as taxes or 'cap and trade' schemes. This is because, in contrast to renewables, CCS does not have a negligible marginal cost of generation and there is no incentive to run even existing capture equipment unless operating costs can be covered. To cope with an undersupply a buy-out price is required, as with the ROCs market. If generators choose to pay the buy-out, their payment will contribute to a buy-out fund that is distributed amongst those who did achieve compliance or who produced traded compliance units. This increase in the value of a compliance unit should tend to correct the undersupply.

Compliance unit market price would, however, fall to zero once market-wide carbon prices rose to levels that covered CCS, except possibly if very high levels of capture (around 95% and above) were required by a performance standard and these proved to be difficult to attain with the technology then in place. It is worth noting, however, that an environmentally sound way of achieving the equivalent of even a 100% capture level is to use biomass that has been produced with reasonably low life cycle emissions in power plants with CCS (*i.e.* Class 3B CCS). Any sectoral (or plant) emissions performance scheme should, therefore, allow for the appropriate treatment of biomass. Also, CO_2 stored by Class 3B CCS should be rewarded in emission trading and carbon tax schemes, even though some, if not all, current schemes ignore biomass use without CCS since it assumed to be 'carbon neutral'. The reward for Class 3B CCS (with similar principles applied for any equivalent mechanism for Class 3A projects) should take the form of additional credits to be generated for the biomass carbon captured and stored. Only if the life cycle emissions of the biomass are taken into account in the emission trading or carbon tax scheme should they then be subtracted from this additional credit, however, since the purpose of this incentive measure for CCS with biomass is to ensure that Class 3B projects get the same additional benefit as other CCS projects for CO_2 stored. A consistent treatment of the life cycle emissions between projects with and without CCS is required to avoid perverse incentives.

5 Scope for Retrofitting CCS and the Role for Carbon Capture Ready (CCR) Plants

As noted in Section 2.1, it is likely that retrofits could have an important role to play in demonstration and rollout. In fact, any rapid overall reduction of emissions from power generation using CCS will require either that existing fossil fuel plants are replaced with new power plants using CCS, or that CCS is retrofitted to existing plants. It is not possible to cut CO_2 emissions simply by building additional capacity with CCS. Particularly in developing economies such as China, which have a recently-built fleet of new fossil power plants without capture that is still increasing, CCS retrofit appears likely to be an effective way to minimise carbon lock-in [see ref. 37 for a discussion of the concept of carbon lock-in, where the diffusion of carbon-saving technologies is inhibited by various system failures].

In addition, in any country, retrofitting CCS to existing plants is likely to be a very suitable way to implement CCS is its earlier stages. Compared to the planning, permitting and financing procedure for a new plant, capture equipment can be added relatively quickly and the overall investment cost is much lower. The residual lifetime of an existing plant is also likely to be shorter than for a new plant. Although this could be viewed as a disadvantage for investors, since it reduces the time available to recover capital expenditure, this may not be a significant problem in this case. If the remaining life of the plant is around 10–15 years, then it is likely that there will be sufficient time for investors to receive a reasonable return on their investment. Additionally, it is likely that any early capture equipment will have become obsolete within a decade or so, given the expected rapid increase in development activity that a CCS deployment programme would trigger. A limited operating lifetime before capture plant upgrade or closure due to competition from power plants with improved capture equipment would, therefore, be expected for early CCS projects anyway. For accurate economic analysis of retrofit projects, it is also important to consider the counterfactual for investors. For example, for a typical electric utility, if the alternative to CCS retrofit is to close an existing plant and build a new one then the cost of capital for the base plant that could be retrofitted is essentially zero. The overall levelised cost of electricity for a retrofit is then quite likely to be lower than for a completely new plant in many cases.

The 'obvious' retrofit technology for most industry-standard fossil-fired power plants, pulverised coal and natural gas combined cycle (NGCC) respectively, is post-combustion capture.[13] One reason for this is that the operation of the base plant can remain essentially unchanged, with less technical and commercial risk, and probably with a shorter down time for the retrofit. Retrofitting oxyfuel technology to existing coal plants is, however, also considered to be feasible, and a small oxyfuel retrofit has been undertaken for research purposes in Australia.[38] Oxyfuel retrofits could offer advantages since the large additional plant required, in this case an air separation unit and the CO_2 purification and compression block, can be more easily sited away from the actual power plant than the large vessels used to remove CO_2 from the flue gas in post-combustion capture processes. In the latter case, siting adjacent to the existing stack is preferred to avoid ducting the flue gases for long distances. Proximity to the power plant is also desirable to facilitate the use of low pressure steam from the steam cycle, to provide the main heat input for solvent regeneration in the capture process. Heat integration is likely to be less critical for oxyfuel plants, although some use of 'waste heat' from the capture process to heat steam cycle condensate is typically proposed for both post-combustion and oxyfuel plants.[39,40]

An alternative CCS retrofit for NGCC plants, provided they have gas turbines that can be adapted for hydrogen firing, would be to supply hydrogen-rich fuel gas, either from an adjacent unit or possibly from a more remote facility by pipeline.[41] Particularly in the latter case, it is possible that the original fuel would also be changed, from natural gas to coal, since pre-combustion capture from natural gas is generally reported to be less competitive than post-combustion capture at an NGCC plant. A remote hydrogen supply allows both coal deliveries and CO_2 transport to be located at a chosen site that is likely to be more suitable than an NGCC site originally selected against different criteria. Pre-combustion capture retrofits to IGCC plants are also technically feasible but their number is expected to limited since only a handful of prototype IGCC plants are currently operating. Some of these units are, however, already developing plans for adding CO_2 capture. For example, the Shell gasifier plant at Buggenum, is planning to develop a pre-combustion capture test facility.[42]

It is worth noting that the cost per tonne of CO_2 avoided from CCS retrofits to existing plants is, in principle, the same as for equivalent capture technology fitted to new plants. CCS costs may be lower if a more efficient capture technology can be used or a lower capital cost option is available for new build, but not retrofit. Although the efficiency of the base plant is not a factor in determining the cost per tonne of CO_2 avoided, a larger volume of CO_2 is produced per MWh of electricity dispatched by lower efficiency plants. Thus the total cost of achieving a given capture rate is higher for lower efficiency base plants, but more CO_2 is captured and stored.

As already noted, retrofits can become less attractive compared to new build CCS projects if significant additional expenditure is required to build the capture plant – for example if other equipment has to be relocated first to make space – or if the capture process cannot be integrated as efficiently with the power cycle and a higher energy penalty in incurred. It is also likely that at least some existing plants will not be sited with convenient access to CO_2 storage. All of these factors that would make retrofit less attractive can be addressed for current and future new build plants that are expected to be retrofitted with CO_2 capture in the future, by designing them to be 'carbon capture ready' (CCR).

A comprehensive study on CCR design modifications has been undertaken for the International Energy Agency (IEA) on behalf of the G8 through the IEA Greenhouse Gas R&D Programme.[16] Practical CCR applications in India have subsequently been studied and reported;[17] additional capital costs are estimated at no more than 1%, with no additional running costs before capture is added. A number of CCR NGCC power plants have been permitted in the UK, and CCR coal plants have been announced in a few countries. Anecdotal reports suggest that other recent coal fired power plants have been built to be CCR, but without any explicit statement to this effect. Making new coal plants CCR has attracted some criticism from environmental NGOs, presumably because of a general antipathy towards coal power coupled with a lack of conviction that it will actually result in a timely CCS retrofit.[43] Making new plants CCR with an identified retrofit timetable is, however, suggested in the draft 'American Clean Energy and Security Act of 2009'.[32] Pragmatically, making new plants CCR in countries such as China and India from as soon as possible could make a significant contribution to facilitating CCS deployment in these countries, since it is unlikely that continued deployment of coal-fired power plants can be halted until CCS is ready for global rollout.

6 Conclusions

Carbon capture and storage (CCS) has been identified as a potentially significant contributor to rapid electricity decarbonisation in the next one or two decades. Although it is expected that CCS will be technically feasible, it is likely that constructing initial commercial-scale integrated CCS schemes quickly enough to allow two tranches of deployment before widespread roll-out will have some significant benefits. It may also be necessary for two tranches to be built before normal commercial guarantees can be provided for some aspects of CCS projects and for regulators to be willing to rely on CCS to meet challenging, but also possible, global CO_2 (and, more generally, greenhouse gas) emissions reductions targets.

In this context, there is a strong case for suggesting that additional mechanisms to support CCS development and deployment should be considered, above and beyond any generally-applicable CO_2 pricing *via* a carbon tax or emissions trading scheme. As incentive mechanisms (and regulations) are developed, various CCS technology characteristics should be considered. For example, CCS projects can be classified according to whether they are producing carbon-containing products (with downstream CO_2 emissions to atmosphere unless CCS is also applied when the product is used) or carbon-free energy vectors such as electricity, hydrogen or heat. There could be a strong case for applying different incentives and regulations to projects in different CCS classes.

At the time of writing, a number of commercial-scale integrated CCS projects are under development and financing mechanisms for a first tranche of plants are under discussion. A range of characteristics can be identified that are desirable for a mechanism to support these, and second tranche and probably subsequent projects. These include the ability to trade compliance in any performance standard approach so that flexible operation and early participation from all fossil fuel users is possible. A number of site- and project-specific approaches are being considered for first tranche plants. It seems likely that this approach will be appropriate for rapid deployment of the limited number of projects required to form a first tranche. For a second tranche and beyond, it is likely that a more general, although CCS-specific, mechanism would be more appropriate until CCS is justified by a generally-applicable carbon price alone.

Retrofitting CCS to existing plants is very likely to be needed for rapid reductions in CO_2 emissions from the existing fleet, and may also be a useful option for early CCS demonstration. To avoid unnecessary obstacles and costs when retrofitting CCS, new plants that are built without capture from the outset should be made 'carbon capture ready' (CCR) as an interim measure. This is a relatively simple approach that is expected to add no more than 1% to capital costs and incur no operating penalties before capture.

Acknowledgements

The authors are grateful for helpful discussions on the topics presented in this chapter with many colleagues, especially through the TSEC UK Carbon Capture and Storage Consortium (UKCCSC) and the Imperial College Centre for CCS (IC⁴S).

Appendix A Carbon Dioxide Capture Technologies Closest to Commercial Deployment

Post-Combustion Capture

In post-combustion capture, CO_2 is removed from power plant flue (waste) gases after a normal combustion process. There is relatively little change to the base plant operation for producing power, although there is some integration with the CO_2 capture plant. Post-combustion separation technologies closest to commercial deployment use a chemical cleaning process based on amines or ammonia.[39] Carbon dioxide is removed from the flue gas in one column and then transferred to a second column within the chemical solvent. In the second column, CO_2 is released from the solvent, typically using a temperature swing process in a reboiler. It has been shown and, is now generally accepted, that the most efficient source of heat for this process is steam diverted from the power plant steam cycle before the low pressure turbine.[44] Once the CO_2 has been removed from the solvent, the solvent is returned to the first column to be reused, while the CO_2 is dehydrated and compressed so that it can be transported to safe geological storage (or use).

Pre-Combustion Capture

In pre-combustion capture, a more fundamental change to the power generation process occurs since the primary fuel does not combust. Instead it undergoes partial oxidation (*e.g.* gasification for coal) to produce a mixture of carbon monoxide and hydrogen. Shift reactions are then used to cover carbon monoxide to more hydrogen plus CO_2, when water (steam) is added. In typical pre-combustion capture processes, a physical solvent, such as Selexol[45] or Rectisol,[46] is used to separate CO_2 from hydrogen using a pressure swing process, before compression for transport to safe storage or use. The produced hydrogen can be used for electricity generation, normally in a combined cycle power plant, or used in other applications.

Oxy-Combustion or Oxyfuel

In oxy-combustion the combustion process is modified so that fuel is burned in oxygen rather than in air. Flame temperatures must be moderated due to material constraints, and most technologies closest to commercial deployment achieve this by recycling some of the flue gas produced by the boiler. Although there is an energy penalty associated with producing oxygen from air, this altered combustion process allows a less energy intensive process to be used to treat the flue gases, since little or no nitrogen should be present in the flue gas. It is, therefore, possible to avoid the chemical separation required for post-combustion capture, although it is necessary to include some CO_2 cleaning within the compression process.

References

1. N. Stern, *The Economics of Climate Change: The Stern Review*, Cambridge University Press, Cambridge, UK, 2007. http://www.hm-treasury.gov.uk/sternreview_index.htm
2. Intergovernmental Panel on Climate Change (IPCC), *IPCC Fourth Assessment Report: Climate Change 2007*, IPCC, 2007. http://www.ipcc.ch/ipccreports/assessments-reports.htm
3. Committee on Climate Change, *Building a Low Carbon Economy – The UK's Contribution to Tackling Climate Change*, TSO, London, 2008. http://www.theccc.org.uk/reports/
4. International Energy Agency (IEA), *World Energy Outlook 2008*, IEA, Paris, 2008.
5. Intergovernmental Panel on Climate Change (IPCC), *Carbon Dioxide Capture and Storage*, IPCC, 2005. http://www.ipcc.ch/pdf/special-reports/srccs/srccs_wholereport.pdf
6. J. Davison, *Energy*, 2007, **32**, 1163.
7. N. Riley in Chapter 6 of this book.
8. World Resources Institute (WRI), *CCS Guidelines: Guidelines for Carbon Dioxide Capture, Transport and Storage*, World Resources Institute, Washington D.C., 2008. http://pdf.wri.org/ccs_guidelines.pdf
9. CCSReg Project, *Carbon Capture and Sequestration: Framing the Issues for Regulation, An Interim Report from the CCSReg Project*, Department of Engineering and Public Policy, Carnegie Mellon University, Pittsburgh, PA, 2009. http://www.ccsreg.org/pdf/CCSReg_3_9.pdf
10. Department of Trade and Industry (DTI), *Strategy for Developing Carbon Abatement Technologies for Fossil Fuel Use*, URN 05/844, Crown Copyright, 2005. http://www.berr.gov.uk/files/file19827.pdf

11. International Energy Agency Greenhouse Gas R&D Programme (IEAGHG), *Estimating the Future Trends in the Cost of CO_2 Capture Technologies*, Report 2006/6, IEAGHG, Cheltenham, UK, 2006.
12. J. Gibbins and H. Chalmers, *Energy Policy*, 2008, **36**, 501.
13. H. Chalmers, J. Gibbins and M. Lucquiaud, *Retrofitting $CO_2\setminus$ Capture to Existing Power Plants as a Fast Track Climate Mitigation Strategy*, Submitted to ASME 3rd International Conference on Energy Sustainability, San Francisco, California.
14. D. Simbeck and W. Roekpooritat, *Near-Term Technologies for Retrofit CO_2 Capture and Storage of Existing Coal-fired Power Plants in the United States*, White Paper for the MIT Coal Retrofit Symposium, SRA Pacific, Inc., Mountain View, CA, 2009. www.sfapacific.com
15. M. A. Wise and J. J. Dooley, *Int. J. Greenhouse Gas Control*, 2009, **3**, 39.
16. International Energy Agency Greenhouse Gas R&D Programme (IEAGHG), *CO_2 Capture Ready Plants*, Report 2007/4, IEAGHG, Cheltenham, UK, 2007.
17. Mott MacDonald, *CO_2 Capture-Ready UMPPs, India,* London, 2008. http://www.defra.gov.uk/environment/climatechange/internat/devcountry/pdf/co2-capture-ready.pdf
18. M. Mahmoudkhani, K. R. Heidel, J. C. Ferreira, D. W. Keith and R. S. Cherry, *Energy Procedia*, 2009, **1**, 1535.
19. J. Gibbins and H. Chalmers, *Modern Power Systems*, 2008, Aug 2008, 12.
20. UK Department for Business, Enterprise and Regulatory Reform (BERR), *Reform of the Renewables Obligation: Statutory Consultation of the Renewables Obligation Order 2009*, URN 08/1022, 2008. http://www.berr.gov.uk/files/file46838.pdf
21. UK Department of Trade and Industry (DTI), *Energy White Paper: Meeting the Energy Challenge, URN 07/1006*, TSO (The Stationery Office), London, 2007. http://www.berr.gov.uk/energy/whitepaper/page39534.html
22. J. Gibbins and D. Barnett, *Will Renewables be Cheap Enough to Fix Climate Change?, Public Service Review*, Jan 2008.
23. G8 Energy Ministers, *Joint Statement by G8 Energy Ministers, Amori, Japan on 8 June 2008*, http://www.enecho.meti.go.jp/topics/g8/g8sta_eng.pdf, 2008.
24. Prime Minister of Australia, *Global Carbon Capture and Storage Institute*, Joint media release with the Minister for Resources, Energy and Tourism, 19 September 2008. http://www.pm.gov.au/media/Release/2008/media_release_0484.cfm
25. Australian Embassy Poland, *Australia's Global Carbon Capture and Storage Initiative*, http://www.australia.pl/wsaw/GCCSI.html, 2008.
26. B. Caldecott and T. Sweetman, in *A Last Chance for Coal: Making Carbon Capture and Storage a Reality*, ed. C. Littlecott, Green Alliance, London, 2008, 31. http://www.green-alliance.org.uk/grea_p.aspx?id = 3278

27. California Climate Change Portal, *Assembly Bill 32 – The Global Warming Solutions Act of 2006*, http://www.climatechange.ca.gov/ab32/index.html, 2008.

28. Sourcewatch, *Schwarzenegger clause*, http://www.sourcewatch.org/index. php?title = Schwarzenegger_clause, 2008.

29. Conservation Law Foundation Maine Advocacy, *Tough, New Global Warming Pollution Limits for Coal Power Plants*, Fact Sheet LD 2126, 2008. http://www.clf.org/uploadedFiles/CLF/Programs/Clean_Energy_&_ Climate_Change/Air_Pollution/Wiscasset_Coal_Plant/ CLF%202126%20%20fact%20sheet_FINALrevised.pdf

30. Department of Ecology, *Carbon Dioxide Mitigation Program for Fossil-Fueled Thermal Electric Generating Facilities*, Chapter 173-407 WAC, State of Washington, 2008. http://www.ecy.wa.gov/biblio/wac173407.html

31. Bellona, *CO_2 emission performance standard proposed in the European Parliament*, 5 March 2009. http://www.bellona.org/articles/articles_2009/ EP_EPS

32. Committee on Energy and Commerce, *American Clean Energy and Security Act of 2009*, http://energycommerce.house.gov/Press_111/20090331/acesa_ discussiondraft.pdf, 2009.

33. H. Chalmers and J. Gibbins, *Fuel*, 2007, **86**, 2109.

34. H. Chalmers, M. Leach, M. Lucquiaud and J. Gibbins, *Energy Procedia*, 2009, **1**, 4289.

35. EU Zero Emissions Technology Platform (ZEP), *ZEP Position Paper, Comitology Process, version 10, 30th March 2009*, http://www.zero-emissionplatform.eu/website/docs/ETP%20ZEP/ZEP%20Position%20Paper-%20Comitology.pdf, 2009.

36. Office of Gas and Electricity Markets (Ofgem), *Renewables Obligation: Guidance for licensed electricity suppliers (GB)*, Ref: 282/07, Ofgem, London, 2009. http://www.ofgem.gov.uk/Sustainability/Environment/RenewablObl/Documents1/Supplier_guidance__GB_2008.pdf

37. G. Unruh, *Energy Policy*, 2000, **28**, 817.

38. C. Spero, *Overview of Australian-Japanese Callide-A Oxy-Fuel Demonstration Project*, Presentation to IEA GHG Inaugural Workshop, Oxy-Fuel Combustion Research Network, Cottbus, Germany, 2005.

39. International Energy Agency Greenhouse Gas R&D Programme (IEAGHG), *Improvements in power generation with post-combustion capture of CO_2*, Report PH4/33, IEAGHG, Cheltenham, UK, 2004.

40. International Energy Agency Greenhouse Gas R&D Programme (IEAGHG), *Oxy Combustion Processes for CO_2 Capture from Power Plant*, Report 2005/9, IEAGHG, Cheltenham, UK, 2005.

41. International Energy Agency Greenhouse Gas R&D Programme (IEAGHG), *Retrofit of CO_2 Capture to Natural Gas Combined Cycle Power Plants*, Report 2005/1, IEAGHG, Cheltenham, UK, 2005.

42. C. Hornby and W. Waterman, *Nuon to Start Building CO_2 Capture Test Facility*, Reuters, 9 April 2009. http://www.reuters.com/article/

internal_ReutersNewsRoom_BehindTheScenes_MOLT/idUSTRE5381O72-0090409

43. K. Allot and A. Kaszewski, *Evading Capture: Is the UK Power Sector Ready for Carbon Capture and Storage?*, WWF-UK briefing paper, 2008. http://assets.panda.org/downloads/evading_capture.pdf

44. J. Gibbins and R. Crane, *Proc. Inst. Mech. Eng., Part A*, 2004, **218**, 231.

45. UOP LLC, *Selexol^{TM} Process*, http://www.uop.com/objects/97%20Selexol.pdf, 2002.

46. Linde, *Rectisol Wash*, http://www.linde-le.com/process_plants/hydrogen_syngas_plants/gas_processing/rectisol_wash.php, 2005.

Carbon Capture and Storage (CCS) in Australia

ALLEN LOWE, BURT BEASLEY AND THOMAS BERLY

1 Background

Australia is a country with large land area, relatively small population and high GDP per capita (7.7 million km^2, 20.8 million people and $A50 700, respectively in 2007/08; ref. 1). The population is geographically dispersed around the coast of the country, predominantly in a small number of large cities and urban areas.

Coupled with the high GDP per capita, is a high per-capita annual energy consumption (274 GJ in 2006/07; ref. 2) based on extensive use of private transport and the development of energy-intensive industries dependent on access to historically cheap, primarily coal-based, energy. This has resulted in Australia in 2006 having an annual per-capita equivalent carbon dioxide (CO$_2$e) emission of 28.1 tonnes (including emissions from land-use change and forestry) .[3] This is among the highest in the developed world. However, because of its small population, Australia makes a relatively minor contribution (1.5%) to total global emissions of greenhouse gases.

The country has a high level of education and actively contributes to the world research effort in many areas, including global-warming science. This has resulted in a population that is broadly conscious of global-warming issues and of Australia's position as a high per-capita CO$_2$ emitter. There is also a strong environmental movement in the country, where Green groups hold seats in both Commonwealth and State Governments. There is strong opposition to many large projects on environmental grounds. Coal-based power, mining and infrastructure projects, particularly, are subject to opposition from environment action groups.

Issues in Environmental Science and Technology, 29
Carbon Capture: Sequestration and Storage
Edited by R.E. Hester and R.M. Harrison
© Royal Society of Chemistry 2010
Published by the Royal Society of Chemistry, www.rsc.org

Australia is energy rich with large reserves of coal that are both relatively cheap to mine and located close to major load centres. Resources of natural gas and uranium are also substantial, but are mostly located far from major domestic markets or, in the case of uranium, not used for energy supply locally due to political and environmental opposition. The nation's oil reserves are significant, but reserves are declining in the face of continued production and the absence of new discoveries. However, Australia's plentiful resources of oil shale, coal and gas could support conversion to liquids if required, albeit at increased CO_2-emission levels.

Access to cheap energy resources has provided Australia with competitively priced energy domestically. In particular, coal now provides approximately 83% of Australia's electricity, with Australia having among the lowest electricity prices in the developed world. This low-cost energy has also allowed the development of large energy-intensive industries such as aluminium smelting and alumina production.

Energy-based commodities provide a major source of income to Australia, with coal (thermal plus coking) being the largest single export-income earner and comprising approximately 16% of total Australian export income for 2005/06. Crude petroleum, aluminium, alumina and natural gas are also all within the top 10 exports by value (both crude and refined petroleum products are also among Australia's largest imports). Australia's high GDP per capita presently depends strongly on access to cheap energy resources.

Australia has a market economy based primarily around service and mineral extraction industries, while the Australian manufacturing sector is relatively small. Australia's small population and geographical isolation do not support a large heavy-manufacturing industry and the majority of power generation and carbon-capture technology probably will be imported.

Historically, utilities such as electricity and water were provided by State governments. However, progressive market reform has led to the breakup and privatisation or corporatisation of many of these former State-owned entities. The economic reforms have also resulted in a fully competitive market for electricity. The states of Queensland, New South Wales, Victoria, South Australia and Tasmania are all interconnected *via* the south-east grid and power on this grid is actively traded *via* the National Electricity Market (NEM).

The electricity industry is a major emitter of greenhouse gases, being responsible for 198 million tonnes of CO_2e, or about 35% of the nation's total CO_2e emissions, in 2006. Following the market reforms, the industry is comprised of a mix of small (by international scale) local utilities and larger multinational companies where power stations in Australia are part only of their broader international portfolio.

That Australia's economic well-being will continue to be based on low-cost coal for electricity generation, energy security and export income for some time to come, has been well recognised by successive governments and also some key environmental action groups. However, global warming is also projected to have a serious impact in Australia, with increased drought and bushfire risk, decreased food production and increased risk of loss of native fauna and flora.

There is therefore also a recognition that, because of Australia's position as a carbon-intensive economy and as a developed and scientifically advanced nation, it has a responsibility to take strong action to develop CO_2-mitigation strategies for fossil fuels, to apply these strategies domestically and to assist our neighbours and energy customers do likewise. To this end, Australia has now set long-term targets for CO_2e emissions at 60% of year 2000 emissions, to be achieved by 2050.

Carbon capture and storage (CCS) is seen as a key strategy for achieving this target. The implementation of CCS in Australia will require at least the following prerequisites to be in place:

- Technology: CCS technology must be available at commercial scale from reputable suppliers able to offer performance guarantees.
- Commercial: the CCS project must be financially viable in a competitive market place.
- Regulatory: the regulatory environment in which the projects operate must be clear and transparent.

Numerous initiatives have been taken in Australia over some years to put these prerequisites in place. While no commercial CCS project is yet in operation, substantial progress has been made in each of these areas. This chapter briefly reviews the status of CCS in Australia as at early 2009 with reference to each of these prerequisite areas.

It is seen that the implementation of CCS is a major undertaking, requiring many years of effort at all levels within industry, commerce, education and government. However, despite these challenges, CCS remains one of the key avenues available to Australia to achieve its long-term greenhouse gas (GHG) reduction targets.

2 CCS Programs and Strategies

2.1 General Policy

Successive Australian Governments have supported a wide range of climate-change initiatives, including CCS. At a national level, Australia is a signatory of the United Nations Framework Convention on Climate Change (UNFCCC), ratifying in 1992. Australia also participated in the UNFCCC Kyoto Conference of Parties that established the Kyoto protocol with its binding commitments to the reduction of greenhouse gases (GHGs).

However, while signing the protocol and setting policies to achieve the agreed targets, the Australian Government initially refused to ratify, on the basis that doing so would compromise jobs and that actions on GHG reduction required the inclusion of the developing economies. Ultimately, following a change in government in 2007, the Kyoto Protocol was ratified and came into effect with respect to Australia in 2008.

The initial approach taken by both Commonwealth and State governments to CCS was to support the development of demonstration projects by private organisations. This was achieved by contributing to the capital cost of the projects and through strong support of CCS R&D. R&D was seen as necessary to optimise overseas technologies to Australian conditions, to facilitate significant cost reductions in technologies and to develop an appropriate skill base in Australia.

Australia has also been instrumental in fostering international initiatives to achieve reductions in greenhouse-gas emissions. Key to these initiatives has been a perceived need to include the rapidly developing economies in the Asia Pacific region in carbon-reduction programs. These initiatives included the Carbon Sequestration Leadership Forum (CSLF) in which Australia was a founding member. Also, Australia, in 2006, hosted the launch of the Asia Pacific Partnership on Clean Development and Climate (APP). APP now includes USA, China, India, Japan, Canada, South Korea and Australia, countries representing over 50% of both global GHG emissions and population.

While continuing with, and expanding on these initiatives, the Australian Government has now introduced the new Carbon Pollution Reduction Scheme (CPRS) as the primary policy approach to achieving domestic greenhouse gas emission reduction targets. The CPRS, in effect a carbon 'cap and trade' scheme, is expected to be implemented from mid 2011.

These strategies are discussed in more detail in the next section.

2.1.1   Carbon Pollution Reduction Scheme (CPRS)

The Australian Government is introducing a market-based scheme to provide a price for carbon. This is seen as the primary mechanism by which low-carbon technologies, such as CCS, can achieve take-up in Australia's market-based economy.

In pursuit of this objective, the Government in June 2008 released the discussion paper *Carbon Pollution Reduction Scheme-Green Paper.*[4] This was followed in December 2008 with a policy white paper titled *Carbon Pollution Reduction Scheme – Australia's Low Pollution Future*[5] and, in March 2009, with an exposure draft of the corresponding legislation.[6]

While some modification to the legislation may still be expected as a result of review of the exposure draft, key features of the Carbon Pollution Reduction Scheme include:

Mechanism. The pollution-reduction scheme will employ a 'cap and trade', emission-control mechanism, based on the creation of Australian Emission Units (AEU) equal in number to the annual amount of carbon pollution permitted under the scheme. The cap will, for a given year, be set equal to the indicative national emissions trajectory for that year, less projected emissions from sources not included in the Scheme. The Government will announce scheme caps for five years into the future, and will provide indicative gateway ranges for up to ten years into the future to assist business planning.

Obligations. The core obligations of liable entities are to report emissions, and to surrender AEUs equal in number to the tonnes of emissions reported. Reported emissions will include direct emissions for which the entity is responsible, with adjustments, where appropriate, for any emission liability transferred to an upstream fuel supplier and for amounts of gases imported, manufactured or supplied.

In the event that insufficient emission units are surrendered, then the entity will incur an administrative penalty and a 'make-good number' of emission units. The penalty is determined at 110% of the average auction price for AEUs for the year times the number of units shortfall. The make-good units must be included in the reported emission calculations for the following financial years.

Coverage. The Scheme will cover all six GHG gases listed under the Kyoto protocol. It has been designed to achieve the maximum practicable coverage of greenhouse-gas emissions and will initially include stationary energy, transport, industrial process, waste, forestry and fugitive emissions. Scheme obligations will apply to facilities with direct emission of 25 000 tonnes of CO_2e a year or more. This is expected to include entities responsible for about 75% of Australia's greenhouse gas emissions.

Targets. The Government has set national targets of from 5 to 15% below year 2000 greenhouse-gas emission levels by 2020, and at 60% below 2000 levels by 2050. The year 2020 minimum 5% reduction level is seen as an unconditional target, irrespective of what actions may be taken by other countries around the world. The 15% reduction below 2000 levels by 2020 would be committed to in the context of agreement by all major economies to commit to substantially restrain emissions, and for advanced economies to take on reductions compatible with that of Australia.

Recognition of CCS. Carbon that is transferred to carbon capture and storage facilities will not be counted towards the originating entities gross emissions. However, fugitive emissions from carbon capture, transport and storage activities will be imposed on the relevant CCS facility operator.

Price Setting of AEUs. The primary avenue for release of AEUs will be by monthly auction. However a price cap will be set for the first five years of the Scheme, starting at $A40 per AEU (or per tonne CO_2e) and rising at 5% in real terms per year. (Government modelling has suggested AEU prices required to achieve the 5% reduction target rising from $A25 initially to $A35 in the medium term).

The mechanism used to maintain the price cap will be the availability of supplementary AEUs at the cap price. Unlimited banking and limited borrowing of AEUs will also be permitted to assist in stabilising the market price.

Coal Fired Power Generation. An electricity-sector adjustment scheme will be implemented to provide some protection of existing investments in certain high-emission assets. For the first five years of the scheme, a quantity of AEUs will be allocated amongst eligible coal-fired generators on a *pro rata* basis and in accordance with annual assistance factors. These factors will be calculated as the product of the historical energy generation of the asset and the difference betweens its actual emission intensity and 0.86 tonnes CO_2e per MWh. Tests will be applied to recover any windfall gains under this provision.

Transitional Assistance to Industry. Certain energy-intensive industries that primarily market product overseas, such as aluminium smelting, are seen to be unable to pass on to customers additional costs of meeting the CPRS. These industries will be provided with an administrative allocation of AEUs for transitional support and to reduce the likelihood of carbon leakage overseas. Support will be based on the emissions intensity of the activity in terms of value of product exported.

International Linkages. Provision has been made for an entity to surrender certain "Kyoto" emission units to satisfy its liabilities. Initially only emission-reduction units created under the Joint Implementation and Clean Development Mechanisms will be accepted. Trading in emission-reduction units created under other schemes such as the EU Scheme and also international trading of AEUs is expected to be allowed in the future.

2.1.2   Carbon Sequestration Leadership Forum (CSLF)

Australia has been a member of the CSLF[7] since its formation in 2003, presently holds the position of Vice-Chair of the Policy Group, and is a key member of the Technology Group. The CSLF includes 21 developed and emerging nations plus the European Commission. This group of countries produces about 75 percent of the world's CO_2 emissions.

The Forum's purpose is cooperation and collaboration in development, demonstration and deployment of more cost-effective carbon dioxide capture and storage technologies with a focus on electric-power production and other industrial activity. This is achieved *via* collaborative efforts that address key technical, economic and environmental obstacles. The CSLF also seeks to promote awareness and champion legal, regulatory, financial and institutional environments conducive to CCS technologies.

Australia's Monash Energy and CO2CRC Otway Pilot CO_2 Storage projects are presently recognised within the CSLF. These projects are described in more detail below.

2.1.3   Asia Pacific Partnership on Clean Development and Climate (APP)

APP is a partnership between countries representing over 50% of the world's emissions, energy use, GDP and population.[8] The seven member countries are Australia, Canada, China, India, Japan, Republic of Korea and the United States. APP was set up with the objective of accelerating the development and deployment of technologies including CCS through collaborative research and on-going development.

The Commonwealth Government has to date committed $A93 million in support of projects through APP. Australia is also Chair of the Cleaner Fossil Energy and Aluminium Task Forces. Many of the Australian-based CCS projects described later in this chapter are recognised as APP projects, allowing member countries to participate in and to share learning from these projects.

2.1.4   Global Institute for Carbon Capture

The Commonwealth Government in September 2008 committed $A100 million to the establishment of a global institute for carbon capture. The initiative was taken in response to the 2008 Toyako G8 Summit Leaders' Declaration expressing strong support for the launching of 20 full-scale CCS demonstrations by 2010. The Institute has been mandated to action all activities leading to the acceleration of the deployment of CCS demonstration projects globally.

It is expected that, when operating, the Institute will provide expertise, assist in building the business case for, and identify and commission critical-path R&D in respect of commercial-scale CCS projects. It is intended that the Institute become an international hub for the coordination of public and private-sector funding for CCS. While the details of the structure of the Institute have yet to be released, the Government has foreshadowed annual funding for the Institute of up to $A100 million per annum.

2.2 Governmental CCS Initiatives and Funding

Both Commonwealth and State Governments have recognised that CCS projects require support both in terms of technology development through continued R&D and in financial support to achieve penetration in a market-based economy. This is particularly the case where carbon emission-reduction technologies impose additional costs of business, but emissions of carbon are free, as is presently the case, or allowed at relatively low cost.

To assist in improving the competitiveness of CCS technologies, the following initiatives have been taken.

2.2.1 Support for Cooperative Research Centres

Both Commonwealth and State Governments have supported research into CCS through a number of Cooperative Research Centres (CRCs). Those that have addressed CCS issues include the CRC for Cleaner Power from Lignite (CRCCPL), the CRC for Coal in Sustainable Development (CCSD) and the CRC for Greenhouse Gas Technologies (CO2CRC).

Both CRCCPL (1999–2006) and CCSD, as well as its predecessor, (1995–2008) primarily sought improvements to power-generation technologies, with CRCCPL focusing on low-rank Victorian lignite while CCSD addressed black coal-related technologies. The CO2CRC has strong programs in both geological storage issues and capture technology. The on-going CO2CRC program is discussed in Section 3.2.

These three centres, initially funded to around $A250 million, have provided Australia with a strong base of scientists skilled in advanced coal technologies and particularly in CCS issues. An important outcome has also been the rapid diffusion of CCS knowledge to industry leaders.

2.2.2 Low Emissions Technology Demonstration Fund

The Commonwealth Government in 2004 committed $A500 million to the Low Emission Technology Demonstration Fund (LETDF). The objective of the LETDF was to help Australian firms demonstrate the commercial potential of new energy technologies or processes, or for the application of overseas technologies and processes to Australian circumstances, within the context of a requirement for long-term, large-scale greenhouse gas emission reductions.

CCS-related projects funded through LETDF include:

- Gorgon ($A60 million): injection of CO_2 separated from natural gas during processing into a deep saline aquifer with low permeability.
- Callide Oxy-Fuel ($A50 million): demonstration of oxy-fuel firing at semi-commercial scale and storage of a portion of the CO_2 collected.
- HRL IDGCC ($A100 million): demonstration of new technology with integrated drying and gasification of high-moisture, low-rank coals.
- Hazelwood 2030 ($A30 million): demonstration of brown-coal drying and post-combustion capture (PCC), with the CO_2 separated from the flue gas used to neutralise alkaline, fly-ash transport waters.

Descriptions of these projects are provided in following sections. Certain other projects which were initially granted LETDF funding have not proceeded. Further commitments are not expected under this program.

2.2.3 National Low Emission Coal Initiative

In July 2008, the Commonwealth Government established the National Low Emission Coal Initiative (NLECI) to support the development and deployment of clean coal technologies. A National Low Emission Coal Council (NLECC) was also set up to advise the Minister with regard to the Fund. The Council will recommend strategies and actions to reduced emissions from coal-fired power generation through deployment of low-emission coal technologies, including CCS.

As part of the Initiative, funding of $A500 million over 8 years has been allocated by the Government. It is noted that Australian coal producers have committed to co-investing more than $A1000 million in funding through COAL21 in low-emission coal technologies.

Development of programs and strategies under NLECI continues. Steps taken to date, include the establishment of a Carbon Storage Taskforce and of a Carbon Capture and Storage Research Centre.

2.2.4 Victorian Government Energy Technology Innovation Strategy (ETIS)

Victoria has extensive reserves of low-cost, but high-moisture, brown coal. While capable of providing electricity at very low cost, this fuel also has high CO_2 emissions relative to black coal. This has been recognised in the Victorian Government Energy Technology Innovation Strategy (ETIS) which seeks to support low emission technologies where a market gap can be identified.

In 2005, ETIS provided $A109 million for low-carbon emission technology development, including $A50 million for the HRL IDGCC project, $A30 million for the Hazelwood 2030 project and $A29 million for R&D with Victorian universities.

A second Request for Proposal was released in December 2008 with applications to close in mid-2009. This initiative is funded to an amount of $A110 million and is intended to bring forward large-scale, pre-commercial demonstration projects. Selection criteria include that the technology can deliver electricity with maximum emissions of 0.3 tonnes CO_2 per MWh, that the technology be scalable to commercial scale and capable of delivering electricity at a competitive sent-out cost. It is also expected that, at the end of the demonstration, the technology will be shown to be commercially and technically 'bankable'.

2.2.5 Queensland Government

In 2006, the Queensland Government allocated $A300 million from the Queensland Future Growth Fund to support the development of clean coal technologies. The Queensland Clean Coal Council, with joint membership from Government and the black-coal industry, was also created to provide advice regarding funding priorities and projects to accelerate deployment of clean coal technology.

The Queensland Government is also presently contributing directly to research at cLET, the University of Queensland and to the ZeroGen feasibility study.

In 2007, the ClimateSmart Strategy was launched, which includes an allocation of $A10 million to the Geological Survey of Queensland (GSQ) for the identification of geosequestration sites. The initiative will identify, evaluate and categorise geological sites in Queensland that have the potential for long-term, safe and secure storage of carbon dioxide emissions. These sites may store emissions from large-scale, clean coal technology power plants, or other stationary energy emissions sources in the future.

2.2.6 New South Wales

The NSW Government in 2008 enacted the Clean Coal Administration Bill which established the Clean Coal Fund and the associated NSW Clean Coal Council. An initial allocation of $A100 million was made to the fund to be sourced from a levy on electricity distributors.

The NSW Government has to date committed some $A22 million to clean-coal projects including:

- Munmorah PCC demonstration project.
- A state-wide survey to identify geosequestration sites in NSW. Drilling of a deep well to provide stratigraphic data and targeting potential storage sites adjacent the Munmorah power station is now underway.
- Funding of basic research at the University of Newcastle.

2.3 Black Coal Mining Industry Initiatives

The Australian black-coal mining industry has taken a leading role in initiating and pursuing coal-related CCS opportunities in Australia, initially through its contribution to R&D and more recently though moving to support commercial-scale demonstration projects. Since 1992, all Australian black-coal producers have contributed $A0.05 per tonne of coal mined to fund the Australian Coal Association Research Program (ACARP). ACARP has supported research and scoping studies into clean coal technologies, including CCS, both directly and through contribution from 1995 through 2008 to the Cooperative Research Centres for Black Coal Utilisation and for Coal in Sustainable Development.

These CRCs were joint initiatives of the coal and power industries, researchers and government, with the coal industry providing funds both directly and *via* ACARP. However, during the course of these centres it became clear that more direct action was required. This resulted in the coal industry initiating the establishment in 2003 of the COAL21 program.

COAL21 is a partnership between the coal and electricity industries, unions, Commonwealth and State governments and the research community. Its objectives are to first identify and then realise the potential for reducing or eliminating greenhouse gas emissions from coal-based electricity generation in Australia. COAL21 developed a national action plan.[9] The plan pointed to the need for direct support of first-of-a-kind commercial-scale demonstration projects for a number of technologies.

In 2006, the Australian black-coal mining industry then moved to establish the COAL21 Fund, a world-leading whole-of-industry commitment for R&D of low-emission coal technologies. The Fund is raised by a voluntary levy on coal producers, the amount contributed being based on each company's production levels. It is expected to raise approximately $A1 billion over a 10-year period from commencement in 2007.

The Australian Coal Association has now established ACA Low Emissions Technologies Limited (ACALET) to administer the COAL21 Fund. ACALET assesses eligible projects for funding support *via* an industry project-assessment committee and provides public information on low-emission coal technology progress.[10]

To date over $A500 million has been committed by ACALET to CCS based projects including:

- $A46 million to the ZeroGen feasibility study, with additional funding up to a total of $A300 million available for a Queensland-based Integrated Gasification Combined Cycle (IGCC) project.
- $A68 million to the Callide Oxy-fuel project.
- $A50 million for a post-combustion capture and storage project in New South Wales, to follow on from the current Munmorah PCC project.
- $A20 million for Queensland geosequestration initiatives.
- $A75 million commitment to the research and development program of the National Low Emissions Coal Council, matching the $75 million commitment of the Commonwealth government.

3   CCS R&D Activities in Australia

CCS Research in Australia may be broadly considered as a three-level hierarchy comprising funding organisations, special purpose research vehicles and research providers.

Funding organisations include Commonwealth and State governments and industry associations seeking to achieve specific policy objectives *via* the funding support. Commonly a call for expressions of interest is released specifying objectives and funding criteria. Applications for funding are then assessed against the criteria and the preferred projects supported.

Australia has also created a number of Special Purpose Research Vehicles (SPRV), including those within the Commonwealth Government's Cooperative Research Centre program. These organisations, comprising partnerships of research providers, relevant State Government and Industry, develop coordinated research programs seeking to address specific issues.

The primary research output in Australia is provided either by University researchers or the Australian Commonwealth Scientific and Research Organisation (CSIRO). The researchers are funded primarily from either government or SPRV sources, or from specific-purpose industry contracts.

The National Low Emissions Coal Research Ltd has recently been established to coordinate CCS R&D in Australia and to facilitate individuals and groups working closely together.

This approach to R&D has resulted in a broad program of work with few significant gaps and little duplication. The broad thrust of this program is reviewed below.

3.1 Australia's Commonwealth Scientific and Research Organisation (CSIRO)

The CSIRO is Australia's national science agency and, with a current budget of over $A1 billion, is one of the larger and more diverse research agencies in the world. It plays a major role in many of the major scientific endeavours being undertaken in Australia and is funded by a combination of government appropriations and external contract revenue. Since internal funding is significant, CSIRO is able to operate as both a funder of research and a deliverer of research.

CSIRO is a key research provider to a number of Australia's CCS technology programs, undertaking R&D on behalf of the CO2CRC in geological storage issues and to cLET on syngas processing. It maintains a strong in-house R&D program on post-combustion capture technology and also operates a pressure-entrained flow reactor for coal-gasification studies. CSIRO has previously carried out a major experimental study of the gasification performance of Australian coals in association with the CRC for Coal in Sustainable Development and ACARP and continues to perform work in this area.

CSIRO has initiated a comprehensive R&D program to improve the performance of PCC on coal-fired power stations, with both laboratory and pilot-plant-scale research being involved.[11] The program seeks improvements in solvent performance, process flow-sheet design, the physical design of equipment, such as absorbers, and thermal integration with the power-station process flow chart.

The solvent research includes screening of advanced amine formulations, ionic liquids and phase-change solvents. The process modifications being investigated include heat exchange integration in the stripper, integration of the compressor and split-flow absorbers. Novel process components include membrane contactors and heat pumps to transfer energy. The application of enzymes to improve solvent processes is also being investigated.

CSIRO undertakes laboratory measurement of CO_2-absorption rates and capacities, supported by modelling of the chemical interactions between CO_2 and the amine. Following screening, the more successful solvents are tested for reaction kinetics and energy requirements, vapour and liquid-phase equilibrium and solution phases. This work allows laboratory-scale, pilot-plant testing of more prospective candidates and, in turn, pilot testing in the field to assess factors such as corrosion, oxidation and impact of typical flue-gas contaminants such as sulfur dioxide (SO_2) and nitrogen oxides (NO_x).

Within this program CSIRO has developed and installed a number of PCC pilot plants at power stations. Projects are presently running at Munmorah, Loy Yang, Tarong and Gaobeidian (Beijing) power stations, as reviewed in Section 4.3.

The goal of these projects is to provide proof of the PCC concept, evaluation of various absorbents, assist in the scale-up to demonstration and commercial-size plants, demonstrate further development potential and provide the science underpinnings for future policy options for CO_2 capture.

3.2   Cooperative Research Centre for Greenhouse Gas Technologies (CO2CRC)

The CO2CRC was set up specifically to research and demonstrate CO_2 capture and storage technologies in Australia. It has wide participation including the Commonwealth and State Governments of NSW, Queensland, Victoria and Western Australia, key Australian Universities including Curtain, Monash, Adelaide, Melbourne and the University of New South Wales, as well as major industrial partners from the petroleum, mining, manufacturing and power industries.

The CO2CRC is the successor to the Australian Petroleum CRC. It was established in 2003 under the Commonwealth Government's CRC program for an initial seven-year program and with a budget of about $A150 million. The budget was allocated between research into storage (25–35%), capture (20–30%) and demonstration and regional assessment (35–40%). While the initial term is scheduled to complete in mid 2010, an application for an additional term is presently being prepared.

The CO2CRC functions primarily to plan, direct, fund and manage research, with the majority of R&D effort being carried out by either the participant Universities, CSIRO or under sub-contract to specialist firms. Funding is contributed by the Commonwealth Government's CRC Program and the participants as 'cash and in kind' under the terms of a joint venture agreement.

The CO2CRC has given a major stimulus to Australian R&D effort in CCS by consolidating contributions from a wide range of participants to fund a much more diverse and capable effort than would otherwise be possible, by providing high-quality scientific leadership with improved focus and cooperation between the researchers and by improving international visibility, credibility and access.

The CO2CRC has published extensively with much of the research documented on the organisation's website at http://www.co2crc.com.au. The major CO2CRC program themes include:

Capture Research. This program seeks to reduce the high cost of current CO_2 capture processes. In general, the R&D is aimed at technologies seen to have potential to achieve step-change reductions in cost.

The program primarily includes laboratory-scale work on a wide variety of current and second generation, pre- and post-combustion capture technologies. Projects include:

- Improvements to equipment design and application of gas-liquid membrane contactors to enhance solvent-based systems.
- Investigation of a variety of polymer- and nano-material-based membrane, gas-separation approaches involving developing and assessing performance, including effects of fuel-gas contaminants.
- Improvements to pressure-swing adsorption systems.
- The use of cryogenics for gas separation, including the characterisation of hydrate formation.

The laboratory work is supported by both economic and system-integration modelling that is used to provide indicative costings of potential storage sites and technologies, and help vet alternative technologies. The more successful technologies are moving to pilot trials using live raw-gas streams as described later.

Storage Program. The storage program includes research into selection and characterisation of sites, in terms of their storage capacity, the physical and chemical processes taking place during injection, monitoring techniques and potential risks and uncertainties involved. Specific sub-programs include regional geology, reservoir and seal characterisation, geomechanics and petrophysics, hydrodynamics and geochemistry, reservoir modelling, geophysics and CO_2 storage in coal seams. Many of the outcomes of the storage program have been applied in both the CO2CRC's regional study publications and in the Otway pilot project. A key outcome from this program is a methodology for assessment of reservoir capacity.[12]

Regional Studies. The CO2CRC has published a number of regional studies of CO_2 storage capacity in Australia and New Zealand. A preliminary Australia-wide survey titled "Geodisc" was initiated by the Australian Petroleum Cooperative Research Centre and completed within the CO2CRC. This has been followed by more detailed studies of specific regions. These studies have included many of the key basins in Australian states with large CO_2-emission sources and also in New Zealand.

Of note is the Latrobe Valley CO_2 Storage Assessment[13] which addresses the potentially large and high-quality reservoirs in the Gippsland Basin, offshore eastern Victoria. This region is Australia's most prolific but now maturing petroleum production field, and is close to the major CO_2 sources of both the Latrobe Valley power stations and potential new projects, such as the Monash Project. The study concluded that the region has the potential to store up to 2 billion tonnes of CO_2 at low unit cost. Injection rates from two million tonnes per annum up to 50 million tonnes per annum were seen to be possible. Assessments of both infrastructure and containment showed risks to be within accepted safety and containment standards. However, as petroleum extraction continues in the Basin, a need was identified for any CCS project developer to work closely with oil and gas producers to avoid adverse CO_2 interference.

Development Program. The development program presently includes both pre- and post-combustion capture projects and the Otway pilot project. The Otway pilot project constitutes one of the key initiatives in CCS within Australia. It is described in detail in Section 4.4.1.

The pre-combustion capture project is being carried out in association with HRL Technologies Ltd and uses raw syngas from the HRL brown-coal research gasifier to provide syngas feed to test rigs. The CO2CRC test rigs include solvent and membrane technology, developed in conjunction with Melbourne University, and adsorption technology, developed at Monash University. The trials will be used to identify the most cost-effective technologies for further development.

The post-combustion project will test similar technologies to that deployed in the pre-combustion pilot projects. However, in this case they will be trialled on the 25 tonne per day PCC unit at the brown-coal-fired Hazelwood power station.

3.3   Centre for Low Emission Technology (cLET)

The Centre of Low Emission Technology is researching coal gasification and syngas cleaning technologies as a prerequisite to moving to an IGCC-based, low-carbon coal cycle using hydrogen as an energy carrier. The Centre is supported by the Queensland Government, local power-generating companies, ACARP, CSIRO and the University of Queensland. CSIRO and the University of Queensland are the primary research providers for cLET. It was set up in 2003 and is presently funded through to mid-2009 with a total budget of about $A26 million.

cLET was created on the premise that improved syngas processing and cleaning technologies are key to enabling competitive cost and performance of IGCC with carbon capture. The program has five main research areas as follows:

Gasification Core Facility. This research seeks to provide access to high-quality, coal-specific, gasification-performance data; objectives include to obtain pilot-scale test data on a selection of Australian coals, to develop a 5 MW thermal-scale national gasification test facility and to develop a syngas generator for research purposes. A detailed design and costing for both pilot-scale gasification facility and syngas generator were completed. However, higher cost and longer delivery times than initially projected have prevented the realisation of these within the current cLET program. Gasification trials of four coals were carried out by the CRC for Coal in Sustainable Development at a test facility in Germany.[14]

Gas Cleaning Program. This program seeks to improve performance of commercially procured candle filters in hot dry syngas through the redesign of cleaning systems, the use of sorbents and guard beds to trap syngas contaminants. This work, carried out within CSIRO, has had success with pulse-less cleaning of candle filters in the laboratory.

Gas Processing Program. This program investigates water–gas shift reactions and the development of catalysts specific to processing of coal syngas. The work is carried out within the CSIRO.

Gas Separation Program. This research investigates catalytic membrane reactors and thin-film membranes for hydrogen separation. This project at the University of Queensland, using catalysts developed in the gas-processing program, has been successful in achieving conversion beyond initial equilibrium values and good separation of hydrogen in chemical vapour-deposited silica-membrane reactors.

Social and Economic Integration. This program addresses the social issues involved in educating the public on CCS technologies, their risks and benefits. Surveys found that people, when presented with the facts, tended to draw positive conclusions with regard to CCS.

3.4 University Research Activities

3.4.1 University of Newcastle

The University of Newcastle has been a leader in oxy-fuel research in Australia and has carried out research into furnace heat transfer, boiler-design issues and coal reactivity and slagging under oxy-fuel conditions.[15] The University contributed strongly to the Callide oxy-fuel feasibility study, including participating in pilot-scale combustion tests in Japan and their analysis. Oxy-fuel research is currently addressing potential issues with corrosion.

The University is also supporting the CSIRO PCC program with fundamental studies of PCC process chemistry.

3.4.2 University of New South Wales

The University of New South Wales' School of Petroleum Engineering contributes to the CO2CRC through the economic modelling of CCS activities.[16]

3.4.3 Monash University

Monash University is researching the application of oxy-fuel combustion, pre-combustion and post-combustion capture technologies to Victorian brown coal. The work is funded primarily through the CO2CRC and the ETIS programs.

3.4.4 University of Melbourne

The University of Melbourne is a key node of the CO2CRC, housing the carbon capture program and contributing R&D effort on gas absorption, gas separation and gas-absorption membranes and technologies.

3.4.5   University of Adelaide
The University of Adelaide is another key node of the CO2CRC, housing the carbon storage node. Research undertaken at the University addresses the selection, characterisation and monitoring of CO_2-storage sites and of the processes taking place during and after injection. Many of the results of this work have been applied in the CO2CRC regional assessment reports.

3.4.6   Western Australian Energy Research Alliance. (WA:ERA)
WA:ERA is an association of the Curtain and Western Australia Universities, CSIRO and major oil companies, set up to carry out research into geological sciences. WA:ERA contributes to the CO2CRC storage program and performs much of the analysis work for the Otway Basin Pilot Project.

3.4.7   University of Queensland
The Centre for Low Emissions from Coal within the University of Queensland is researching storage of CO_2 in coal seams.[17] This may have particular application to Australia due to the proximity of large coal reserves to major power stations, but generally poor access of those power stations to high-quality conventional geological-storage structures.

The Centre has developed specialised equipment to measure coal properties, including absorption isotherms and permeability under *in situ* stress conditions. The potential of injecting flue gas immediately adjacent to power stations with a view to maintaining permeability, purging coal-seam gas for sale with the included nitrogen, while sequestering the CO_2 in the degassed coal is presently being investigated.

The University's Centre for Functional Nanomaterials researches applications for nanomaterial-based membranes as part of the cLET program.

4   CCS Projects in Australia

The recognition that CCS offers a potential solution to Australia's high CO_2 emissions from industrial activities has led to the consideration of a wide range of projects at both commercial and demonstration scale. However, in the absence of strong cost drivers and the only-recent enactment of appropriate enabling legislation, these presently remain either relatively small-scale or at pre-feasibility level. The following projects are currently active.

4.1   Commercial-Scale Projects Incorporating CCS

4.1.1   ZeroGen
ZeroGen is a Queensland Government project in partnership with the Australian Coal Association.[18] The Government has enrolled a number of service providers, with technical and engineering support from various local and overseas consulting firms.

The project envisages a two-stage development of IGCC with CCS. The first stage will be an 80 MW net coal-gasification plant with carbon capture and storage, to be developed adjacent to the Stanwell Power Station in Queensland. The plant will capture up to 75% of the CO_2, which will then be transported approximately 200 km for injection and storage in the Northern Dennison Trough.

In parallel with Stage 1, ZeroGen will develop a 300 MW net project with CCS for a site yet to be selected in Queensland. The location of the plant will ultimately be determined on the basis of access to coal, water, storage sites and transmission links.

This two-stage approach has been adopted to 'de-risk' the project by proving the technologies, in particular, the integration aspects of the IGCC and the CCS processes at pilot scale, and then applying these learnings to the larger commercial-scale project.

At present Stage 1 of the project is a feasibility study, with funding being contributed by the Queensland Government ($A102.5 million) and the Australian Coal Association ($A26 million). In parallel with the feasibility study, ZeroGen is drilling to demonstrate suitable reservoir capacity in the Denison Trough. Trial injections of CO_2 are planned to allow optimisation of the injection process and a better understanding of monitoring and verification techniques.

The feasibility study is expected to be completed by mid-2009 and, assuming a positive outcome, construction would start by late 2012. Total project cost is unknown, but Stage 1 is expected to cost around $A1.7 billion. The Queensland Government has earmarked $A300 million for the project, with the remainder expected to be raised through normal financing channels.

4.1.2   Monash Energy

The Monash Energy project proposes the development a large-scale, coal-to-liquids plant with carbon capture and storage. It is a joint development of Anglo American plc. and Shell Gas and Power. The project would capitalise on the large brown-coal resources in the Victorian Latrobe Valley and the nearby depleting Gippsland Basin oil and gas reservoirs for CO_2 storage.

The partners have recently determined that key enabling requirements for coal-to-liquids projects are not yet in place, due to oil price volatility and high and escalating construction costs. As a result, the project has been placed under review.

4.1.3   HRL IDGCC

HRL Pty Ltd proposes to develop a large-scale, brown-coal power-generation demonstration project in the Latrobe Valley, using their Integrated Drying Gasification and Combined Cycle (IDGCC) technology. The project is a 400 MW commercial-scale plant, with two 200 MW gasification systems feeding a gas turbine combined-cycle power block.

The IDGCC technology has been designed specifically for very high-moisture, low-rank brown coals. It is an air-blown design and uses the hot syngas from the gasification plant to dry the brown coal, which is then used as a feedstock for the gasifier. The drying of the coal cools the syngas and adds to its vapour content, hence increasing the mass flow of gas through the combined-cycle power plant. It is estimated that the technology will reduce emissions by about 30% and require about 50% less water than conventional brown coal power-generation technologies.

As this will be the first full-scale demonstration of IDGCC, it will not include CCS. However, the technology will be compatible with CCS processes as would be applied with conventional IGCC technology, and future IDGCC projects can include CCS if appropriate.

HRL have been awarded $A50 million from the Victorian Government and $A100 million from the Commonwealth Government toward the $A750 million project. Harbin, one of the largest power-construction companies in China, will contribute $A500 million to the project. China has more than 10% of the world's brown-coal reserves and this technology could assist China in reducing its greenhouse-gas emissions in a significant way.

4.1.4   Coolimba Power Project

Coolimba is a 400 to 450 MW power project proposed for north of Perth in Western Australia.[19] The project is being developed by Aviva and AES Corporations and envisages two 200 MW, circulating fluid bed (CFB) boiler/turbine units firing high-moisture, high-ash sub-bituminous coal. Commercial operation is scheduled for 2013.

Coolimba is a merchant power project seeking to satisfy demand growth in the Western Australian network. The project also brings increased security to the network as it is located close to a regional growth-centre, but far from current major generation sites. The initial construction will be optimised for maximum efficiency, and designed for ready conversion to oxy-fuel firing at some time in the future. Details of design modifications to enable the plant to be converted to oxy-fuel have not been released.

The Coolimba site is close to a number of potential storage sites including depleting oil and gas reservoirs, deep saline aquifers and deep coal measures. A study has been carried out by the CO2CRC into the CO_2-storage potential of the region immediately north of the proposed power station site. This study indicated capacity for up to 40 million tonnes of CO_2 in depleted gas reservoirs. It also identified deep saline aquifers with storage potential estimated at 500 million tonnes of CO_2. The depleted gas fields are among those owned by oil and gas producer, Australian Worldwide Exploration Ltd, who contributed extensive data from the oil and gas fields to the study.

Conversion of the Coolimba Power Plant to oxy-fuel would depend on there being sufficient financial incentive to do so. However, the project proponents consider that, given a capture-ready plant design and the proximity of the well characterised but depleted oil and gas fields, this project would be among the first to become viable as a cost of carbon is introduced.

4.1.5 FuturGas

FuturGas is developing a coal-to-liquids project to utilise the Kingston lignite deposit in South Australia. It is proposed to use gasification followed by Fischer–Tropsch synthesis to manufacture about 10 000 barrels per day of low-sulfur diesel and naptha. The plant is also proposed to produce about 40 MW of power from surplus synthesis-process gas, plus 100 tonnes per day of elemental sulfur.

The project would also produce approximately 1.6 million tonnes per annum of CO_2 which is extracted from the process during synthesis-gas cleanup. The CO_2 so removed would be piped to the Otway Basin to the south of the Kingston lignite resource for storage.

This project is in the early stage of development.

4.1.6 Gorgon

The Gorgon Joint Venture partners[20] plan to develop the Greater Gorgon gas fields, located approximately 130 km off the north-west coast of Western Australia, to produce liquid natural gas and domestic pipeline gas. These fields contain resources of some 40 000 PJ of gas and are Australia's largest presently known natural-gas resource. The project will produce some 15 million tonnes per annum of LNG, 100 PJ per annum of domestic gas and associated condensates.

The project comprises the development of the gas field with sub-sea wells and pipelines to connect to the processing facility on nearby Barrow Island, the construction of a three-train, 15 million tonnes per annum gas-processing and liquefaction facility on Barrow Island, LNG shipping facilities and injection of CO_2 into deep formations below Barrow Island.

The injected CO_2 will be separated from the raw natural gas using a methyldiethanolamine (MDEA) scrubbing process. It will then be compressed and transported a distance of about 10 km by pipeline and injected through some 8–9 injection wells, directionally drilled from 3–4 drill centres. Associated infrastructure will include 3–4 pressure-management wells (for formation-water extraction) and a similar number of water re-injection wells.

Environmental and planning approvals were, subject to certain requirements, initially granted for a two-train project by the Western Australia Government in 2007. Since that time, a revised submission for a three-train project has been submitted for assessment while the project partners move toward a final investment decision.

4.2 Large-Scale Demonstration Projects

4.2.1 Callide Oxy-Fuel

The Callide oxy-fuel project involves the conversion to oxy-fuel of an existing 30 MW coal-fired boiler at CS Energy's Callide A Power Station.[21] The project scope includes the overhaul and refurbishment of the No. 4 boiler/turbine unit, the construction of a nominal 660 tonnes per day air-separation unit (oxygen at 98% purity), retrofit to the boiler of oxy-combustion and flue-gas recycle plant and equipment, and the installation of a CO_2-compression and purification plant. In a second stage, some 50–75 tonnes per day of liquid CO_2 will be transported to a geological storage site.

The project is being carried out by a joint Japanese/Australian consortium, including an Australian power utility, Japanese boiler manufacturer, a coal miner, drilling-service provider and others. The total project cost is estimated at $A206 million. Significant government support is being received, including from the Commonwealth Government Low-Emission Technology Demonstration Fund ($A50 million), Queensland Government ($A35 million) and the Australian Coal Association COAL21 program ($A68 million).

The project achieved financial closure in July 2008 and has now moved to implementation, with all major plant-supply contracts in place and refurbishment works underway. A two-month test run on air firing was carried out in early 2009 to commission and confirm the reliability of the main plant after the refurbishment. The oxy-fuel conversion will take place during 2010, with operation on oxy-fuel from mid 2011. A three-year period of operation on oxy-fuel is envisaged. During this period it is expected that design and operational requirements can be optimised.

The objectives of the demonstration include the establishment of design and operating requirements for large-scale retrofit and new-build oxy-fuel plants, including geological-storage aspects. It is also expected that realistic data on capital and operating costs for the technology will be produced. It is expected that successful demonstration of the technology at this scale will allow the large existing conventional air-fired boiler fleet in Australia and overseas to be targeted for conversion.

The conventional swirl-burner, wall-fired boiler will be modified to include recycle of flue gas, so as to satisfy boiler heat-transfer requirements under oxy-fuel conditions. Approximately 10% of the boiler combustion products will be bled from the flue-gas stream to a gas processing unit. Here it will be washed, compressed, dehydrated, cleaned of mercury and then liquefied by chilling. The process will produce $0.9\,kg\,s^{-1}$ of 99.9% pure CO_2 with non-condensable gases, including residual oxygen, nitrogen from casing leakage and other minor gases vented to atmosphere.

A number of locations for the geosequestration component of the project have been considered, with sites within the Denison Trough presently being assessed in more detail. The Denison Trough is some 200–250 km from the power-station site and it is envisaged that the liquid CO_2 would be transported by road tanker. The program envisages injection to be carried out for up to two years with a total of about 50 000 tonnes of CO_2 to be injected.

The Denison Trough option contains a number of producing and near-depleted natural-gas fields. As a result, the region is well explored geologically, with gas-exploration and production wells in place and extensive seismic investigation previously done. The region is also tectonically stable and does not contain potable water aquifers.

Preliminary studies have indicated that the region has ample storage capacity from moderate-scale injection activities. However, the permeability of the conventional gas-bearing sands is relatively poor (< 100 mD) and natural-gas production typically requires stimulation to achieve commercial flow rates.

4.2.2 Hazelwood 2030 Project

The Hazelwood 2030 project entails a number of initiatives being carried out by International Power at the Hazelwood Power Station in Victoria. The \$A370-million project includes installation of coal-drying equipment to boost power-station efficiency and of an amine-based PCC plant, with an innovative approach applied to sequestration of the CO_2. The project is supported by the Commonwealth Government LETDF fund (\$A50 million) and the Victorian Government ETIS project (\$A30 million).

The PCC component of the project comprises an amine-based capture process, designed and constructed by the Process Group and the CO2CRC to suit Victorian brown coal. The 25 tonne per day (9000 tonnes per annum) plant treats a flue-gas stream taken from the No. 8 boiler, with 15 tonnes per day of CO_2 being then used to neutralise alkaline-ash transport water at the station. Reaction of the CO_2 with calcium ions in the ash water results in sequestration of the CO_2 as calcium carbonate. The remainder of the CO_2 produced is available for sale.

International Power has also made the facility available to the CO2CRC for field testing of new PCC technologies, including CO2CRC's patented "UNO" solvent and for assessment of process-heat integration alternatives.

4.3 Pilot-Scale Demonstrations

4.3.1 CSIRO Pilot Plant – Loy Yang Power Station

CSIRO in 2008 commissioned a 1000 tonne per annum, amine-based PCC plant, operating on a slip-stream of flue gas from the Loy Yang power station. The project is designed to assess the performance of the amine technology where the brown-coal flue gas is not pre-treated with either flue gas desulfurisation (FGD) or de-NO_x.

The project forms part of the broader Latrobe Valley Post Combustion Capture Project – a joint collaboration between Loy Yang Power, International Power Hazelwood, government, researchers from CSIRO's Energy Transformed Flagship and the CO2CRC (including Monash and Melbourne Universities).

4.3.2 CSIRO Pilot Plant – Munmorah Power Station

CSIRO, in a joint initiative with Delta Electricity, has constructed a \$A5 million, 3000 tonne per annum PCC pilot plant at the Munmorah Power Station in NSW. The plant is trialling a chilled-ammonia-based capture technology under Australian conditions with low-sulfur black coal, but without FGD or catalytic de-NO_x. Successful operation of PCC without FGD or de-NO_x equipment would substantially reduce costs to retrofit PCC to the large fleet of black-coal-fired plant existing in Australia and elsewhere that do not already have FGD and de-NO_x equipment fitted.

The project was commissioned in mid 2008 with test work to be completed in 2009. Results will then be compared with those from amine-based technologies being tested elsewhere, to allow selection of a technology for a 100 000 tonne per annum project, expected to be in operation in NSW by 2013 at a cost of around $A150 million.

4.3.3   Tarong Power Station
CSIRO, in conjunction with Tarong Energy, is constructing an amine-based PCC pilot plant at the Tarong Power station in Queensland. The plant is designed to capture 1500 tonnes per annum of CO_2 and, as with the Munmorah and Loy Yang plants, operates without FGD or de-NO_x applied up-stream of the pilot plant. The trials are expected to be complete by 2011, with results being used to provide a comparison with the ammonia-based PCC technology at Munmorah.

4.3.4   Gaobeidian Power Station
CSIRO, in partnership with China's Huaneng Group and China's Thermal Power Research Institute, has constructed a PCC pilot plant at Gaobeidian Power Station in China. The 3000 tonne per annum, amine-based plant will provide performance data from a black-coal-fired power plant that is fitted with FGD and de-NO_x equipment. This will enable comparison with alternative capture technologies being trialled at Munmorah and Tarong Power Stations in Australia.

4.4   Storage Projects

As noted previously, the preparation of legislation permitting the exploration for, and assessment of, sequestration sites presently remains in the development stage. However, one significant pilot injection project has been approved to date, while a number of scoping studies have been carried out to assess storage potential based on existing geological data.

4.4.1   Otway Pilot Project
The CO2CRC has initiated Australia's first demonstration of geological storage of CO_2 at a site near Warrnambool in south-western Victoria. The project involves the extraction of CO_2-rich natural gas (approx. 80% CO_2, 20% CH_4) from the Buttress 1 well, the dehydration and recompression of this gas, and its transmission by pipeline approximately 2.5 km to the injection well, CRC1. An existing well, Naylor 1, has been re-equipped to allow monitoring of the injected CO_2. The cost of the project is estimated at $A 40 million. A schematic of the project is shown in Figure 1.

CRC1 was drilled to a depth of 2249 m into the Eumeralla Formation. It is fitted with a 194-mm surface casing down to 400 m, and a 114-mm 13Cr

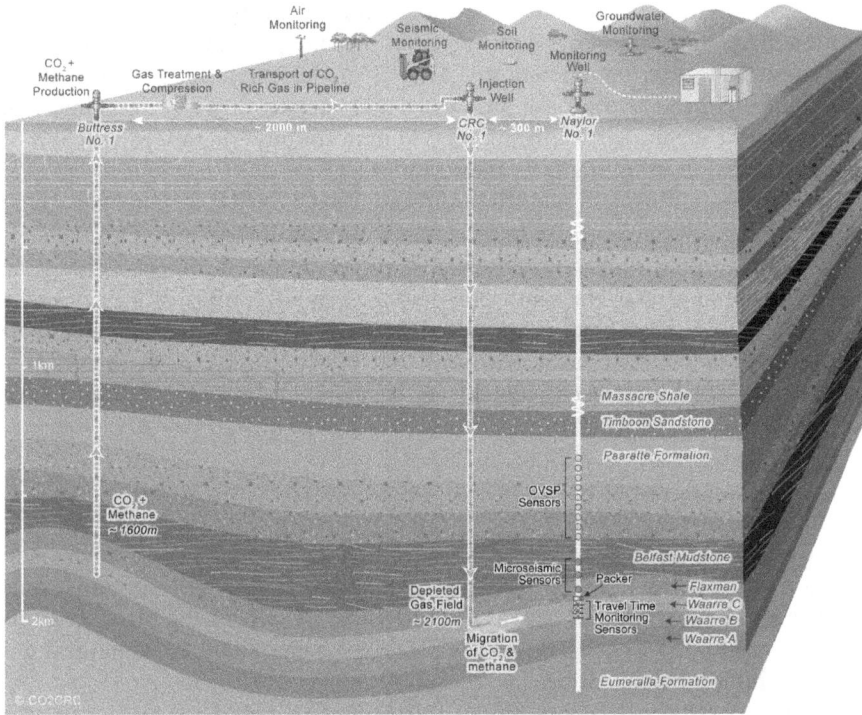

Figure 1 Schematic of Otway Basin Pilot Project (courtesy of Cooperative Research Centre for Greenhouse Gas Technologies).

production casing to total depth. The production casing is perforated over an injection interval from 2039 to 2055 m to allow injection into the Waarre C formation.

The Waarre C formation is a depleted gas producer which has high permeability (0.5–1.0 D) and is capped by the Belfast mudstones to provide a high-quality seal. There are additional seals and saline aquifers between the injection level and the surface, providing multiple barriers to leakage.

Injection commenced on 2 April 2008 and is projected to run for 12–18 months, with up to 100 000 tonnes of CO_2 to be injected. The injection rate is 120 tonnes per day at a maximum pressure of 20 MPa.

The Otway project has multiple objectives. At a research level, it provides an opportunity to test and verify reservoir simulation and modelling approaches, and to develop and verify monitoring techniques. As this is the first such project in Australia, it also provides the opportunity to help shape the development of a satisfactory regulatory framework, and to provide guidance on the steps required to obtain approvals to carry out such a project. Finally, it is intended to provide a clear public demonstration that CO_2 can be safely produced, transported, injected, stored and monitored to help in establishing public credibility for geosequestration.

To achieve these ends, extensive monitoring is being carried out on the project. This includes down-hole vertical seismic profiles to provide high-resolution images in the immediate environs of the borehole. This will detect fine changes in the level of the gas–water interface to verify volume injected. Micro-seismic techniques are being used to monitor any fracture production or reactivation of existing faults in the strata.

Extensive monitoring of wide areas around the CO2CRC tenements is being conducted to detect variations in CO_2 that may indicate leakage. This includes monitoring of ground water, sampling of the unsaturated air zone above the water table and also of the atmosphere above ground level.

A Stage 2 of the project is proposed whereby the CO_2-rich gas would be injected into a poorer-quality reservoir that exists below the Waarre A formation. This work has been included in the scope of a proposed extension of the CO2CRC.

4.4.2 Santos Cooper Basin Hub

The Cooper Basin is Australia's largest on-shore oil and gas producer and supplies natural gas to much of eastern Australia. It presently contains some 160 gas and 75 oil fields, as well as the Moomba gas-processing plant. Over its productive life, the Cooper Basin has produced some 8 000 PJ of natural gas and 250 million barrels of oil.

The Cooper Basin has a number of attributes that make it attractive as a CO_2-storage hub. Exploration in the field began in 1954, with first production in 1969. Extensive exploration over the production history has ensured that the basin geology is very well understood, with over 2300 wells drilled to date, while the hydrocarbon accumulations confirm the existence of strong seals. The Basin is centrally located with respect to, though distant from, a number of major sources of CO_2 emission, including Sydney, the NSW Hunter Valley, Brisbane and Adelaide. The site also contains extensive gas-processing infrastructure, including underground gas-storage facilities.

Santos, a publicly-listed Australian oil and gas company, holds large acreages in the Cooper Basin and is the operator for the Cooper Basin Joint Venture. Santos[22] has proposed the development of a CO_2-storage hub in the Cooper Basin based initially on use of existing, near-depleted, oil and gas reservoirs. Ultimately it may be possible to utilise the extensive saline aquifers that underlie the Basin.

It is proposed that the project be developed in three phases, with initially the collection and injection of all CO_2 presently separated at the site as a result of the natural-gas processing activities. The second stage would link to the development of a clean-coal power project in New South Wales or Queensland and the development of the CO_2-transport infrastructure, while Stage 3 would take CO_2 from the major point sources in New South Wales, Queensland and South Australia. The initial stage would capture approximately one million tonnes per annum of CO_2. It is estimated that the facility could accept up to 20 million tonnes per annum, with an ultimate storage capacity of up to 1 billion tonnes of CO_2.

The cost of Stage 1 of the development is estimated at over $A700 million. While the project should benefit from revenues due to enhanced oil recovery, commercial viability will still require additional financial support. Later stages would depend primarily on revenue from storage activities.

Santos has identified a number of factors key to enabling the project to proceed, including:

- Oil and carbon prices adequate to ensure commercial viability.
- Ability to finance the construction with either public or private monies.
- Legislative rights to store CO_2.
- Right of way to transport CO_2.
- Customers.

The Moomba Carbon Storage project was recently (March 2009) placed under review, as key financial enablers in terms of sufficiently-high oil and carbon prices are not yet in place.

4.4.3 Geodisc

In 1999, the Australian Petroleum Cooperative Research Centre (APCRC) initiated a program to assess the sedimentary basins of Australia for their CO_2-storage potential.[23] The project was completed by the CO2CRC with some 300 sedimentary basins in Australia being assessed. Of these, a short-list of 48 was drawn up, based initially on suitable geological characteristics, such as, thickness, depth, structure, *etc.* Within these basins, some 77 potential injection sites were ranked by risk factors, including storage capacity, injectivity potential, containment, the absence of viable natural resources, economic and technical viability, *etc.*

The potential storage capacity of the different suitable sites was found to be highly variable. Of the sites studied, 43% were hydrodynamic traps which contained about 94% of the probable storage capacity of all sites. Hydro-dynamic traps were found typically to have a large storage capacity relative to other trap types. Here probable capacity is based on considerations of likely storage efficiency, injectivity potential, containment strength, site physical details and the possibility of access to the site being constrained by the existence of valuable resources such as hydrocarbons.

On a national scale the total probable capacity of the 65 sites studied was around 740 billion tonnes of CO_2, equivalent to 1600 years storage at Australia's 1998 CO_2-emission levels.

The Geodisc study also considered major CO_2-source locations in Australia, to assess matching with the storage sites. It was concluded that the top 50 point sources around Australia represented about 90% of the potential storable emissions. These sites were then further collected into emission nodes, with a view to identifying reductions in costs associated with the required infrastructure.

Figure 2 illustrates the major sedimentary basins in Australia, with the main emission nodes and potential sequestration sites identified. It is clear that the

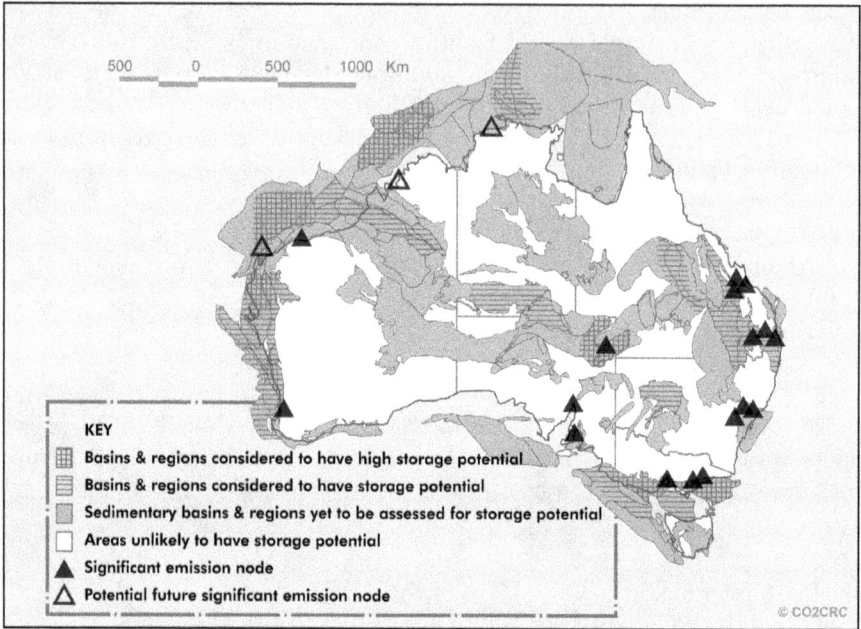

Figure 2   Major sedimentary basins of Australia showing their expected CO_2 storage potential, overlain with the major emission nodes in the country. (Courtesy of the Cooperative Research Centre for Greenhouse Gas Technologies).

north-west area of Australia and Victoria contain most of the high-quality storage potential, but relatively fewer emission sources. By comparison, NSW, Queensland and South Australia have major emission sources, but potential storage sites are generally either remote from the source, expected to be of poorer quality or unexplored. It is therefore clear that while Australia's storage capacity is expected to be high, significant CO_2 transport infrastructure and exploration will be required to capitalise on this opportunity.

An overview study of storage opportunities within coal seams was also carried out by the CO2CRC as part of the Geodisc program.[24] This study concluded that Australia's coal seams have the potential to store almost nine billion tonnes of CO_2, with the added potential benefit for substantial increase in production of coal-bed methane. The study also pointed out that saline aquifers and depleted gas fields in the same coal-bearing basins offer additional potentially viable geological sequestration options.

The study highlighted a number of key risks associated with storage in coal, including the need to compromise between increased depth, which renders the likelihood of future extraction less likely, and the low permeability that tends to characterise deep coals. However, it was noted that the large coal-bearing basins are close to the major CO_2 sources in the eastern states, offering benefits in terms of reduced transport-infrastructure costs.

5 CCS Legislation and Regulation

Responsibility for legislation in Australia is shared between Commonwealth, State and local governments. At the present time, none of these jurisdictions addresses all aspects that would be required for the approval of a CCS project encompassing capture, transport, injection, storage and monitoring. However, regulatory frameworks do exist at both Commonwealth and State levels in relation to the minerals and petroleum-extraction industries that have clear parallels with CCS projects.

Specifically, Commonwealth government legislation regulates activities associated with off-shore petroleum exploration, recovery and production, while State legislation covers petroleum-related operations both on-shore and in coastal waters, and also on-shore mineral-extraction operations. Here the Commonwealth off-shore legislation applies to waters greater than three nautical miles from the coast. In general, the States administer the Commonwealth off-shore legislation on behalf of the Commonwealth and, in most cases, the State coastal waters legislation mirrors the Commonwealth Act.

Within this pre-existing framework, a number of different situations existed in respect of CCS activities. For example, the South Australian Petroleum Act 2000 and the Queensland Petroleum and Gas (Production and Safety) Act 2004 provide for transport by pipeline and storage in natural reservoirs of substances, including carbon dioxide, regardless of the source location or the activity that produced it. The Commonwealth, State and Northern Territory Petroleum (Submerged Lands) Acts provide a mechanism for authorising and regulating the capture and storage by a production licensee of CO_2 separated from the petroleum stream, as part of the integrated petroleum operations of the licensee. However, CCS streams from other sources (*e.g.* from an on-shore power station or off-shore petroleum operations unrelated to the production licence) were not able to be authorised.

No provision existed for the exploration or assessment of reservoirs specifically for geosequestration. It is also likely that geosequestration will involve transport of CO_2 between jurisdictions, including between states and between on-shore capture activities and off-shore storage sites, where differing requirements may be applied. No mechanisms existed for resolution of conflict of interest between different operations, nor were there any regulations in regard to risk assessment, site storage or long-term monitoring of the stored CO_2. There was, therefore, a clear need for both the Commonwealth and the States to amend and update legislation relating to CCS activities.

5.1 Regulatory Guiding Principles

The Commonwealth and State governments initiated actions to update the legislative framework for CCS across Australia, with the establishment by the Ministerial Council on Mineral and Petroleum Resources of a CO_2 Geosequestration Regulatory Working Group in September 2003. This resulted in the release in 2005 of the *Carbon Dioxide Capture and Geological Storage Australian Regulatory Guiding Principles*.[25]

Six key issues were seen as fundamental to a CCS regulatory framework:

Assessment and Approvals Processes. In order to suitably protect rights and responsibilities of all concerned parties, it is important to ensure that assessment and approvals deal with all stages of a project and incorporate best current practice.

Access and Property Rights. CCS projects are expected to be of high cost and require long periods of access to sequestration sites. Project proponents will then require a high degree of certainty about access to a selected injection site, while ownership of the CCS stream at each stage of a CCS project needs to be established with clearly defined rights and responsibilities.

Transportation Issues. Transportation constitutes an integral component of CCS projects and the unique characteristics of CO_2 must be recognised in legislation, particularly the differences between CO_2 and natural gas, such as:

- CO_2 produces acid when dissolved in water.
- CO_2 is heavier than air, odourless and non-flammable.

Monitoring and Verification. Monitoring and verification are required to ensure safe performance of CCS projects, to verify the amount of CO_2 injected and to confirm the continued storage of the CCS stream in its intended location.

Liability and Post-Closure Responsibilities. Clearly defining liabilities and post-closure responsibilities associated with CCS projects is essential.

Financial Issues. The general fiscal system of taxes, insurance and funding of post-closure obligations and liabilities must be clear.

5.2   Commonwealth Offshore Petroleum and Greenhouse Gas Storage Act 2006 (OPA)

Following the development of the Regulatory Guiding Principles, the Commonwealth Government amended the Offshore Petroleum Act 2006 (OPA), with the Offshore Petroleum Amendment (Greenhouse Gas Storage) Act 2008. The amended act has been renamed Offshore Petroleum and Greenhouse Gas Storage Act 2006 (Cth) (OPGGS Act).

The OPA was identified as the most appropriate vehicle to implement a CCS access regime, due to the extensive processing of CO_2 in the petroleum industry, the need to establish determinable rights between these two industries and the similarities in the technologies used by both. As a result, the legislation incorporates a licensing framework for CO_2 broadly similar to the existing regime for petroleum activities.

The Act establishes greenhouse gas titles as follows:

- Greenhouse Gas Assessment Permit: authorisation to explore for storage formations and injection sites.

- Greenhouse Gas Holding Lease: enables lessee to retain tenure over acreage while a commercial source of greenhouse gas for injection is obtained.
- Injection Licence: authorises injection and storage of greenhouse gas to an identified storage formation.
- Greenhouse Gas Search Authority: allows for preliminary assessment of acreages not yet covered by other authority.
- Greenhouse Gas Special Authority: authorises preliminary assessment in regions near to boundaries of permit areas or, subject to conditions, to areas where title is already granted.
- Greenhouse Gas Infrastructure Licence: authorises construction and operation of infrastructure relating to greenhouse gas substances in the licence area.
- Pipeline Licence: expends current pipeline licence to provide for greenhouse gas transport.

The Act also creates categories for gas storage formations including:

- Potential Greenhouse Gas (GHG) Storage Formation: where the title-holder reasonably expects the formation has potential to store GHG.
- Identified Greenhouse Gas Storage Formation: a formation reasonably believed to be able to effectively store GHG.
- Eligible Greenhouse Gas Storage Formation: a formation confirmed as suitable for the permanent storage of a particular amount of GHG injected over a particular time frame.

The Act allows for overlapping petroleum and CCS titles that may be held by different owners. In circumstances where carbon sequestration or petroleum operations could impact negatively on operations under another title, mechanisms are provided to resolve the conflict.

Pre-existing petroleum rights are recognised. Where they exist, the CCS proponent will be required to demonstrate to the regulator that the proposed activities will not pose risk of significant adverse impact on current, or future, petroleum activities. However, where there are no pre-existing rights, the two industries compete on a level playing field.

Once an Injection Licence or Production Licence has been issued, the relevant Licence holder is protected in the same way as a pre-commencement petroleum title holder. This provides greater certainty once major investment decisions have been made.

The assurance of safe and secure storage is managed through a number of mechanisms, including a requirement for an approved site plan that addresses issues such as reservoir integrity, plume-migration modelling, risk assessment, and monitoring and verification to ensure that the injected CO_2 behaves as predicted.

The decommissioning and closing of the site requires the title holder to report to the Minister on:

- Modelling of the behaviour of the stored gas.
- Assessment of the expected migration pathways and short- and long-term consequences of migration.
- Suggestions for monitoring, measurement and verification of the injected-gas behaviour.

The Minister can issue a site-closure certificate, only when it is demonstrated that the CO_2 is behaving as predicted, and does not pose a significant risk to the geological integrity of the area, the environment or human health.

The operator will bear all liability under both statutory and common law for up to 20 years after the completion of the storage project. This term includes up to 5 years following application by the operator for the minister to issue a provisional closure certificate, and a further 15 years for a site-closure certificate to be issued. Following the issuing of the site-closure certificate, all liabilities revert to the Government.

Under the Act, greenhouse gas storage acreage is released in a manner equivalent to petroleum acreage *via* gazette notice, inviting applications for a Greenhouse Gas Assessment Permit on a work-bid or cash-bid basis.

5.2.1   Release of Carbon Sequestration Acreage

Following the passing of the OPGGS Act, the Commonwealth Government in early 2009 announced details of ten off-shore areas designated specifically for the assessment of their greenhouse gas storage potential. The release covers areas across five off-shore basins adjacent to Victoria, South Australia, Western Australia and the Northern Territory.

5.3   State CCS Legislation

Following the release of the Commonwealth Offshore Petroleum and Greenhouse Gas Storage Act, a number of the states have moved to enact legislation to govern geosequestration activities in on-shore areas. Queensland and Victoria have opted to create new Acts specifically for geosequestration activities; South Australia has followed the Commonwealth approach of amending the respective petroleum production act; while Western Australia has created project-specific legislation for the Gorgon project. The other states have yet to release details of their proposed legislation.

A summary of the key State on-shore legislation is given in the following sections.

5.3.1   Queensland

The Petroleum and Gas (Production and Safety) Act 2004 (P&G Act) previously provided for the transportation by pipeline and storage of CO_2 in underground reservoirs by a petroleum lease holder, while the Petroleum and

Gas (Production and Safety) Regulation 2004, as amended in December 2005, enabled the evaluation of natural underground reservoirs for CO_2 storage.

In 2009, the Greenhouse Gas Storage Act 2009 (Qld) (GGSA) was enacted to introduce a new tenure regime, specifically governing the discovery and use of underground reservoirs for the storage of CO_2. The Act follows a similar format and many of the tenure concepts and processes to those in the earlier P&G Act.

The Act regulates "GHG streams", which are defined as a stream of CO_2 or a substance that 'overwhelmingly' consists of CO_2 in a gaseous or liquid state. It does not permit the sequestration of other greenhouse gases.

The Act creates a number of authorities, including:

- GHG injection and storage data-acquisition authority: allows a GHG permit or lease holder to carry out geophysical surveys on land close to that in an existing GHG tenure.
- GHG exploration permit: permits exploration for underground geological structures suitable for injecting and storing GHG streams.
- GHG injection and storage lease: permits the injection, storage and monitoring of GHG streams.

The Act vests ownership of all GHG reservoirs in the state, regardless of the form of land tenure under which the reservoirs reside. If a GHG-exploration permit holder discovers a reservoir, it can apply for an injection and storage lease or for a declaration of a 'potential storage area' where there are currently no GHG streams available to utilise that reservoir.

The grant of a GHG lease is subject to ministerial discretion, and will require that the applicant be able to demonstrate that GHG storage will occur at the site within five years after the grant of the lease.

Surrender of a GHG lease requires that the Government be satisfied that the risks associated with the carbon storage have been reduced "as much as possible". Prior to surrender, the carbon-storage operator is potentially liable for any adverse consequences or impacts on third parties associated with the operations. On surrender of the GHG lease, ownership of the CO_2 passes to the state; however, the Act contains provisions that may leave a GHG-storage operator open to long-term liabilities.

The Act allows for the granting of GHG tenements that overlap existing mining or petroleum tenements, and geothermal-exploration permits. Provisions, including a public benefit test, are made to resolve conflicts of interest that may occur between holders of alternative tenement rights.

GHG-permit and lease holders need to take into account potential groundwater issues. Approval of work programs for GHG permits and development plans for GHG leases, require that the program or plan must be referred to the minister administering the Water Act 2000 (Qld) for approval, to ensure that there is no undue or adverse impact on groundwater. However, in general, environmental requirements in terms of the GHG-storage operations will be detailed in the accompanying environmental authority.

5.3.2 Victoria

The Greenhouse Gas and Geological Sequestration Act 2008 (Vic) (GSGSA Act) was passed in late 2008. This Act is separate from, but establishes a system of greenhouse-gas titles similar to, that of the current Victorian on-shore petroleum titles.

Petroleum and CO_2 storage titles can co-exist; however, under the GCGSA Act, the Minister must not allow certain greenhouse-gas operations, where those operations present a "significant risk of contaminating or sterilising other resources in the permit area". This provision is subject to a public-interest criterion which may allow the CO_2 storage priority, subject to the petroleum-tenement holder receiving due compensation.

The GSGSA Act is silent on long-term liability, meaning that long-term liability appears to stay with the CO_2 storage-site operator indefinitely. However, because the State assumes responsibility for monitoring and verification activities after a licence has been surrendered, it is not yet clear where long-term liability will fall in practice.

5.3.3 South Australia

CCS operations in South Australia will be regulated under the existing Petroleum Act 2000 (SA) which applies to regulated resources, defined to include natural underground reservoirs. The Petroleum Act authorises a system of licences applicable to both GHG and petroleum operations. An application for production licence must demonstrate that the operations are, or will within two years, become commercially feasible, or the licence may be cancelled by the minister.

Ownership of the regulated resource is vested in the Crown to the point where a substance is "produced". In terms of a natural gas storage, "produced" is defined as when a reservoir is used for the storage of the gas. Therefore, ownership of the injected CO_2 should remain with the CCS operator. No provision is made to transfer liability to the State on closure of the site.

The Act provides that a licensee will bear liability for any reasonable costs incurred by the state in rehabilitating serious damage due to activities, and that the operator may be directed to undertake action to prevent or minimise environmental damage.

5.3.4 Western Australia

Western Australia has pursued a project-specific approach to the regulation of CCS operations for the Gorgon project. In that case, the Barrow Island Act was amended to allow ministerial approval of a range of CCS titles.

6 CCS Challenges in Australia

It is clear that substantial progress has been made toward implementation of CCS in Australia. Relevant research has been carried out at numerous locations across the country for over a decade; substantial sums of money have been made available by Governments and industry for R&D and to support pilot and demonstration projects. Furthermore, policies have been formulated at State and Commonwealth level to create business environments compatible with CCS projects and specific legislation has been drawn up to govern storage of CO_2.

Despite this intensive effort, no commercial-scale CCS project has yet been implemented in Australia, although several currently are in the planning stages. It is clear that widespread implementation of CCS is immensely challenging, especially to a country of Australia's size and nature.

As a result of efforts to date, many pre-conditions for commercial-scale CCS projects now exist. These include:

- Legislation enabling CCS activities in a number of jurisdictions.
- Policies and support programs designed to assist CCS projects to overcome the natural hurdles for new technologies seeking market entry.
- A sound knowledge base of CCS technology within government, academia and industry.
- A growing skill base in research and in design and construction of CO_2 capture and storage facilities.

However, a number of key challenges remain to be addressed before CCS can make a significant contribution to CO_2 reduction in Australia. These include:

Commercial. At present, CCS projects are not commercially viable, as the costs of capture must be borne by the project and there is as yet no commensurate return for storing the CO_2. The lack of commercial viability relates to the high cost of the CO_2-capture plant; transport infrastructure and storage operations; the large impact that CO_2 capture has on plant performance, particularly on output and efficiency in electricity generation; and the current electricity market structure, which actively discriminates against high-cost generators while not ascribing a cost to CO_2 emissions.

It is of note that numerous companies are investing heavily in developing projects incorporating CCS in the expectation that CCS is now, or will soon be, a commercially viable proposition and that there will be a competitive advantage in possessing the technology. It is expected that the carbon pollution-reduction scheme to be introduced in 2011 will, in part, redress the issues associated with the financial viability of CCS.

It is clear that large capital grants or subsidies will be required for early CCS projects to be viable, and that such funding will probably be provided by several parties who will each want to participate in the project. This will necessitate the establishment of a range of complex legal agreements which, experience has shown, can take considerable time to negotiate. One way to minimise this delay is to focus initially on the principles of involvement of each participant, to determine why they wish to be involved and what outcomes are being sought.

Storage. Legislation is now in place to enable the active exploration for and proving-up of geological storage reservoirs. However, while nation-wide and, in some instances, basin-wide surveys have been carried out, these are generally preliminary in nature. Proving the capacity and quality of a reservoir for a commercial-scale CCS operation will require a substantial commitment of time and money, probably involving extensive seismic and geological surveys, the drilling of multiple deep exploration wells, extensive measurements of reservoir and seal-rock geophysical and geochemical properties and extensive modelling of the behaviour of the injected fluids.

Currently, the only potential near-term, commercial-scale storage sites in Australia have been characterised as a result of past petroleum production activities. With the exception of the Gippsland Basin oil and gas fields, these are far from the major CO_2 sources in eastern Australia. These regions also contain currently active petroleum-production operations, where CO_2-storage activities may be complicated by petroleum-recovery requirements and pre-existing title. Identification of feasible near-term, large-scale storage sites will require substantial commitment of capital to high-risk exploration operations.

The assessment of storage for a CCS project must reach a bankable level of certainty to allow the project to proceed to a final investment decision. Clearly, a project cannot commit to high levels of capital expenditure on the plant and the capture component, until there is certainty with respect to the storage component. This introduces an additional timing issue which has to be managed by a project, to bring together the capture and storage components which will normally be handled by different groups, with different skill sets.

Industry Convergence. Viable storage options are a make-or-break requirement for the success of CCS. However, resource exploration is characterised by large, up-front and at-risk exploration expenditure, with expectations of high return to justify the risk. Neither of these characteristics is shared by infrastructure businesses, such as, electricity generation or natural gas pipelining. Here, investment requirements are relatively well known, but return on investment is low and may well be regulated. The intimate linking of industries with such different characteristics to create an integrated CCS industry, will require substantial vision, time and effort.

Legal. A basic legal structure is now in place in most states to allow the carrying forward of CCS projects. However, there are significant differences between jurisdictions with regard to key issues, including ownership of the stored CO_2 during operations and following project closure, with respect to long-term liability for any environmental damages or personal loss that may be suffered by others as a result of the CCS operations; for monitoring and verification procedures required to be satisfied for closure certificates to be issued; or for the steps to be followed to achieve environmental and project approvals. All of these factors increase the cost, time and complexity of project development.

Further development of legislation and, in particular, the drafting of regulations is required.

Infrastructure. Large-scale infrastructure for the transport of CO_2 does not presently exist in Australia. The majority of Australian pipeline infrastructure is owned and operated on a commercial basis with some regulatory overview. The longer-term vision of capture nodes linked to storage nodes by pipeline will, under present circumstances, require that a suitable infrastructure operator arise, with sufficient long-term capital to fund the pipeline during construction, and for the initial years of low toll income while capture projects ramp up. The conflicting requirements of keeping initial capital cost at a manageable level (low), but building enough capacity to be economically efficient as demand for pipeline capacity grows, are very difficult for a commercial pipeline-infrastructure company to manage.

Technology Obsolescence. Large-scale applications of CCS technology to the power industry are rare and the technology, while technically viable, is presently the subject of much research. It is, therefore, likely that second generation technologies with substantially improved performance will arise. Technology selection is, therefore, a key issue for some CCS projects; delaying project-design lockdown pending the delivery of new technology may ultimately prove beneficial to the project. A number of Australian CCS projects remain in this technology selection phase.

Timing. Australia has set targets of 5 to 15% reduction below year 2000 emission levels, by 2020. This requires a reduction in emissions below the 'business as usual' case of between 25 and 35% of the year 2000 emissions, to be achieved within the next 11 years. In absolute terms, this equates to emissions in the range of 140 to 190 million tonnes of CO_2e per year to be sequestered or forgone by 2020.

It must also be recognised that design, approval, financing and construction of large projects such as commercial-scale CCS would typically require some 5 to 7 years. Similarly, to prove-up a reservoir to store multi-million tonnes per year of CO_2 will take years, even assuming that the preliminary exploration stage is fast. For CCS to take a significant contribution to the GHG-reductions targets set by Australia will, therefore, require an acceleration of the current rapid build-up of expenditure and effort that has been directed toward CCS technologies to date.

A National Project? The implementation of CCS technologies in Australia has now moved through R&D and early pilot-scale projects, to the serious investigation and planning for commercial-scale demonstrations. To date, this has involved many groups and association of groups working collaboratively. In view of the scale of the challenge ahead there may be a need for the setting up of a national project, with the strength to assemble the required human and financial resources required, minimise duplication and galvanise public interest and support.

In spite of these challenges, much has already been achieved with CCS in Australia. The Commonwealth Government has set ambitious targets for near-term implementation of commercial-scale CCS. CCS therefore remains a key technology for Australia to achieve substantial CO_2 reductions in the future.

References

1. Australian Bureau of Statistics, *Australian Economic Indicators*, Cat No 1350.0, Commonwealth of Australia, Canberra, 2009.
2. Australian Bureau of Agricultural and Resource Economics, *Energy Update 2008*, Commonwealth of Australia, Canberra, 2008.
3. Department of Climate Change, *National Greenhouse Gas Inventory 2006*, Commonwealth of Australia, Canberra, 2006.
4. Department of Climate Change, *Carbon Pollution Reduction Scheme – Green Paper*, Commonwealth of Australia, Canberra, 2008.
5. Department of Climate Change, *Carbon Pollution Reduction Scheme – Australia's Low Pollution Future*, Commonwealth of Australia, Canberra, 2008.
6. Department of Climate Change, *Exposure Draft of the Carbon Pollution Reduction Scheme*, Commonwealth of Australia, Canberra, 2009.
7. http://www.cslforum.org
8. http://www.asiapacificpartnership.org
9. Australian Coal Association, *Reducing Greenhouse Gas Emissions Arising from the Use of Coal in Electricity Generation. A National Plan of Action for Australia*, Canberra, March 2004
10. http://www.newgencoal.com.au
11. P. Feron, *Post Combustion Capture (PCC) R&D and Pilot Plant Operation in Australia*, presented at the 11th IEA GHG Post Combustion Capture Network Meeting, Vienna, May 2008.
12. J. K. Kaldi and C. M. Gibson-Poole, *Storage Capacity Estimation, Site Selection and Characterisation of CO_2 Storage Projects*, Report No RPT08-1001, Cooperative Research Centre for Greenhouse Gas Technology, Canberra, 2008.
13. *Latrobe Valley CO_2 Storage Assessment*, ed. B. Hooper, L. Murray and C. Gibson-Poole, CO2CRC Publication RPT05-0108, Cooperative Research Centre for Greenhouse Gas Technology, Canberra, 2005.
14. D. Harris and D. Roberts, *Summary Report of Pilot Gasification Trials in Australian Coals*, Technology Assessment Report No 70, CRC for Coal in Sustainable Development, Kenmore, Queensland, 2008.
15. B. Buhre, L. Elliott, C. Sheng, R. Gupta and T. Wall, *Prog. Energy Combust. Sci.*, 2005, **31**(4), 283.
16. G. Allinson, N. Nguyen and J. Bradshaw, *APPEA J.*, 2003, **43**(1), 611.
17. P. Massarotto, V. Rudolph and S. Golding, Queensland Government Mining J., 2007, Summer, 44.
18. www.zerogen.com.au
19. http://www.coolimbapower.com.au
20. http://gorgon.com.au
21. http://www.callideoxyfuel.com
22. J. Anderson, *Moomba Carbon Storage (MCS)*, presented at Coal21 Conference, Crowne Plaza, Hunter Valley, Australia, 2007.
23. J. Bradshaw, B. E. Bradshaw, G. Allinson, A. J. Rigg, V. Nguyen and L. Spencer, *APPEA J.*, 2002, **42**(1), 25.

24. B. E. Bradshaw, G. Simon, J. Bradshaw and V. Mackie, *Geodisc Research: Carbon Dioxide Sequestration Potential of Australia's Coal Basins*, CO2CRC Report No: RPT05-0011, Cooperative Research Centre for Greenhouse Gas Technology, Canberra, 2005.

25. Ministerial Council on Mineral and Petroleum Resources, *Carbon Dioxide Capture and Storage, Australian Regulatory Guiding Principles*, Canberra, 2005.

Underground Coal Gasification (UCG) with Carbon Capture and Storage (CCS)

DERMOT RODDY* AND GERARDO GONZÁLEZ

1 Introduction

Underground Coal Gasification (UCG), taken on its own, offers the prospect of increasing the world's usable coal reserves by a factor of at least three,[1] and doing so in a cost-effective manner with the added benefit of avoiding the risk to life and limb that accompanies deep coal mining. Coupled with Carbon Capture and Storage (CCS), it also offers a highly attractive carbon management plan whereby most of the expected CO_2 emissions are permanently sequestered back in a coal seam void that has recently been created. If this can be made to work at a commercial scale, suddenly we have an excellent bridging technology that enables the world to use fossil fuels for longer, in an environmentally acceptable manner, whilst waiting for the full panoply of renewable energy technologies to become fully cost-effective at scale. That is the reason for the intense interest in UCG-CCS.

UCG is a process which can safely exploit the energy of coal without having to mine it using conventional techniques. This is achieved by gasifying the coal, *in situ*, by utilising directional drilling technology developed by the oil and gas industry. Accurately controllable boreholes are drilled from the surface into appropriate coal seams and these are used to introduce hot steam and oxygen to gasify the coal. The resultant hot gas mixture – known as synthesis gas or 'syngas' – is drawn to the surface *via* neighbouring boreholes, where it can be used for a wide range of purposes, such as driving turbines to generate electricity, or for manufacturing products ranging from plastics, to gas and liquid transport fuels.

*Corresponding author

Issues in Environmental Science and Technology, 29
Carbon Capture: Sequestration and Storage
Edited by R.E. Hester and R.M. Harrison
© Royal Society of Chemistry 2010
Published by the Royal Society of Chemistry, www.rsc.org

The UCG process creates voids deep underground following gasification of the coal. Furthermore, as most UCG processes are oxygen-fuelled, carbon dioxide and water vapour are the only gaseous exhaust streams produced after gasification, thus making separation and capture of the carbon dioxide relatively simple and cheap. The process is, therefore, particularly compatible with CCS. In most deep UCG scenarios, the proposition is that the captured carbon dioxide is consigned to storage within the UCG voids and surrounding strata *via* the same boreholes used for injection and extraction. A combined UCG-CCS project aims to achieve a reduction in CO_2 emissions of up to 85% compared with conventional coal-fired power generation. Such a project therefore offers a very attractive solution and is the only process yet devised, that offers integrated energy recovery from coal and storage of CO_2 at the same site.

UCG-CCS can sit happily alongside some other CCS approaches described in other chapters. In particular, where CO_2 collection and transmission pipelines can be linked together, new degrees of freedom for carbon management emerge.[2]

This chapter begins by outlining the history of UCG since its first inception in 1912 and explaining why it is becoming economically viable now, almost a hundred years later. With regard to technology, we cover the key areas of drilling; controlled underground gasification; environmental monitoring and risk assessment; the mechanism for storing supercritical CO_2 in coal seam voids; and the gas processing technologies that pertain to the various end uses that can be envisaged for the syngas arising from gasification – with the notable exception of CO_2 separation technology which is covered elsewhere in the book. References to the open literature are included where appropriate. There is also an extensive body of knowledge available through the Underground Coal Gasification Partnership (UCGP) of member companies.

2   A Brief History of UCG

The world's first UCG experiments were carried out by Sir William Ramsay in 1912 in County Durham, North East England. Although these experiments were successful, further progress was halted by the First World War. Limitations in the technology and relatively low prices for oil and gas have meant that for many years the technology has been largely unexploited. Nevertheless, a number of countries around the world have developed UCG operations. Most notably these have included the former Soviet Union and China, where commercial-scale operations have been conducted. Feasibility studies or trial operations have also been conducted in Australia, the USA, Spain, South Africa, India and the UK.

The focus in early trials, such as those carried out in the former Soviet Union from 1930 and in China, was on finding ways of controlling the underground gasification process; developing effective ways of drilling the injection and production wells; and developing effective ways of linking the two wells as a precursor to gasifying the coal. Most of this work can only be done at a significant scale in real coal seams, so trials tended to be expensive. In time the trials became bigger and longer in duration, up to a point where it became sensible to build power generation facilities, to run off the syngas being produced from an ongoing programme of coal seam gasification. At least one power generation facility from the 1970s is still operating today in Uzbekistan.

Europe started to look at UCG during the 1950s and 1960s. Meanwhile, trials in the USA developed new variants on UCG technology and also explored more carefully the potential environmental impacts.[3] During the 1970s and 1980s, thirty two separate tests were carried out, along with a large supporting development programme. The main centre of interest was the Powder River Basin in Wyoming. One outcome from all of this was the development of a new control technique called "moveable injection".[4] Some work was also done on routes to chemicals production from syngas as an alternative to power generation. India started looking at UCG in the 1980s.

By the early 1990s, UCG was considered to be largely technically proven, but unfortunately that coincided with the start of the era of low-cost natural gas. Consequently, interest in UCG diminished and much of the development activity stalled, except in China and also in Europe, where UCG came to be seen as an alternative to mining in respect of deeper, thinner coal seams.

Trials were carried out at depths in excess of 500 metres by a European consortium (UK, Spain and Belgium) between 1992 and 1998 at Teruel in Spain.[5] These trials demonstrated the efficacy of a new technology termed "CRIP" (controlled retractable injection point), in which the nozzle releasing the steam and oxygen into the coal is gradually drawn back out of a horizontal stretch of borehole as the coal surrounding it is gasified. The trials demonstrated that UCG in deep seams is feasible with minimal environmental impact at surface level. They also found that the gas produced had a calorific value similar to that achievable with surface-level gasification of coal, and showed that at the higher operating pressure involved there were significant volumes of methane produced, in addition to the normal syngas components. Following on from these trials, a review of the feasibility of UCG in the UK was carried out, leading to a trial under the Firth of Forth in Scotland.[6]

The costs of performing UCG using a CRIP are dominated by the costs of geological exploration and drilling. However, parallel developments in the oil and gas industry over the last few decades have led to the opportunity to use effective forms of directional drilling (first used in the Spanish trial) to access coal for UCG. Many countries with indigenous coal resources are now re-examining the opportunities offered by UCG, including South Africa and Australia. The driving forces include: security of energy supply; the cost of syngas relative to natural gas and crude oil; and the option of using the syngas for power generation in a combined-cycle power plant.

The Underground Coal Gasification Partnership (UCGP) has estimated that around 20 billion cubic metres of syngas has been produced to date from UCG activities across the world, equivalent to about 15 million tonnes of coal. To date, the largest power generation plant based on UCG is a 100 MW steam turbine plant at Angren in Uzbekistan.

To date, there are no examples of integrated UCG-CCS projects being constructed anywhere in the world. CCS projects collect CO_2, pressurise it and store as a supercritical fluid, *i.e.* a fluid with the density of a liquid, but the compressibility, viscosity and diffusivity of a gas. In order to store CO_2 within UCG voids, the voids need to be more than 800 metres underground, where the hydrostatic pressure and temperature are such that they maintain the CO_2 in its supercritical form. Most UCG projects around the world to date have targeted coal seams which are far shallower than this.

Further details on the recent history of UCG are included in Section 5, as part of the review of the current state of deployment of the technology worldwide.

3   The Economic Case for UCG

Even the most optimistic projections for renewable energy point to the need for one or more "bridging technologies", to close the world's growing energy gap during the next few decades whilst development work on renewable energy technology continues. The most optimistic estimates for build time on new nuclear technology, similarly point to at least a transitional role for large-scale use of fossil fuels. Most estimates for conventional oil and gas suggest a peak in production in the next 10 to 15 years, with reserves being substantially depleted by 2050 for oil and 2070 for gas. Compared with these, coal reserves are usually estimated at more than 200 years. Concerns about reliance on oil and gas extend also into areas such as political stability, national energy security and price volatility.

Against this background, if UCG technology can effectively increase usable coal resources three-fold, by releasing the energy from coals that are inaccessible using conventional mining techniques, the economic case for UCG as a bridging technology becomes compelling. It has been estimated by UCGP that the total coal resource in the world is around 8 trillion tonnes, of which nearly half could be suitable for UCG. The estimated coal resource for the United States alone is 1 trillion tonnes.

In broad terms, the economic case for UCG is a balance between positive and negative factors. On the positive side, UCG offers a low-cost route to emissions reduction; the cost is lower than for surface gasification plants because there is no need to mine, store or transport coal, there are no solid residues to dispose of, and there is no need to purchase a gasifier; it converts an abundant natural resource into a secure, economic supply of gas; it enables stranded coal resources (*e.g.* deep or offshore) to be converted into commercial reserves; there is a range of potential end uses and markets, *e.g.* power generation, heating, synthetic fuels, chemicals and hydrogen; it is largely immune to crude oil price swings (unlike conventional coal mining which relies on diesel-fuelled equipment and transportation); and it is cheaper than natural gas for power generation.

The negatives are: technical and commercial uncertainty, since the technology has not yet been widely deployed; syngas production rates and composition are variable compared with pipeline-delivered natural gas; open-cast coal mining (where acceptable) is cheap; there is a risk of ground subsidence and a risk of aquifer contamination (especially freshwater aquifers); trials and prospective site evaluation are expensive; there can be significant costs in transporting the syngas to the point of use; and planning approval processes are still under review in various countries. This is the background against which prospective investors have to make a decision.

Wide-scale proliferation of commercial UCG projects has not yet been seen. Historically, relatively high technological risk and the availability of comparatively cheap supplies of crude oil and natural gas have been the major barriers to the proliferation of UCG. Looking at 2008 data, against a natural gas price (in the USA) of $9 per million Btu (1 British Thermal Unit = 1.055 kJ) raw syngas can be produced *via* UCG in the USA for $1.8 per million Btu based on air gasification.[7] Using oxygen-blown UCG in Europe, the cost of syngas becomes $3.8 per million Btu. These figures are sufficiently low for UCG to look commercially attractive when oil and gas prices are reasonably high.

The economic case for UCG syngas displacing natural gas or coal for power generation is relatively straightforward. Alternative uses, such as conversion of syngas into liquid fuels, chemical intermediates or hydrogen, are more difficult because, whilst the added value is well known (and much higher than for power generation), there is a tighter requirement for syngas cleanup. Technologies for cleaning up UCG syngas to chemical feedstock standard are still under development and so the costs are less well known. There are several such projects underway at present, which should help elucidate the figures in due course.

Other factors beyond straight economics then come into play to tip the balance. The main considerations are: CO_2 emissions and climate change; air quality and power station emissions; a desire for some protection against volatile and rising oil and gas prices; and considerations of national energy and fuel security in a politically unstable world. In broad terms, the distribution of coal around the world is different from the distribution of oil and gas.

There is an additional factor to consider – and one that is difficult to place a value on in economic terms. More than 5000 deaths a year occur in the coal mines of China: four deaths for every million tonnes of coal mined. In Ukraine, the death rate is even worse: seven deaths per million tonnes. To put these figures into perspective, the last time death rates in UK coal mines were as high as they currently are in China was back in the 1920s; in the case of Ukraine the parallel figures occur way back in the 1880s. Much of this mining is linked to energy provision in support of the manufacturing of goods for export to

Western countries. UCG could provide an ethically acceptable way of enabling this economically driven, low-cost manufacturing activity to continue.

4   An Introduction to UCG Technology

The basic idea is that energy can be recovered from deeply buried coal seams by gasification of the coal *in situ*. This is readily achieved by introducing hot steam and oxygen or air to the coal *via* injection boreholes. The syngas produced from UCG is usable at the surface in gas turbines for power generation and other purposes. This section explores the technology that pertains to producing syngas by gasifying coal *in situ*.

In a sense, the uncontrolled combustion of coal underground is well known, as a result of the many coal fires that have occurred around the world. The controlled gasification of underground coal is a different matter. Here the challenge is to produce a syngas of consistent quality and with a high calorific value, by establishing optimised gasification conditions within a zone in the coal seam which moves along the coal seam in a controlled manner.[8] This requires a good understanding of the reaction kinetics and of fluid transport phenomena. To achieve it in practice also requires a good knowledge of the underlying geology. This includes such considerations as, changes in seam gradient, thickness and quality, as well as location of faults that interrupt the coal seam. The locations of permeable strata and aquifers relative to the target seam are also important. Much of this relies on a combination of data obtained from boreholes, from coring and logging, and from pilot holes, and on seismic data (2D or 3D).

UCG has been approached in many different ways. One approach (still favoured by the Chinese today) involves using mined tunnels or roadways to connect the injection wells to the production well. This is known as the "Long and Large (cross-section) Tunnel gasification method" (LLT).[9] Since it operates at a low pressure (atmospheric pressure), gas production rates tend to be low. The Chinese also use a two-stage UCG method in which steam and air are injected alternately, in order to keep the gasification temperature high. Another approach involves sinking a series of vertical wells (the "Linked Vertical Well method", LVW), and moving the injection point along to a new well whenever the current stretch of coal seam has been exhausted. The LVW method is used in South Africa. Specific methods have been developed for steeply dipping coal seams. A popular approach, especially in Europe and the USA, is the use of CRIP technology. However, even with CRIP technology there are variants for enabling several channels to be gasified in parallel – within the same overall seam or in seams that lie above one another. Since CRIP-based technology is becoming so popular now, it will be used to provide a focus for the remainder of this section.

The key enabling technologies for UCG are: gasification systems (including gasifier control techniques), directional drilling, monitoring and measurement systems, modelling and surface plant requirements. Gasification and directional drilling technologies are covered here. The key requirements for monitoring, measurement and modelling are picked up in Section 7, and surface plant requirements will be addressed in Section 9.

4.1 Gasification Configuration and Control

The important things to achieve are a continuous gasification channel through the coal seam and tight control of the gasification reactions. The CRIP system involves a burner attached to retractable coiled tubing which is used to ignite the coal. It operates by moving the injection system to a location within the target coal seam close to the production well, and igniting the coal to start the gasification reaction. The injection point is then gradually retracted away from the production well as the rate of gas production begins to fall off. A new gasifier can be started at any chosen location, as the CRIP is drawn back through the coal seam. Given the variability of coal seam thickness and its surrounding heat transfer parameters, the challenge of maintaining steady gasification conditions in order to generate consistent syngas at a steady rate is considerable. Gasification can be stopped by cutting off the oxygen supply, and re-started using the CRIP. Other gasifier control systems are theoretically possible, and include: the advancing gasifier system trialled in Liuzhuang, China, and the hydro-gasification process proposed by the Brookhaven National Laboratory in the US.

4.2 Directional Drilling

The target coal seam can be on-shore, near-shore or off-shore. In all three cases, a fundamental requirement is the ability to accurately and remotely direct drilling equipment to create the network of gasification channels, injection wells and production wells for a UCG operation. The current tendency is for the industry to standardise on a 4.75 inch hole size. Directional drilling is a proven technology in the oil and gas industry. The in-seam drilling of coal seams has been part of coal exploitation since at least the 1950s. It was common practice to drill horizontal boreholes from the coal galleries for exploration, degassing and de-watering of the coal ahead. Early trials of UCG all used "directed in-seam holes" between parallel galleries as the primary gasification path for the UCG process.

Underground steering of boreholes made its commercial entrance in the oil and gas industry around 1990, when operators established the benefits of lateral drilling for extending the life of wells and fixed drilling platforms, and for reaching inaccessible locations. British Petroleum plc, for example, drilled more than 9 km of lateral well at Wytch Farm, Dorset. Directional drilling has been applied successfully to date on distances of up to 14 km. Transfer of these steerable techniques from oil and gas exploration to the conventional coal mining industry was very limited. Exceptions were the few occasions when coal fields were encountered *en route* to the oil reservoirs, as in the Southern

North Sea and the UCG trial in Spain, which deliberately brought in oil and gas expertise. The coal industry, meanwhile, had little use for directional drilling, until it was discovered that coal bed methane (CBM) production could be enhanced by the drilling of in-seam boreholes from the surface. There is now considerable interest in injecting CO_2 in this manner, resulting in preferential adsorption of CO_2 and therefore release of methane. The current state of the art of directional drilling technology is summarised below.

Directional drilling using steerable assemblies is a widespread and growing activity for the extraction of fossil fuels. It is used extensively in the oil and gas industry, to extend the life of reservoirs, and most drilling companies serving this sector have a range of sensing and steering options to meet the requirements of the extraction industry. One common technology is termed "Measuring While Drilling" (MWD), which uses electronic sensors to identify variations in the local magnetic field and transmit the data to the surface. The control system reacts to the data by sending a signal to the positive displacement motor in the steerable assembly, thereby keeping the drill within the coal seam. Another common technology is called "Logging While Drilling" (LWD), which detects variations in gamma ray emissions. Coal tends to have a low gamma ray count while clays have high gamma ray count. A variety of homing devices are available for the accurate intersection of wells, usually based on magnetic sensors. They have proven to be satisfactory for coal, where the challenge is to ensure that the pilot hole through the coal seam meets up with the vertical production well that has already been drilled. A highly relevant development is the down-hole motor (DHM), which incorporates bends of fixed angles behind the drilling assembly to allow deflection of the drill head in a chosen direction. Rotary surface steering systems allow surface control of the direction of drilling, without the need to withdraw the drill. Such steering systems are compatible with DHMs to achieve rapid and accurate drilling.

Now that directional drilling has become common for CBM and enhanced CBM applications, there are specialist drilling companies around who supply services to CBM operators. The focus to date has been on reducing costs. UCG has a tighter requirement on accuracy. The ability of directional drilling to meet these requirements at an affordable cost is still under review.

There is an important requirement to be able to drill branch wells off a main vertical well, in order that an efficient network of gasification channels can be created. This branching technology has already been demonstrated in CBM applications. The difference with UCG is that UCG wells have to be lined using metal casings, and these are exposed to an aggressive thermal and chemical environment. Ideas for handling this have been developed, but have not, as yet, been extensively proven.

A number of features in the underground geology can potentially complicate the operation. For example, a fault may be encountered in the coal seam, or there may be a need to drill through a live aquifer without introducing a pathway for contamination, or it may be necessary to drill through old mine workings. Techniques have been developed for handling this using grouting, injection of concrete, insertion of secondary liners, *etc.*

5   Current Status of UCG Deployment Worldwide

As a measure of the spread of international activity, the figures for the number of UCG trials conducted up as far as July 2008 are: 200 in the former Soviet Union, 33 in the United States, and 40 elsewhere. Looking at it a different way, the figures for the number of tonnes of underground coal gasified to date, stand at: 50 million for the former Soviet Union, 50 thousand for the United States, 32 thousand for Australia and 10 thousand for Europe. It is clear that no other country is close to catching up the early former Soviet Union lead in terms of sheer scale, but some countries are moving very quickly now on projects which deploy advanced technology. A summary of key recent developments is given below on a region-by-region basis.

5.1   UK and Europe

The UK Department of Trade and Industry (DTI) produced a report which reviewed the most promising locations in the UK for UCG, and came to the conclusion that UCG has the potential to contribute to the UK's energy requirements, but placed question marks on its economic viability and environmental impact.[10] UCGP have estimated that the UK has about 17 billion tonnes of coal resources, much of which is suitable for UCG. Within England, the areas of potential UCG interest are North East England, East Yorkshire, North Wales and South Wales. A follow-on study has been carried out by Heriot-Watt University to look at the feasibility of UCG in the Firth of Forth.[6] Based on the available data, the report concluded that the potential project was economically viable, especially when it included CCS.

Interest in UCG continues in Russia, with Promgaz operating six UCG installations – three based on brown coal and three on hard coal. Several East European nations are known to be investigating UCG, *e.g.* Ukraine, Romania, Poland (which is leading the HUGE project – see Section 9) and the Czech Republic.

5.2   North America

The largest UCG facility in the USA is at Hoe Creek[4]. Growing concern over security of supply has rekindled US interest in UCG. It remains to be seen whether the new US administration will focus efforts on UCG, although it would appear reasonable to speculate that, given the country's enormous coal reserves (estimated at 250 years), UCG will become increasingly significant. Four UCG projects are currently under development in North America with commercial partners: two in Alberta, one in Wyoming and one in North Dakota. Interestingly, all of them are non-power projects (liquid fuels, hydrogen or Substitute Natural Gas).

5.3 Asia

China has an urgent need to grow its energy markets to support its expanding industrial base, and increased requirements for energy and chemical feedstocks. Currently, more than 70% of China's energy is derived from coal. The Chinese authorities have long recognised the potential of UCG, and China currently has the world's largest commercial UCG programme underway with over 16 trials having been conducted since the late 1980s.[9] There is also an environmental cleanup agenda, with nine out of the world's ten most polluted cities being in China. In addition to power generation, syngas is also used for industrial boilers and for ammonia synthesis.

In India, a government co-ordinated programme of UCG development has been initiated as of 2005. Over 30% of India's proven coal reserves are deep and the ash content is high – both of which make UCG attractive. The Indian government is developing a regulatory and fiscal regime for UCG. Coal India, the largest coal company in the world, has earmarked vast areas of its coal reserves for exploitation *via* UCG.

5.4 Australia

Following on from the UCG trial conducted during 2000–2001 by Linc Energy in Chinchilla, in which gas turbine manufacturers confirmed that the syngas produced was suitable for use on their equipment, plans are now well advanced to build a full-scale plant.[11] Linc Energy's primary focus is on combining UCG with gas-to-liquids technology (Fischer–Tropsch) to produce clean liquid fuels.[12] Carbon Energy are active in Queensland looking at coal deposits at a depth of 200–400 metres. Cougar Energy are also active in Australia and have looked at UCG for a 420 MW combined cycle power plant at Kingaroy.

5.5 Africa

Sixty eight per cent of Africa's primary energy comes from coal. South Africa is the fifth largest coal producer in the world. In South Africa, Eskom Ltd have undertaken a UCG feasibility study in the Majuba coal field north of Johannesburg, looking at 300 metre deep coal, and a feasibility study is underway in Swaziland. There are plans to develop a 1.2 GW UCG project, with a parallel integrated gasification combined cycle (IGCC) power plant.[13] Sasol have a long-standing interest in coal-to-liquids (CTL) technology. UCG has the potential to reduce the overall carbon footprint of CTL technology.[14]

6 Mechanism for Carbon Dioxide Storage in Gasified Coal Seam Voids

Sub-surface injection of gases is being successfully accomplished world-wide for different purposes and in different scenarios. This includes oil and gas operations, temporary storage and permanent disposal. As some examples of this, since the 1970s, the oil industry has been practising Enhanced Oil Recovery (EOR), which involves the injection of CO_2 into the oil reservoir, and more recently EGR for gas reservoirs. For almost 100 years natural gas storage in salt caverns has been practised to allow supply flexibility against a fluctuating demand, and acid gas has been injected underground since the 1990s as waste in Canada.

With regard to carbon dioxide geological sequestration, the Intergovernmental Panel on Climate Change[15] proposed the following main scenarios for underground storage of CO_2: active and depleted oil and gas fields, deep saline aquifers, deep unmineable coal seams and (marginally) caverns or basalts. Based on the expected storage capacity and current experience, most of the efforts in research and all of the commercial-scale operations have been directed to storage in oil and gas operations, depleted hydrocarbon fields and associated deep saline aquifers. That is the case with Sleipner, Weyburn, In-Salah and more recently, Snohvit. Their individual annual injection rates are in a range of 0.7 to 2 million tonnes of CO_2, and their total storage will amount to 17 to 20 million tonnes of CO_2 each. Injection into deep unmineable coal seams has been tested in laboratory and field, with disparate results. The Recopol project in Poland found major problems in the injection of the CO_2, due to the plasticisation and swelling of coal when the carbon dioxide is adsorbed in the coal matrix and displaces the methane. However, one option that has not been widely considered yet, and could be of great interest due to a combination of economic and technical aspects, is the storage of the carbon dioxide in the void created by UCG.

Carbon sequestration in a UCG operation is a serendipitous association of a source of CO_2 and storage site. As with the other major alternatives, it takes place in a sedimentary basin with particular geological features that are most appropriate for geological storage. The general requirements of a site for carbon geological storage are:

- Proximity to a source of carbon dioxide, to guarantee the supply of CO_2 and improve the economics of the operation by avoiding long transportation routes.
- Injectivity: the formation needs a certain permeability that allows the injection of the fluid.
- Storage capacity: sufficient to store the CO_2 produced during the plant lifetime.
- Containment: some trapping mechanism has to guarantee the permanence of the CO_2 for a considerable amount of time, *ca.* 1000 years.[15]

In addition to the generic site requirements, it is important to note the effect of the characteristics of the CO_2 stream in the constraints set on the storage site.

The first of the four requirements is fully achieved by the UCG-CCS configuration. The plant and CO_2 injection infrastructures, geological and geophysical studies will have already been developed for the UCG operation when the time comes for CCS. Though capture is the main component of the cost of CCS (70–80%), the cost reduction in the remaining 20–30% is very significant.

For the other three requirements, extensive experience and knowledge in underground coal mining, especially the mining methods of longwall and room-and-pillars, provides the insights for the assessment and preliminary prediction of CO_2 storage capability. Though there are different possible lay-outs for a UCG operation, one configuration is a chamber with a length of 500–600 m, 30–40 m wide and with the height equal to the thickness of the coal seam (1–3 m in the case of North East England). A longitudinal pillar would separate the gasification chambers. Compared to typical longwall panels,[16,17] which are 1–2 km long, 250–350 m wide and 1–3 m high, the above-mentioned UCG void would have an intermediate size between the cavity obtained with the longwall panel method and that of the room-and-pillar method.

Regarding the second requirement, injectivity, the absolute rock permeability is the most significant parameter to consider, the injection pressure is the second most significant parameter, and porosity is not as relevant. Thus, the injected amount varies almost linearly with permeability, while a 20% increase in injection pressure results in a 50% increase in injected CO_2 and a 100% increase in porosity shows only a 7% increase in injected CO_2. It is also important to note that the existence of a high permeability area (or "sweet zone") close to the injection point and a high contrast with the regional permeability, act as enhancing factors for injectivity.[18]

Permeability changes in the surrounding strata of underground coal mining operations have been thoroughly studied in the past to avoid water inflow and gas leakage.[16,17,19] The advance of the working face in a longwall panel causes the collapse of the roof behind and the formation of four vertical distinct zones:

- A caved zone, with broken blocks that have come off the roof. This broken material is referred to as "goaf". The zone extends vertically to between three and six times the coal seam thickness. The final permeability of this zone will depend on the grade of re-compaction of the goaf. In the case of longitudinal pillars along the cavity, these would help to decrease the compaction, resulting in a higher permeability of the goaf. As a general approach, permeability of the goaf can be estimated in a range of 1–20 darcy.[20] (A medium with a permeability of 1 darcy permits a flow of $1\,cm^3\,s^{-1}$ of a fluid with viscosity 1 cP under a pressure gradient of $1\,atm\,cm^{-1}$ acting across an area of $1\,cm^2$).
- A fractured zone with continuous fractures, joint opening and low stress. It extends to between 15 and 60 times the extraction height. Water and gas can drain directly to the void, as permeability in this zone can be up to forty times the original permeability.
- A bending zone where horizontal bed separation and joint opening takes place, increasing horizontal hydraulic conductivity. This can extend to 60 m ahead of the longwall face.
- A zone of intact rock which can suffer tensile fracture at the ground surface.

In longwall mining under the North Sea, changes in permeability of three orders of magnitude due to mining have been reported.[21] The usual figures assumed are an increase of 100 millidarcy in vertical permeability and 50 millidarcy in horizontal. Predictions with improved modeling techniques show increases of up to 35 times in vertical permeability, and 1000 to 2000 in overall permeability 50 meters above the mining void.[16]

In contrast, oil and gas reservoirs and deep saline aquifers show permeability values in the range of 0.001 to 1 darcy.[20] In the Alberta basin the average regional permeability is 0.006 darcy and the maximum is 0.4 darcy,[22] while in Teapot Dome the average is 0.03 darcy and the range is 0.01–0.1 darcy.[23] Therefore, it can be concluded that the UCG voids provide an excellent combination of enhanced permeability and size of the "sweet" zone so convenient for the practical injectivity of the CO_2. However, there are still some gaps in knowledge on the thermal effects on the overlying strata of the cavity, the presence of ashes and coal which can swell, and the effect of the injection pressure if it takes place before collapse. These have not been addressed, and they could affect injectivity.

The third requirement to consider is the storage capacity. Obviously, a remarkable advantage of UCG is the creation of a void that was not previously present. However, a rough estimation shows that the volume needed at 800 metres depth to store the CO_2 produced from the syngas, can be four or five times the volume occupied by the extracted coal. As with depleted hydrocarbon fields and deep saline aquifers, the storage capacity will depend on the specific storage, i.e. the compressibility of the fluid in the strata without exceeding the fracturing limit of the rock. Regulations for sub-surface injection of waste gases in Alberta set the injection pressure limit at 90% of the rock fracturing pressure.[18]

The last requirement of the site for CO_2 geological sequestration is that it can provide containment for a considerable period of (say) 1000 years. The trapping mechanisms have been described in detail for deep saline aquifers and depleted hydrocarbon fields.[15,22,24] First of all, a capping rock is required which acts as a structural seal for the buoyant supercritical CO_2. Then, with increasing temporal and spatial scale, the rest of mechanisms start to work: hydrodynamic trapping, residual gas dissolution in formation water, and mineral precipitation. All these processes will occur accordingly to the physical and chemical characteristics of the site and the CO_2 stream. In addition, in the case of UCG, the organic and inorganic by-products will certainly be mobilised with the CO_2 due to its solvent power. This will affect the chemical reactions in the water and the rock. It is also possible that the CO_2 is adsorbed in the coal matrix due to its higher affinity compared to other elements. In such a case, any resulting methane emissions would need to be factored into the overall greenhouse gas sequestration calculation.

As with permeability issues the experience of underground coal mining is helpful in quantification of containment in a coal basin. For many years, underground coal mining has been carried out under the sea in North East England. Ellington Colliery progressed several kilometres offshore, and no incident of sea water in-rush was ever reported.[25] Coal regulations in Britain state that the minimum distance from the seabed to mine works is 105 m, of which 60 m must lie within the coal measures.[26] However, it can be argued that the viscosity of water is higher than that of supercritical CO_2. In a more similar example, Whittles, Lowndes *et al.*[17] simulated the leakage of methane (with a viscosity of $1.75 \times 10^{-5}\,\mathrm{N\,s\,m^{-2}}$) into the mine workings in a UK colliery. They found that potential sources beyond 20 m of the working face would not leak into the roadway, due to the reduced permeability resulting from increased confining stress.

In conclusion, it can be expected that the trapping mechanisms described in the literature will work in the case of UCG, and that specific sites which meet the requirements can provide the necessary confinement. However, caution has to be exercised, as there is no experience of the large scale that CO_2 storage would imply. The cumulative effects of multi-seam extraction could also be important.

Another critical aspect which influences the mechanisms and requirements for the CO_2 storage site is the characteristic of the CO_2 stream to be injected. Anthropogenic CO_2 contains impurities which depend on the combustion process and the capture method. Some of these impurities are: water, sulfur dioxide, nitric oxide, hydrogen sulfide, oxygen, methane, hydrogen cyanide, argon, nitrogen, hydrogen, and particulates,[27] and they will affect the thermodynamics (density, viscosity, critical point) compared with pure CO_2.[28] In general, the presence of impurities decreases the critical temperature and increases the critical pressure.[29] As a matter of storage efficiency, CO_2 has to be injected in its supercritical state. A stream emanating from a post-combustion process shows the smallest difference compared to pure CO_2; but in the case of pre-combustion or oxy-fuel processes, the supercritical pressure can reach 83 or 93 bar while the critical temperature decreases to 29 °C or 27 °C, respectively.[29] This means that in the case of the North Sea with a sea depth of 50 m, a sea-floor temperature of 5 °C and a thermal gradient of 33 °C km^{-1} (ref. 30), the minimum depth below the sea floor required for maintaining the CO_2 in supercritical state is 800 m. It could be deeper than 900 m if an oxy-fuel combustion stream is to be stored.

7 Approaches to Environmental Risk Assessment

An Environmental Risk Assessment (ERA) should answer four questions: what can occur that causes adverse consequences; what is the probability of occurrence of these consequences; how severe can they be; and how can they be reduced? This last issue is often referred to as Risk Management, though it differs from the more general Risk Management which takes into account economic and social considerations. The steps in performing an ERA to address these questions are:

- Hazard identification that will reveal the contaminants and adverse situations that can be expected.
- Exposure assessment to describe the intensity, frequency and duration of exposure, routes of exposure and the nature of the population exposed.

- Effect assessment to describe the response of the receptors.
- Risk characterisation to provide an estimate of the likelihood for adverse impacts, with end-points definition and a qualitative or quantitative approach.
- Risk management, which includes monitoring and mitigation options.

The ERA is a critical aspect in the development of both UCG and CCS and so far the two aspects have tended to be addressed separately. However, they share some common ground. Usually, when the ERAs are undertaken for UCG and for CCS, each is split into surface operations and what happens underground. The handling of gases like syngas from UCG or CO_2 is common practice in industry, and environmental, health and safety, and other standards and regulations are well established. Also the engineering design is controlled, resulting in low failure rates. For example, CO_2 pipeline failures in USA in the period of 1990–2001 had a frequency of 3.2×10^{-4} incidents per km per year, and the frequency of oil-well blowouts in the Gulf of Mexico and the North Sea was of 10^{-4} incidents per well per year.[31] However, when it comes to the underground side, lack of previous experience and large uncertainties in geology and in hydrogeological, chemical and geomechanical behaviour appear. This is the area considered here.

The environmental risks associated with UCG were one of the reasons why UCG was not further developed in the 1970s and 80s in the US. Despite its indisputable benefits, initial trials at shallow depth (less than 200 m below surface) which produced groundwater contamination and even severe surface subsidence (as in Hoe Creek), discouraged the administration from pursuing new experiments. Nevertheless, more recent projects (like Chinchilla in Australia) have proved that with good site selection and operation control, groundwater contamination can be avoided.

The main risks to be considered in UCG are groundwater depletion, groundwater contamination, gas leakage and subsidence. Set against these, its outstanding environmental advantages are: the elimination of coal stock piles and coal transport and much of the disturbance at surface; low dust and noise levels; the absence of health and safety concerns relating to underground workers; the avoidance of ash handling at power stations; and the elimination of sulfur dioxide (SO_2) and nitrogen oxide (NO_x) emissions.

Most of the contaminants produced in coal gasification are included in the *List I* of the *Water Framework Directive* (2000/60/EC), which forbids release into a water body. Consequently, for a UCG operation to be permitted in the EU, any potential water contamination would almost certainly have to be restricted to water which had been previously classified as Permanently Unusable (PU).

The potential contaminants have been described and are well known.[32,33] They include organic compounds (phenols, benzene, PAHs and heterocyclics) and inorganics (calcium, sodium, sulfate, bicarbonate, aluminium, arsenic, boron, iron, zinc, selenium, hydroxide and uranium).

The approach that is proposed for environmental assessment of UCG is a risk-based approach with a source-pathway-receptor scheme.[34]

For CO_2 storage, the main risks identified are divided into three groups:[35] leakage, dissolution in formation water, and displacement. At a local scale, leakage into the atmosphere or the shallow sub-surface can cause asphyxiation to animals or humans, or affect plants and underground ecosystems. If the leakage is off-shore, it can affect the living organisms in the water column and the sea-bed and interfere with other legitimate uses of the sea. It is also important to make a distinction between sudden large releases and continuous small ones. Large releases from the storage site, however, are not expected unless there is a secondary accumulation close to the surface. The CO_2 injection could also initiate the mobilisation of methane which is potentially explosive. On a global scale, the leakage of carbon dioxide or methane would hinder the ultimate aim of the sequestration, which is to reduce the concentration of greenhouse gases in the atmosphere. If a leak contains some other contaminants, they can pose an additional threat if exposure times and concentrations are toxic or carcinogenic.

Regarding the risks resulting from dissolution in other fluids, the variation in pH of water caused by carbon dioxide can lead to the mobilisation of metals. As CO_2 is a very good solvent, it can also transport other organic contaminants and contaminate potable water.

The displacement of the CO_2 plume can induce seismicity or ground heave or subsidence. The brines pushed away can contact and contaminate potable aquifers or damage other mineral or energetic resources.

It is also important to note that, although a single specific site should not pose a high risk, there is a cumulative effect as the number of storage sites increases in response to the large-scale opportunity for global warming mitigation.

The coupling of UCG-CCS alters the hazards and the risks inherent in UCG or CCS on its own.[3] On the one hand, the operation takes place at a much greater depth than conventional UCG (so that the conditions for CO_2 in its supercritical state are met). This certainly decreases the risk of potable aquifer contamination and of subsidence effects on the surface. In addition, the physical response of the surrounding coal to the CO_2 injection can help reduce the migration pathways by swelling. On the other hand, the pressurisation of the cavity with the injection of the CO_2 can increase the risk of fracture propagation. Under these circumstances, the organic and inorganic by-products of gasification are forced out of the reaction chamber as the CO_2 is injected and pressurised in the void. The transport of organic and inorganic contaminants dissolved in the carbon dioxide through a fractured and porous medium is an area that has not been studied yet. Their concentration in the CO_2 plume and the changes in flow and chemical reactions between the CO_2, tars, ash, coal, brine and the formation rocks are unknown.

Although it is desirable to have a quantitative risk assessment, this has not always been possible in the ERAs performed for CO_2 storage projects. The reason for this is the lack of available data at this stage. Consequently, some studies have been performed using a deterministic analysis based on well-proven numerical simulation tools and examining different scenarios.[36]

There have been various projects around the world to perform the ERA of CO_2 geological storage using several methodologies to characterise risk.[37] These methodologies include Structured What If Technique (SWIFT) for qualitative risk assessment, Fault Tree and FEP (Feature-Event-Process) Database, deterministic modelling, FEP and Scenario development, and the expert-based Delphi process. Under the CO2STORE project, a performance assessment was employed in Schwarze Pumpe (Germany).[38] In Weyburn (Canada), a comprehensive deterministic risk assessment numerical simulation approach has been used. This was based on the Eclipse 3000 reservoir numerical model program, followed by a simplified stochastic simulation. The code CQUESTRA incorporated a link to a program (Crystalball) for Monte Carlo simulation. The comparison of both approaches, deterministic and probabilistic, illustrates the strengths and limitations of both.[39]

Two databases have been developed for the identification of the most important FEPs: Quintessa and TNO-NITG. The first one was created as part of the Weyburn monitoring project as a general reference source, while the latter aimed to be a tool for assessing risk.

Therefore, it seems likely that the most effective approach for UCG-CCS risk characterisation in the future will have to combine deterministic numerical models with probabilistic analysis. These models will have to be able to couple thermal, geomechanical, transport and chemical processes.

Monitoring is especially important at the local scale to be able to detect and mitigate any threat to safety. On a global scale, monitoring is important for dealing with issues related to credits for emissions reduction. Monitoring of the UCG operation is based on measurement of temperature, pressure and mass balance, as well as water chemistry and water level in monitoring wells. In the carbon dioxide injection phase, CO_2 stream injection rates, composition, temperature and pressure have to be monitored. The migration of the CO_2 can be checked with periodic sampling of air, water and soil; with pressure and logs in wells; with CO_2 flux chambers or using eddy covariance; and indirect techniques, such as geophysics or remote sensing. Since sealed wells represent the preferential pathway for leakage, their integrity has to be assured. That can be done with cement bond logs.[3,15,35] The mitigation options in the event of leakage are: recapping of leaking wells, reducing injection pressure, stopping injection, sealing the fracture, or transferring the CO_2.[40]

8   Linking UCG to CCS

UCG offers exceptional opportunities for CCS. From a technical point of view, the enhanced permeability and the presence of a "sweet" zone of injection offer a definite advantage over deep saline aquifers and depleted hydrocarbon fields, since the operational costs of injection will be reduced. In addition, whereas CO_2 transportation can account for 5–15% of a conventional CCS budget, no piping or shipping are required in the case of a self-contained UCG-CCS project. Regarding storage cost (which can be 10–30% of the total), most of this cost is usually related to geological and geophysical studies and drilling of injection wells. The UCG phase will have undertaken all of this already. Consequently, the cost reduction obtained by sequestering the CO_2 in the gasification reactor zone compared to storage in a deep saline aquifer could be very significant, involving little more than the capture, compression and subsequent monitoring of the injected carbon dioxide.

On a European level, the HUGE project (Hydrogen Oriented Underground Coal Gasification for Europe) is being carried out by a consortium which spans Spain, Belgium, the UK, Poland and Ukraine.[41] The scope of the technical investigation includes: hard coal, lignite, deep coal, shallow coal, and various options for CO_2 sequestration, such as linking into CBM technology, or much longer-term options like mineralisation. The project hopes to demonstrate minimal environmental impact and a potential link to Hydrogen Economy interests.

Some of the environmental technology issues associated with CO_2 storage have been covered previously in Sections 6 and 7. To some extent, however, the nature of the issues depends on the use to which the syngas is to be put, and hence the point at which the CO_2 is stored. This section therefore explores a range of such uses and draws out the implications for surface-level equipment in order to complete the picture. However, in general terms, the compositions and outlet pressures of UCG syngas at the surface are broadly comparable with those from surface gasifiers. The costs and methodologies for pre-combustion separation are therefore also comparable.

A commercial UCG-CCS project requires infrastructure above the ground, both on the injection side to process the gases fed into the underground gasifier and on the production side to cool, clean and dry the syngas produced. Injection and production facilities could be located on the same site or on different sites, depending on the well configuration. The major equipment items on the injection side are an air separation unit (ASU), a water treatment plant, gas compressors, and gas and water storage. Power and other utility connections are also required, along with a control facility.

The injection side facilities represent those required to supply the materials injected into the underground gasifier and the systems to control both the injection and the underground processes. The major equipment item on the injection side is an ASU, to extract the oxygen required for the gasification process. The ASU will be relatively large, so a cryogenic unit is likely to be the appropriate technology approach. There is a choice between air-blown gasification and oxygen-blown gasification. The oxygen route is preferred as it produces a higher calorific value syngas. Oxygen storage and a compression station are required to supply the gas to the injection well at the appropriate pressure. Nitrogen storage and compression are also required, assuming that nitrogen is used to control the combustion and to extinguish it when necessary.

The facilities required on the production side depend very much on the intended use of the syngas. The minimum requirement is likely to include gas cooling, cleaning and drying, as well as compressors to allow subsequent piping of the gas to customers and a flare stack for emergency disposal of gas. The expected composition of the syngas is roughly: 32% hydrogen, 16% carbon monoxide, 35% CO_2 and 17% methane, but the gas cleanup system also needs to be able to handle contaminants, such as, hydrogen sulfide, hydrogen chloride, ammonia, carbonyl sulfide, arsenic, cadmium, mercury, selenium, tars and suspended dust. It is expected that the syngas will be produced at a high temperature. Before it can be used, it may be necessary to cool the syngas and to clean it to remove dust and some higher hydrocarbons. A lot of work is being done in the field of coal gasification, generally to develop ways of cleaning and processing syngas without repeatedly cooling it down (*e.g.* from 250 °C to 40 °C) and heating it up again. The complexity of the cooling and cleaning process will depend to some extent on the future use of the syngas. The main options are outlined below.

The simplest option for use of the syngas, is to use it as a supplementary fuel in an existing power station after it has undergone some basic cleaning and drying. In this way, variations in syngas production rate can be tolerated since other fuel sources are available to make up any shortfall. Then there are the obvious advantages of low capital cost, existing grid connections, ability to handle the interruptions to be expected with a relatively new technology, *etc.* Note that this approach does not allow for any pre-combustion capture of CO_2. Therefore the carbon management plan would be to retro-fit post-combustion CO_2 capture to the existing power plant based on, for example, the well-established amine absorption process.

A second option is to use the syngas in a new power station. In a power station designed to use the syngas as its principal fuel, combined cycle gas turbine (CCGT) technology could be used, with the detailed design being adapted to the specific properties of the syngas. Compared to Option 1, this route offers greater efficiency in converting the energy in the syngas into electricity, since it is specifically designed for the purpose and is based on the latest CCGT technology. It should give an efficiency of about 55–56%, depending on the cooling system used, compared with a much lower figure of around 40% that might be obtained on a modified older power station. There is likely to be, however, less flexibility to adapt to changes in syngas volumes and composition compared to Option 1, unless the plant is designed as a number of smaller units (which would result in a higher capital cost) or natural gas is used as a supplementary or back-up fuel (increasing both capital and operating cost). The efficiency figures presented assume post-combustion capture of CO_2 designed into the new CCGT plant.

Besides power generation there are several other potential uses for the syngas, such as, hydrogen production or a gas-to-liquids process. These can be pursued by inserting an additional process step: a Water–Gas Shift reaction using steam over a catalyst to convert CO into CO_2 and additional hydrogen. This then allows pre-combustion separation of CO_2 for subsequent compression and storage (as described previously) without the need for handling large gas volumes. The hydrogen (which will also contain methane) has a number of potential uses, as described below.

Depending on the relative demand and prices for methane and hydrogen, there may be some advantage in separating the methane from the hydrogen after the shift reaction and CO_2 removal. The methane could be sufficiently pure to feed into the local natural gas grid, while the hydrogen is used as a low-carbon feedstock for power generation, a low-carbon feedstock for chemical processing, or a low-carbon transport fuel. Methane and hydrogen tend normally to be priced based on natural gas, so at times of high gas prices there could be attractive margins available – and perhaps a "green premium" under certain circumstances in the light of the low carbon footprint.

In all of the above cases, the CO_2 is captured either before combustion or after combustion and becomes available for storage as part of the carbon management plan. The preferred mode of storage for the purposes of this chapter is back in the original coal seam void as described at the beginning of this section.

9   North East England Case Study

Despite the long history of industrial coal mining in the region (starting in 1585), huge reserves of coal remain in North East England. Only about 25% of the total coal resources have been extracted. There is coal in abandoned mines that is technically mineable – perhaps as much as 500 million tonnes – but because the mines have not been maintained there would be expensive problems to overcome. Typically it is found that roadways deteriorate, and electrical and mechanical equipment are destroyed as a result of flooding once water pumping operations cease. There is also the possibility of trapped water leading to sudden in-rushes, which would present a risk to personnel in the event of underground mining being re-established. Some of this coal lies under land, some under the sea.

Then there is the coal that lies at depths considered uneconomic for conventional mining – both under the land and under the sea. This is particularly attractive for UCG when linked to CCS for the reasons given previously, and could amount to another 500 million tonnes. Previous estimates for the UK have suggested that between 7 and 16 billion tonnes of coal suitable for UCG could be available – and that ignores all coal below a depth of 1200 metres.

Project Ramsay (named after Sir William Ramsay) was established to assess the opportunity for UCG-CCS in North East England. As part of the project, specialists were commissioned to undertake a thorough review of all available data to determine the quantity and accessibility of coal suitable for UCG and for UCG-CCS, in near-shore areas of the North East coast. The study examined available data from a number of sources, including the Coal Authority, the British Geological Survey and BERR as well as data held by others, including Newcastle University.

Suitability of the area was considered for UCG and UCG-CCS taking into account coal seam thickness, depth of cover between the top of target coal seams and the sea bed where relevant, permeability of the relevant strata and, where relevant, stand-off distances from old workings. For UCG, a depth of 100 metres or greater was used, and for UCG-CCS, the minimum depth was increased to 800 metres to achieve the storage pressures necessary for CO_2 in its supercritical state.

The project considered both near-shore coal seams (<2 km) and off-shore coal seams (up to 10 km) at a few locations. The primary difference in approach between near-shore areas and further off-shore, is in the ability to reach the coal reserves from a wholly shore-based enterprise using directional drilling *versus* the need to utilise off-shore rigs. Cost analysis has shown that there is not a significant cost advantage in one approach over the other for the coal reserves under consideration. The initial high cost of off-shore rigs is broadly offset by the more expensive long-reach drilling costs associated with a near-shore project. The project found some very interesting coal seams and concluded that previous estimates of UCG-compatible coal resource had been conservative.

However, generating syngas from coal is only part of Project Ramsay. The region also provides ready energy and chemicals markets for syngas and its derivates, and therefore offers a genuine prospect for a commercial UCG-CCS operation. Geography is important. These markets need to be sufficiently close to the chosen UCG base to be serviceable economically. The siting of a UCG production operation in North East England, allows ready access to the process industry markets on Teesside for syngas and for derived gas products of methane and hydrogen. Equally, there are a number of existing power users and potential new investments in power generation plant at a scale that could make syngas a viable fuel. These options were all reviewed as part of the feasibility study.

From its inception, Project Ramsay always considered CCS as being an essential element of a successful UCG project. Consequently, detailed consideration has been given to those coal targets that are at sufficient depth to provide the option for CO_2 storage, and where significant revenues can be generated by providing a long term storage site for CO_2. Note, however, that CO_2 is also generated in large quantities by the same process and by power industries that provide a potential market for the syngas and its derivatives. Increasingly there is a business opportunity in CO_2 collection, transmission and storage. There is, therefore, the option of extending the envelope to take in CO_2 from other industrial sources and offer additional storage capacity. The voids created through the UCG process in deep coal seams provide a storage option for CO_2, whether that CO_2 was produced through use of UCG syngas or from other industrial activities.

The general options listed in Section 9 have all been identified for the target locations[2] and assessed. The study took account of specific local factors such as: the location of the most suitable coal seams relative to existing power plants and potential new power plants; the existence of pipeline corridors; the location of the most suitable coal seams relative to large industrial users of syngas and hydrogen; the potential for linking into other sources of CO_2 and CO_2-collection systems; the potential for connecting the UCG facility to the proposed new CO_2 pipeline linking the Eston Grange IGCC/CCS plant to storage locations under the North Sea, and so on.

The broad conclusions were that: previous estimates for UCG-compatible coal had been conservative; there are coal seams that appear to be usable for CO_2 storage following UCG; and some of the end uses for syngas are potentially attractive. The most attractive options in financial terms are: (1) to sell syngas, take back captured CO_2 and store it for a fee, and (2) to sell decarbonised hydrogen and methane. It was concluded that a project could be done in three phases, ramping up the scale over time in order to minimise technical risk and investor exposure. Such a project could deliver a profit before year 10 and therefore might warrant follow-on discussion on a more commercial basis with interested parties. If developed on a broader scale, it could act as a source of investment funds for the renewable energy sector, and thereby go beyond the ambition of being a bridging technology on the road to a sustainable energy future.

10 Concluding Remarks on Scale of Opportunity and Challenges

Recent estimates of the total remaining coal resource in the world quote a figure of 18 trillion tonnes.[42] Compared with the figures usually quoted for accessible coal reserves (typically tens of billions of tonnes), there is a huge gap between reserves and resource. UCG offers the tantalising prospect of closing that gap quite considerably. If the UCG opportunity can be linked successfully to emerging CCS technology, then the implications for addressing the twin challenges of climate change and finite fossil fuel reserves is truly game-changing.

There are particular attractions in developing a "self-contained" solution whereby clean use of coal and carbon dioxide sequestration are combined in the same location, without a need for material transfer. From a different perspective, there is an attraction in extending the envelope to include syngas export and CO_2 import/export. The former opens up the prospect of linking into lucrative opportunities beyond the power generation sector; the latter offers contingency plans on a number of fronts.

The main environmental challenges lie in guarding against: (1) aquifer contamination which can impact on potable water supplies, and (2) surface-level subsidence. The main economic challenges relate to the up-front costs associated with evaluating specific sites from a commercial perspective and from an environmental perspective, largely because of the drilling costs associated with characterising deep coal seams.

The pace is tending to be set by those countries and regions that are blessed with significant coal resources and are concerned about the greenhouse gas emissions agenda.

References

1. R. McCracken, *Energy Economist*, 2008, March, 16.
2. D. J. Roddy, *Proceedings of the International Coal Conference, Pittsburgh, USA*, CD ROM, 2008.

3. E. Burton, J. Friedmann and R. Upadhye, *Best practices in Underground Coal Gasification*, Lawrence Livermore National Laboratory, 2006, available on https://co2.llnl.gov/pdf/BestPracticesinUCG-draft.pdf

4. J. Friedmann, *Proceedings of the International Coal Conference, Pittsburgh, USA*, CD ROM, 2008.

5. DTI, *Underground Coal Gasification-Joint European Field Trial in Spain*, Project Summary No. 017 1999, 1999.

6. DTI, *The Feasibility of UCG under the Firth of Forth*, Project Summary No. 382 DTI/PUB URN 06/885, 2006.

7. M. Green, *Proceedings of the International Coal Conference, Pittsburgh, USA*, CD ROM, 2008.

8. M. Green, *Energy World*, 2007, April, 16.

9. J. Liang and S. Shimada, *Proceedings of the International Coal Conference, Pittsburgh, USA*, CD ROM, 2008.

10. DTI, *Review of the Feasibility of Underground Coal Gasification in the UK*, DTI/PUB URN 04/1643, 2004.

11. M. Blindermann and R. Jones, *Gasification Technologies Conference, San Francisco, USA*, 2002.

12. J. Peters, *3rd International UCG Partnership Conference, London, UK*, 2008.

13. J. Varley, *Modern Power Systems*, 2008, January, 23.

14. J. Brand, *Proceedings of the International Coal Conference, Pittsburgh, USA*, CD ROM, 2008.

15. IPCC, *IPCC Special Report on Carbon Dioxide, Capture and Storage*, Cambridge University Press, Cambridge, UK and New York, USA, 2005, pp. 195– 264,.

16. H. Guo, D. P. Adhikary and M. S. Craig, *Rock Mechanics Rock Eng.*, 2009, **42**, 25–51.

17. D. N. Whittles, I. S. Lowndes, S. W. Kingman, C. Yates and S. Jobling, *Int. J. Rock Mechanics Mining Sci.*, 2006, **43**, 369–387.

18. D. H. S. Law and S. Bachu, *Energy Convers. Management*, 1996, **37**, 1167–1174.

19. G. S. Esterhuizen and C. O. Karacan, *40th U.S. Rock Mechanics Symposium, June 25–29, 2005, Anchorage, AK, USA*, 2005.

20. P. L. Younger, D. J. Roddy and G. J. González, *7th Petroleum Geology Conference*, 2009.

21. C. J. Neate and B. N. Whittaker, *U.S. Symposium on Rock Mechanics*, 1979, 217–224.

22. B. Hitchon, W. D. Gunter, T. Gentzis and R. T. Bailey, *Energy Convers. Management*, 1999, **40**, 825–843.

23. L. Chiaramonte, M. Zoback, J. Friedmann and V. Stamp, *Environ. Geol.*, 2008, **54**, 1667–1675.

24. J. Bradshaw, S. Bachu, D. Bonijoly, R. Burruss, S. Holloway, N. P. Christensen and O. M. Mathiassen, *Int. J. Greenhouse Gas Control*, 2007, **1**, 62–68.

25. N. Bicer, Newcastle University, Ph.D. Thesis, 1987.

26. National Coal Board, NCB Mining Department Instruction for working under the sea, Production Department Instruction PI/1968/8 (revised 1971), R. S. Orchard (ed).
27. M. Anheden, A. Andersson, C. Bernstone, S. Eriksson, J. Yan, S. Liljemark, C. Wall, E. S. Rubin, D. W. Keith, C. F. Gilboy, M. Wilson, T. Morris, J. Gale and K. Thambimuthu, in *Greenhouse Gas Control Technologies 7,* Elsevier Science Ltd, Oxford, 2005, pp. 2559–2564.
28. H. Li, J. Yan, J. Yan and M. Anheden, *Applied Energy*, 2009, **86**, 202–213.
29. P. N. Seevam, J. M. Race, M. J. Downie and P. Hopkins, *7th International Pipeline Conference, Calgary, Alberta, Canada*, 2008.
30. S. L. Nooner, O. Eiken, C. Hermanrud, G. S. Sasagawa, T. Stenvold and M. A. Zumberge, *Int. J. Greenhouse Gas Control*, 2007, **1**, 198–214.
31. IEA, *Risk Assessment Workshop. Report PH4/31*, IEA GHG R&D Programme, London, 2004.
32. M. J. Humenick and C. F. Mattox, *Water Res.*, 1978, **12**, 463–469.
33. S. -Q. Liu, J. -G. Li, M. Mei and D. -l. Dong, *J. China Univ. Mining Technol.*, 2007, **17**, 467–472.
34. WS Atkins Consultants, *Review of Environmental Issues of Underground Coal Gasification-Best Practice Guide*, DTI, Birmingham, 2004.
35. A. Chadwick, R. Arts, C. Bernstone, B. Mayer, S. Thibeau and P. Zweigel, *British Geological Survey*, 2008, p. 267.
36. T. Espie, E. S. Rubin, D. W. Keith, C. F. Gilboy, M. Wilson, T. Morris, J. Gale and K. Thambimuthu, in *Greenhouse Gas Control Technologies 7,* Elsevier Science Ltd, Oxford, 2005, pp. 1277–1282.
37. K. Damen, A. Faaij and W. Turkenburg, *Climatic Change*, 2006, **74**, 289–318.
38. E. Kreft, C. Bernstone, R. Meyer, F. May, R. Arts, A. Obdam, R. Svensson, S. Eriksson, P. Durst, I. Gaus, B. van der Meer and C. Geel, *Int. J. Greenhouse Gas Control*, 2007, **1**, 69–74.
39. C. Preston, M. Monea, W. Jazrawi, K. Brown, S. Whittaker, D. White, D. Law, R. Chalaturnyk and B. Rostron, *Fuel Process. Technol.*, 2005, **86**, 1547–1568.
40. L. Protocol, *28th Consultative Meeting of Contracting Parties under the London Convention and 1st Meeting of Contracting Parties under the London Protocol*, 2006.
41. J. Rogut and M. Stein, *Proceedings of the International Coal Conference, Pittsburgh, USA*, CD ROM, 2008.
42. G. Couch, Underground Coal Gasification (to be published), IEA, 2009.

Towards Zero Emission Production – Potential of Carbon Capture in Energy Intensive Industry

DAVID POCKLINGTON* AND RICHARD LEESE

1 Overview

1.1 Greenhouse Gas Reduction/Issues for Energy Intensive Industry

Acknowledgement of the importance of climate change and the need for a global solution gathered momentum in the Earth Summit in Rio in 1992, where the developed countries agreed to voluntary reductions in their emissions of greenhouse gases to 1990 levels. Subsequently, the Kyoto Protocol – a Treaty of the United Nations Framework Convention on Climate Change (UNFCCC)- laid down legally binding reductions in the emissions of a basket of six greenhouse gases[i] (GHG), to be achieved over the period 2008-2012, on the basis of 1990 emissions.

The Treaty commits industrialised countries to cap their emissions and reduce their collective emissions of greenhouse gases by 5.2% compared to the year 1990, by 2008-2012. The European Community agreed to reduce its emissions by 8% and the UK by 12.5% within the burden sharing agreement.[1] Early ratification of Kyoto was slow and resulted in only small reductions in GHG emissions. By October 2008, 180 nations had ratified

---

*Corresponding author. On 2 March 2009 the British Cement Association merged with the Quarry Products Asssociation and The Concrete Centre to become the Mineral Products Association (MPA). The work referred to in this chapter was undertaken by the British Cement Association.
[i] Carbon dioxide, CO_2; methane, CH_4; nitrous oxide, N_2O; hydrofluorocarbons, HFCs; per-fluorocarbons, PFCs; sulfur hexafluoride, SF_6.

Issues in Environmental Science and Technology, 29
Carbon Capture: Sequestration and Storage
Edited by R.E. Hester and R.M. Harrison
© Royal Society of Chemistry 2010
Published by the Royal Society of Chemistry, www.rsc.org

the Kyoto Protocol, including Russia, but not the United States, China, or India.

The UK has been in the forefront of climate reduction legislation, introducing the Climate Change Levy in 2001,[2] the UK Emissions Trading Scheme in 2002,[3] and the Climate Act in 2008.[4] These measures were augmented, and to some extent duplicated, by the EU Emissions Trading Scheme,[5] the first phase of which was from 2005 to 2007. Phase II is from 2008 to 2012, to coincide with the first period of commitment under the Kyoto Protocol, and Phase III from 2013 to 2020.

The energy-intensive sectors of industry have played a significant role in the development of this legislation through:

- Detailed technical discussions between industry and government on the UK and EU legislation through the UK Emissions Trading Group (ETG), other government/industry groups,[ii] and bilateral meetings with government departments and their consultants.
- Supporting other initiatives to engage in emissions trading, through commitment to the "UK Manifesto on EU ETS" initiated by the Secretary of State for the Environment, speaking alongside government representatives at conferences in Prague, Berlin and Paris, and meeting delegations from the US Senate, California, Australia.
- Advocacy through National Associations, such as the British Cement Association and the CBI, and European bodies, such as CEMBUREAU (European Cement Association) and Business Europe.

In addition to the above legislative measures, the Treasury commissioned Sir Nicholas Stern to undertake an economic review of the potential impacts of climate change, and the resulting influential report was published in 2007.[6] The Stern Report stated that if annual emissions remained at current levels, then greenhouse gas levels would reach around 550 ppm CO_2e by 2050. However, the worst impacts of climate change could be avoided if atmospheric GHG levels were stabilised between 450–550 ppm CO_2e at a cost of *ca.* 1% of global GDP. Significantly, the report concluded that immediate action would cost substantially less than action at a later date.

A recent analysis by IPCC[7] indicated likely temperature rises:[iii] 1.8 °C rise (range is 1.1 to 2.9 °C) in a low scenario, and 4.0 °C (range is 2.4 to 6.4 °C) in a high scenario. Defra's Chief Scientist has stated[8] that whereas plans for *mitigation* should be based upon on a 2 °C rise in global temperatures, those for *adaptation* should assume a 4 °C temperature rise.

---

ii *e. g.* the Energy Intensive Users Group (EIUG), *Ad Hoc* Environmental Taxation Steering Group (AHETSG) and the Manufacturers' Climate Change Group (MCCG).
iii Global average surface air warming, °C, at 2090–2099 relative to 1980–1999.

The UNFCCC 15th Conference of the Parties, (COP-15), held in Copenhagen in December 2009, is providing a stimulus for political action with a view to achieving an international agreement in the *post*-Kyoto period:

- *26th November 2008:* the UK Climate Change Act received Royal Assent, thus introducing the world's first long-term legally-binding framework to tackle climate change.
- *1st December 2008:* the UK's Climate Change Committee was established and published its first report detailing, *inter alia*, plans to achieve an 80% reduction in greenhouse gas emissions. The cost of achieving this by 2050 is estimated to be 1–2% GDP (slightly higher than the estimate provided by Stern).
  - The Committee has proposed[iv] a GHG budget for the traded[v] sector of 1233 Mt CO_2e (2008–2012) dropping to 800 Mt CO_2e (2018–2022), which equates to a 35% reduction compared to a reduction of 19% by the non-traded sector.
- *17th December 2008*: the European Parliament approved the EU's Energy package, aimed at reducing EU greenhouse gas emissions by 20% in 2020 relative to 1990, providing 20% of the Community's energy from renewables and cutting primary energy use by 20%.

This is expected to cost up to €90 billion to 2020 and reduce the EU's GDP by 0.35–0.5% by that year. The Package is underpinned by four Directives on Renewables, Emissions Trading, Fuel Quality, and Carbon Capture and Storage,[9] and other measures on new car emissions, energy efficient buildings and energy labelling.

Each of these initiatives gives prominence to the role of carbon capture and storage (CCS). Of the options available for the mitigation of emissions in cement manufacturing, carbon capture and storage has been identified by the International Energy Agency (IEA) as the only technique that is likely to be feasible, and scenarios have been developed where by 2050 CCS is incorporated at *ca.* 50% of the manufacturing capacity of the developed world. Other commentators have suggested that since each element of carbon capture and storage has been demonstrated at a non-trivial scale, CCS is *"clearly feasible and no fundamental research breakthrough* [is] *required"*.[10]

Whilst, at one level, this statement presents an accurate general summary of the situation, it is not sufficiently focused to provide a sound basis for the development of government policy or industrial strategy. These demand a rigorous examination of many interrelated issues, and this chapter provides an overview of the factors that an energy-intensive industry must take into consideration when determining whether to install plant for the capture of carbon dioxide emissions.

[iv] The intended budget reflects a successful international climate change agreement at the United Nations Framework Convention (UNFCCC) Conference of Parties (COP 15) in Copenhagen 2009.
[v] Energy-intensive firms regulated by the EU Emissions Trading Scheme.

As such, the primary focus will be on capture rather than the subsequent transport and storage. Although these are important components of the abatement equation, the main initiatives in these areas are being undertaken by major emitters such as the power sector, and smaller sources such as the steel and cement industries are unlikely to take the lead on the development of a collection infrastructure or storage facilities.

2 Carbon Dioxide Emissions in Cement Manufacture

2.1 Cement Manufacture

There are two basic types of process for the production of Portland cement,[vi] using a variety of kiln types. Cement is produced by either "wet" or "dry" process, depending on the water content of the material feedstock, itself a function of the local geology. The wet process was the original rotary kiln process and is used to process raw materials with high moisture content. However, it has a higher energy requirement due to the slurry water that must be evaporated before calcination can take place.

The dry process avoids the use of slurry material and, as a result, uses less energy. Between these two extremes are "semi-wet" and "semi-dry" processes, although for each process the basic principles are the same and involve the following steps:

- Raw material preparation.
- Production of clinker in the kiln.
- Production of cement.

2.1.1 Raw Material Preparation (Quarrying and Grinding)

Following extraction, limestone/chalk, marl, and clay/shale are crushed at the quarry site and transported to the cement plant, after which the raw material mix is homogenised. If the local rock does not meet the designated raw material requirements in terms of calcium oxide, silicon oxide, aluminium oxide, ferric oxide, and magnesium oxide, then bauxite, iron ore or sand may be required to adjust the chemical composition of the raw mix to the requirements of the process and product specifications. In some cases waste-derived (or "alternative") raw materials replace naturally quarried material.

For the dry (and semi-dry) process, the raw materials are dried and ground together to produce a "raw meal" in the raw mill. In the wet (and semi-wet) process, the materials are ground with water to produce raw material slurry. Additional steps in the variant processes may be required, such as preparing raw meal "pellets" from dry meal (semi-dry process) or "filter cake" by dewatering of the slurry in filter presses (semi-wet process).

[vi] The most common grey cement has been given its name because of its resemblance to Portland Stone.

The resulting intermediate raw meal or raw slurry is then fed into the kiln in its homogenised state, the "kiln feed".

2.1.2 Production of Clinker

Typically, cement kilns are fired with fossil fuels; however, increasingly, waste-derived alternative fuels, such as solvents, tyres, biomass and refuse-derived fuels, are being used to replace coal and petcoke (petroleum coke). As the "kiln feed" enters the kiln system it is subjected to a thermal treatment process, consisting of the consecutive steps of drying/preheating, calcination, and sintering (formation of clinker).

Modern kilns have 4–5 stages of preheating/precalcining at around 850 °C before the material enters the rotary kiln. In the lower-temperature part of the kiln, calcium carbonate (limestone) decomposes to calcium oxide and carbon dioxide:

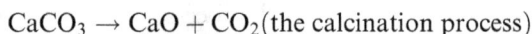

$$CaCO_3 \rightarrow CaO + CO_2 \text{(the calcination process)}$$

The material travels down the kiln towards the flame due to its slight inclination and rotation. In the high-temperature part of the kiln (up to 1450 °C), calcium oxides and silicates react to form dicalcium silicate (Ca_2SiO_4; belite). The dicalcium silicate reacts with calcium oxide to form tricalcium silicate (Ca_3SiO_5; alite). Small amounts of tricalcium aluminate ($Ca_3Al_2O_6$) and tetracalcium aluminoferrite ($4CaO \cdot Al_2O_3Fe_2O_3$) are also formed. The resulting material is referred to as clinker (sintered but not fused lumps). The clinker is cooled with air to around 60 °C in the clinker cooler, and the hot air is recovered for raw material and fuel drying.

Cement production is an energy-intensive process. Experience in the EU[11] confirms that the specific thermal energy demand ranges from 3000 to <4000 MJ tonne^{-1} clinker, for modern dry process multi-stage pre-heater/pre-calciner kilns (variations depend on variables such as raw material moisture, fuel mix and operating periods). Consequently, with its high energy requirements and emissions of carbon dioxide, and in view of its widespread use – globally, concrete is the second most widely used commodity after water – its consumption is a significant contributor to anthropogenic greenhouse gas emissions.

However, 60% of the CO_2 directly generated by the process results from calcinations, and only 40% from the combustion of fuels in the kiln. Further 'indirect' carbon dioxide is emitted *via* the use of electricity and material transport.

Recent estimates[12] indicate that the cement industry contributes around 3.8% to global GHG and around 5% to global CO_2 emission. In the UK, the cement industry emits around 9.7 Mt CO_2 per year (about 1.7% of the UK total emission).[vii] However, significant improvements have already been made; the 2007 direct emissions of CO_2 from the UK cement industry were 27% below

---

[vii] Calculated using National Air Emissions Inventory and British Cement Association Data.

the 1990 level, a year-on-year difference of 3.7 Mt CO_2. The improvements in CO_2 emissions relate to: energy efficiency improvements, plant rationalisation (the closure of old wet kilns and replacement with new pre-calciner kilns), fuel switching (replacing traditional fossil fuels with waste-derived alternatives) and a fall in production.

2.1.3 Production of Cement from Clinker

Portland cement is produced by inter-grinding cement clinker with natural or industrial gypsum (mainly anhydrite) in a cement mill. About 5% gypsum is added to adjust the setting time of the finished cement, and other minor additional constituents are also allowable within limits specified in the European standard.[13] Other cements, known as blended or composite cements, may be produced for specific applications and generally contain proportionally less clinker (and therefore less CO_2 is emitted in their production).

Composite cements contain other 'cementitious' (cement-like) constituents in addition, such as blast-furnace slag, natural or industrial pozzolans, *e.g.* volcanic ash or fly ash from power stations, where available. However, although using slag as a clinker replacement may reduce the CO_2 emitted in cement's production, it should be noted that blast-furnace slag originates from steel production which is in itself an energy and CO_2-intensive process.

Discussion is taking place regarding 'novel' cements (low carbon/low energy) that may be based on different chemistry and raw materials.[14] However, although their credentials concerning CO_2 emission reduction are promising, their commercial scale, technical appropriateness and availability in the UK are largely questionable.

2.2 Incentives for Carbon Reduction

2.2.1 Environmental

Until recently, the general public could be excused for its lack of awareness of climate change, but now that it is at the top of the media and political agenda it is hard to avoid references to its possible impact. A number of climate change sceptics still remain, yet theirs is a message that is finding less and less acceptance – even the (English) courts have acknowledged the scientific basis of global warming.[viii]

There is, however, a significant difference between commentators who express their views on global warming but are not required to put these into practice, and sectors of industry that have not been afforded this luxury for

---

[viii] In *Dimmock vs. Secretary of State for Education and Skills (2007) EWHC 288*, the plaintiff sought to prevent the educational use of Al Gore's film, *An Inconvenient Truth*, on the grounds that schools are legally required to provide a balanced presentation of political issues. The court ruled that the film was substantially founded upon scientific research and fact, and could continue to be shown, but it had a degree of political bias such that teachers would be required to explain the context *via* guidance notes issued to schools along with the film. The court also identified nine 'errors' in the film, and ruled that the guidance notes must address these errors specifically.

some time, and have needed to modify their business model to respond to the ever-changing legal demands in this area.[ix]

Although in the past, industry (and government) has been resistant to the introduction of environmental legislation,[15] within the UK, relatively few sectors have campaigned on the basis that global warming is *not* occurring, and most consider that this is an issue that is not simply restricted to current legislative requirements.

This is particularly so for the so-called "energy intensive industries" – power generation, steel, aluminium, cement, lime, glass, paper – where a substantial percentage of the costs of production are associated with energy usage. It has been suggested that financial issues alone would provide the drivers for emissions reduction/energy efficiency legislation improvement, and, in terms of emissions trading, each these industries differs significantly with regard to the ratio of carbon-to-product cost, and the extent to which emissions reduction costs can be passed through to the customer.

Global warming is unlikely to be resolved in the short- to medium-term, and companies must look beyond the current, direct effects and existing climate-related legislation, towards the controls that are likely to be in operation in 2050 and beyond. In this aspect, the cement industry has been one of the leading sectors. In the UK, the sustainability initiative (see Section 2.2.2) that was drawn up in 2005, contained a carbon strategy that sought to develop a carbon trajectory for the sector that would achieve a 60% reduction in its greenhouse gas emissions by 2050, based upon the level of reduction identified in the 22nd Report of the Royal Commission on Environmental Pollution, *Energy – The Changing Climate*.

In 2008, the UK government set a goal of 80% reduction in CO_2 for 2050, based on the recommendation of the Committee on Climate Change, a body established under Part 2 of the Climate Change Act 2008, Ch. 27. Nevertheless the principle remains the same – industry must identify the means by which it can make the meaningful reductions in its greenhouse gas emissions that will be necessary for it to continue in business.

In the case of the European cement industry, a substantial part of the additional reductions necessary to meet such goals will result from the application of carbon capture and storage. The IPCC[16] has said that CCS for cement is a key mitigation technology that is projected to be commercialised before 2030, and the International Energy Agency (IEA) has highlighted CCS as the *only* low-carbon solution for coal, cement, and iron & steel sectors,[17] and indicated that by 2050 about 50% of cement manufacturing capacity would include carbon capture and storage.[x]

Against this background, the British Cement Association[xi] has been working with the UK-based Carbon Capture and Storage Association (CCSA) in

---

[ix] *i.e.* since the introduction of the UK Climate Change Levy in 2001. See reference 2.

[x] Usage by other sectors was estimated as: iron and steel, 75%; ammonia, 100%; pulp and paper, 30%.

[xi] On 2nd March 2009, the British Cement Association merged with the Quarry Products Association and The Concrete Centre to become the Mineral Products Association (MPA).

relation to the development of CCS-related legislation; the International Energy Agency GHG R&D Programme (IEA GHG) on the application of CCS to cement manufacture; and other bodies, including the World Business Council on Sustainable Development (WBCSD). The findings of the IEA[18] work formed the basis of the development programme undertaken by the German trade association representing the cement sector, VDZ, and its research arm, ECRA (European Cement Research Academy).

With regard to the developing world, the Lafarge Conservation Partnership in conjunction with WWF has produced a report[19] indicating how reductions in greenhouse gas emissions can be made in China. The report considers the options for reducing CO_2 emissions in the face of significant increases in cement manufacture.

2.2.2 Corporate Social Responsibility

The UK cement industry has been in the vanguard of industry sectors that have advanced the sustainable development agenda. Following the pioneering work of the Society of Motor Manufacturers and Traders in 2000, the British Cement Association[xii] worked with its members to develop a sectoral approach, building upon their own corporate initiatives and involvement in the World Business Council for Sustainable Development Cement Sustainability Initiative (WBCSD CSI). Unlike the SMMT, which based its key performance indicators on those of the initial participants in the scheme, the BCA scheme began with a highly-focussed approach.

Commencing with a two-day Masterclass for company MDs and their senior advisors facilitated by Jonathan Porritt, a one-year Task Force was established under the leadership of the MD of a BCA Member Company, to deliver four major objectives: a business case for sustainable development within the cement industry; a carbon strategy; a programme of stakeholder engagement; and a "cement makers' code".

The development of a carbon strategy and CO_2 emissions profile to 2050 was identified as a key objective, and a number of options considered for reducing the sector's emissions by 60% in this time frame. Subsequently, UK government has adopted a target reduction of 80% by 2050, and this has been incorporated into the targets.

Overall, the Task Force's initiatives were underpinned by the development of the business case for the sector. Building on other sustainability accounting work carried out with other sectors, Forum for the Future provided an estimate of the costs to society of cement manufacture and how to allocate these costs.

Its analysis combined the social, environmental and economic benefits and costs of cement manufacturing, and attempted to place a monetary valuation onto these. In collaboration with Forum for the Future, BCA made a valuation of these benefits and costs using the latest and most reliable sources of

---

[xii] Now incorporated in the Mineral Products Association (MPA).

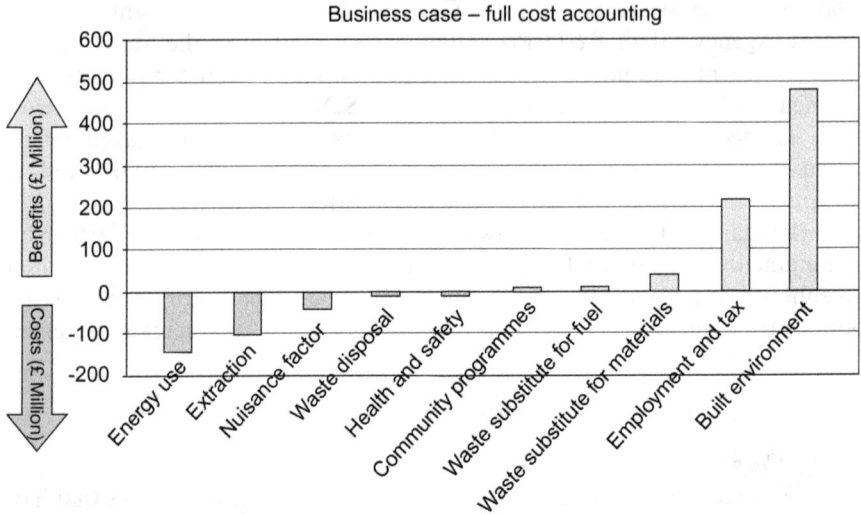

Figure 1 Cost and Benefits of Cement Manufacture. (Source: BCA Sustainable Development Task Force – work undertaken by Forum for the Future).[50]

information available. Values were only been put on the most significant costs and benefits. Forum determined the areas of significance in discussions with the BCA Sustainable Development Task Force with reference to environmental management systems, sector data reports and external publications.

The principal aim of the analysis was to capture the most significant impacts and aspects of the cement manufacturing, which considered five stages in the product supply chain, *viz.*, raw material extraction, cement manufacture, construction (concrete and precast), in use (buildings and infrastructure), and end-of-life (disposal or recycling). The output from this work (see Figure 1) gives a semi-quantitative indication of the benefits and costs of cement manufacture.

On aggregate, there is a strong business case for cement manufacture, but a significant part of the positive component results from the "in use" phase of cement-based materials, primarily concrete. Significant energy-saving reductions can be made from well-designed concrete buildings which compensates for the carbon dioxide emissions during manufacture, using properties of 'thermal mass'.[20]

Nevertheless, carbon-focused legislation seldom, if ever, takes such an holistic view, and manufacturers are required to reduce emissions from manufacturing, even if these are balanced in the "in use phase" of the product. Furthermore, processes such as iron making and cement manufacturing are required to reduce "process CO_2", that results from the decomposition of limestone, in addition to "combustion CO_2", produced from the burning of fuel. Process CO_2 is governed by the chemistry of the reaction and, as such, is essentially irreducible at source.

2.3 Costs Associated with Carbon Emissions

The decision to construct an installation with the potential to capture the CO_2 that is generated, or to retrofit carbon capture equipment to an existing plant, will be determined by a number of interrelated factors:

- Legislation that *requires* CCS to be installed, or influences the *cost* of emitting carbon.
- Cost of alternative carbon abatement options (taking into consideration the degree of their technical development and associated operational logistics).
- The logistics of transporting and storing the carbon that has been collected.
- The long-term commercial strategy of the organisation, globally, as well as domestically.

2.3.1 Legislation

The European Commission has stated, "*Whether CCS is taken up in practice will be determined by the carbon price and the cost of the technology. It will be up to each operator to decide whether it makes commercial sense to deploy CCS*".[21] Furthermore, it envisages that through providing a stable carbon price, the EU Emission Trading Scheme, and any global counterpart, will provide the main incentive for CCS deployment.

Despite the overwhelming environmental requirement for abatement, recent legislative changes relating to the EU Emissions Trading Scheme and CCS have not included a requirement for *mandatory* CCS for power generators. An early Commission Communication[22] envisaged all new *post*-2020 plants would require CCS, and *pre*-2020 plants to be capture-ready to enable a rapid retrofit *post*-2020.

However, the CCS Directive introduces a modification[23] into the Large Combustion Plants Directive (2001/80/EC), whereby new plants of greater than 300 MW capacity must include sufficient space for collection and compression of CO_2, where CCS is deemed to be technically and economically feasible[xiii] by the competent authority.

Such a provision was made early in 2009, when planning permission was granted to three new gas-fired power stations in the UK.[24] Many might regard this approach to "capture ready" as mere tokenism, demanding very little commitment from the firms involved, but the 2015 review of the CCS Directive will assess the need and practicability of introducing mandatory requirements on new large combustion installations generating electricity.

In England and Wales, the permitting of large industrial plant and ensuring Best Available Techniques (BAT) of the IPPC regime is regulated by the Environment Agency. Its Chairman, Chris Smith, has described any attempt to build coal-fired power stations without carbon capture and storage technology as unsustainable. Even ensuring that such plants are "capture ready" would be

---

[xiii] *i.e.* where suitable storage sites are available, transport facilities are technically and economically feasible, and it is technically and economically feasible to retrofit for CO_2 capture (new Article 9a of the *Large Plants Combustion Directive*).

wrong, because of the scale of the challenge the world faces in bringing down emissions.[25]

In terms of the effectiveness of abatement and the logistics of the transport, injection and storage of the collected carbon dioxide, the "higher volume CCS users" such as coal-fired power stations will lead the way, rather than cement and steel manufacture whose CO_2 emissions are substantially less – a single 2 GW coal-fired power station produces more carbon dioxide than all of the UK's thirteen cement works.

However, it would be wrong to assume that carbon capture and storage will first be introduced by the "higher volume CCS users", and only then by the "lower volume CCS users". The existence of infrastructure for CO_2 transport is a critical factor in any decision to install carbon capture, and clearly its development will take into account all potential sources of carbon dioxide.

Capture is the major component of the overall cost of CCS, but there are clearly commercial advantages to parties of making provision for the collection of CO_2 from all sources within a given region. Whilst a "National Grid" of CO_2 collection and transport would be neither practicable nor cost effective, many of the larger emitters of CO_2 are clustered in localised regions and the largest concentration is around the Humber and East Midlands (see Section 3.3.8). Fortuitously, the CO_2 large sinks – gas fields and saline aquifers – are located nearby in the southern North Sea.

2.3.2 Alternative Abatement Options

An analysis undertaken by the World Business Council for Sustainable Development (WBCSD)[26] indicated a *world-wide* potential of the cement industry to reduce CO_2 emissions by *ca.* 30% by 2020 using conventional approaches. In the UK, small, progressive reductions in CO_2 emissions have been achieved through: improved kiln control systems; high efficiency motors and drives; improved energy management procedures; higher efficiency crushing and grinding techniques; and optimisation of raw material chemistry.[27] Larger 'step changes' in fossil fuel CO_2 emissions require substantial investment.

To achieve $>60\%$ CO_2 reductions by 2050, demands new technological solutions and the WBCSD report identified a number of "advanced CO_2 management approaches", including: the use of non-limestone-based binders; production of cement and electrical energy on hybrid cement-energy facilities; employment of carbon capture and sequestration.

As early as 2003 CCS has been recognised as a low-carbon solution for coal, refineries, cement, and iron & steel sectors,[xiv] and the carbon trajectory

---

[xiv] *"Carbon dioxide capture would be most efficiently applied to large 'point sources' in order to gain economies of scale both in the capture process itself and in subsequent transportation and storage. Examples of such sources include fossil-fuelled power stations, oil refineries, petrochemical plant, cement works and iron and steel plant,"* G. Marsh, *Carbon Dioxide Capture and Storage – A Win-Win Option?* Future Energy Solutions, Oxon, Report Number ED 01806012 for DTI, May 2003.

developed by BCA envisages that further CO_2 reductions will occur in three phases:

- *Short- to medium-term*: further energy efficiency improvements and increasing use of biomass waste-derived fuels.
- *Medium-term*: incorporation of greater quantities of pre-calcined, waste-derived materials with cement clinker.
- *Long-term*: carbon capture.

Phase III of the EU Emission Trading Scheme will impose the requirement of a further 21% reduction in emissions, which for the UK cement industry will yield a total of 50% reduction on 1990 levels by 2020.

2.3.3   Logistics of Transporting and Storing Collected Carbon

The technical and economic feasibility of carbon capture must be considered holistically, and factors relating to transport to the repository for the CO_2 are essential considerations. A site's location in relation to both a distribution pipeline/CO_2 sink and other CO_2 sources may have a significant influence on the feasibility of its use of CCS, particularly for lower volume emitters, such as steel and cement.

IEA recognises the need to build regional CO_2 pipeline transport infrastructure, and indicates that near-term Enhanced Oil Recovery (EOR) can jump-start CO_2 pipeline infrastructures. However, it believes that saline formations will provide the most likely long-term solution.

Whilst the location of CO_2 storage sites is determined by geology, it also determines the applicable legislation (see Figure 2) under the United Nations Convention on the Law of the Seas (UNCLOS): the territorial sea, the Exclusive Economic Zone and the high seas.

- Within the Territorial Sea, up to twelve miles from the nation's shores, its sovereignty is determined by international law.
- The Exclusive Economic Zone (EEZ) extends from the end of the Territorial Sea out to 200 miles from a country's coast (*i.e.* 188 miles from the end of the territorial sea), and coastal states have sovereign rights to explore and exploit the natural resources of the sea bed and sub-soil of the continental shelf (land which is usually contained within the EEZ).
- The high seas are beyond the EEZ and are open to all states. However, they may complain if activities of others cause undue harm to their interests.

2.3.4   Corporate Commercial Strategy

Many manufacturers within the energy-intensive industries are part of European or global groups whose domestic investment programmes and investment policies are determined in a wider context than domestic production. An important feature of the third phase of the EU Emissions Trading Scheme was

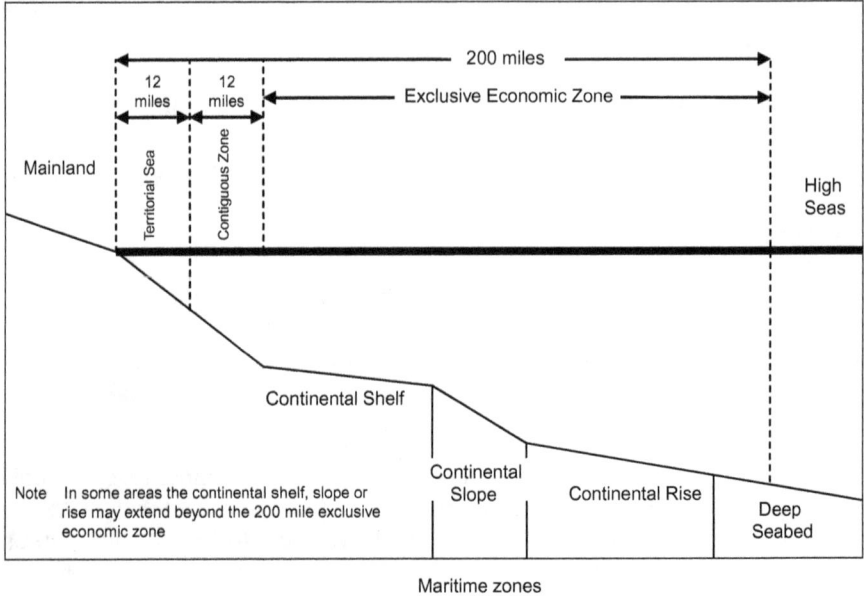

Figure 2   Legal Zones of the Sea. (Source: © OECD/IEA, 2005).[51]

the recognition that a number of industries would be at risk of "carbon leakage" – *i.e.* production being relocated outside the EU to no-carbon constrained counties – until a system of global carbon trading had been established.

Whilst this is an important short term issue, if by 2050 carbon capture and storage is to be a component of manufacturing in the developed world – up to 75% of iron and steel production; 50% cement; 100% ammonia; 30% pulp and paper (see Figure 3) – then companies must develop the CCS technologies for worldwide application. Under a scheme of global emissions trading, the proximity of storage sites will assume a much greater importance and as indicated by the sources and sinks research carried out by the British Geological Survey.[28]

Just as electricity from renewable sources is frequently marketed on the basis of its "green" credential, it has been suggested that products manufactured from low-carbon sources might be placed to receive a similar commercial advantage. This is a component of responsible sourcing initiatives and, although complex and in their early stages, could become a persuasive factor in purchasing choice.

3   Options for Mitigation

3.1   Mitigation in Cement Manufacture

As described in Section 2.2, 60% of the CO_2 emitted from clinker manufacture originates from the calcination process and the remaining 40% from fuel combustion. The thermodynamics of the process mean that opportunities only

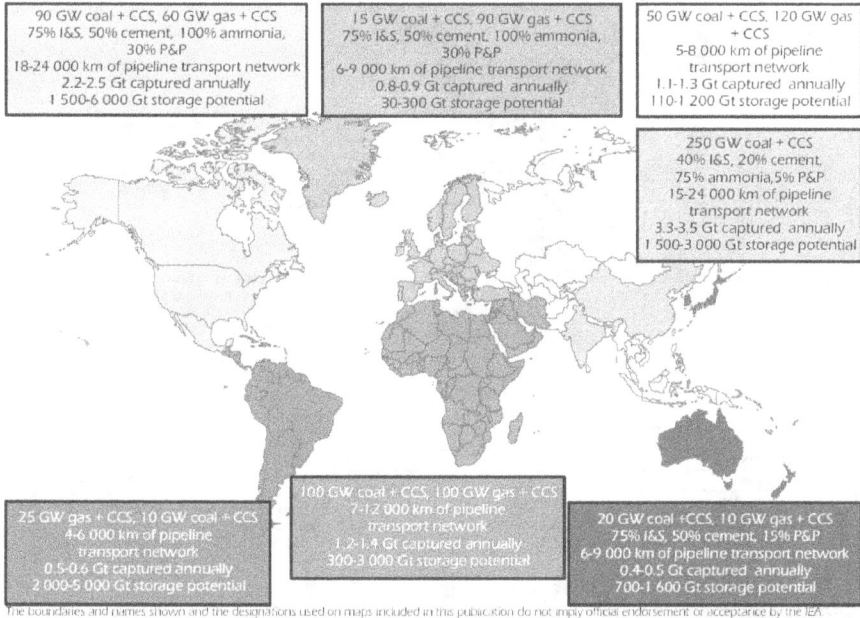

Figure 3   Carbon Capture and Storage in 2050. (Source: Presentation entitled *"CO$_2$ Capture and Storage Legal & Regulatory Update,"* given at COP 14 in Poznan, Poland © OECD/IEA, 2008, slide 14).[52]

exist to reduce CO$_2$ from 40% of the total emission. This has significant implications on any regulatory enforced reduction on the total emission, *e.g.* a 20% reduction target on the *total* emission translates into a 50% reduction on the 'reducible' portion. As such, fuel efficiency and renewable fuel measures only provide a partial solution to CO$_2$ reduction in the cement industry and thus illustrates the attractiveness of carbon capture and storage.

Further improvements in energy efficiency by replacing inefficient kilns with newer kilns will only be partially effective in addressing the UK contribution to climate change, because much of the investment has already been made. Figure 4 illustrates some of the abatement options available to the industry against the sources of emission from cement production.

There is, however, further scope to fuel switch from finite fossil fuels to 'regenerable' waste-derived alternatives – in particular, biomass. In 2007, the thermal replacement of kiln fuel with waste-derived fuels (WDF) was 18.6% in the UK (around 333 kilotonnes of coal equivalent), of these WDF, 4.2% comprised 100% biomass. The waste-derived fuels comprised used solvents, waste tyres, paper and packaging waste, refuse-derived fuel (RDF), processed sewage pellets (PSP) and meat and bone meal (MBM). This means that, in a life-cycle context, around 790 000 tonnes of CO$_2$ emission[xv] has been avoided if those wastes were land-filled and/or incinerated.

[xv] Based on a coal equivalent emission factor.

Carbon Dioxide Emissions and 'Step Change' Abatement Options Associated with Cement
Production

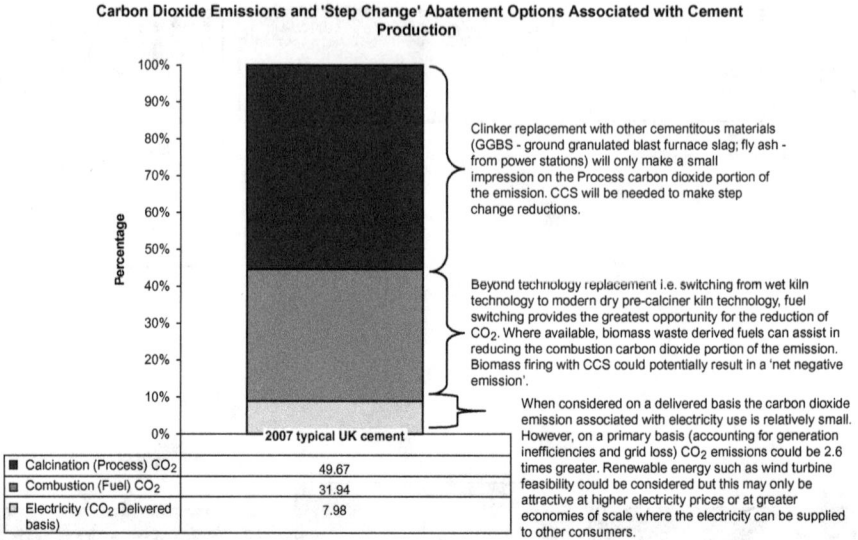

Clinker replacement with other cementitous materials
(GGBS - ground granulated blast furnace slag; fly ash -
from power stations) will only make a small
impression on the Process carbon dioxide portion of
the emission. CCS will be needed to make step
change reductions.

Beyond technology replacement i.e. switching from wet kiln
technology to modern dry pre-calciner kiln technology, fuel
switching provides the greatest opportunity for the reduction of
CO_2. Where available, biomass waste derived fuels can assist in
reducing the combustion carbon dioxide portion of the emission.
Biomass firing with CCS could potentially result in a 'net negative
emission'.

When considered on a delivered basis the carbon dioxide
emission associated with electricity use is relatively small.
However, on a primary basis (accounting for generation

	2007 typical UK cement
■ Calcination (Process) CO_2	49.67
▨ Combustion (Fuel) CO_2	31.94
☐ Electricity (CO_2 Delivered basis)	7.98

inefficiencies and grid loss) CO_2 emissions could be 2.6
times greater. Renewable energy such as wind turbine
feasibility could be considered but this may only be
attractive at higher electricity prices or at greater
economies of scale where the electricity can be supplied
to other consumers.

Figure 4 Abatement options.

However, even if extensive waste-derived fuel supplies were available, fuel
switching would still only address less than half of the direct CO_2 impact of
cement manufacture.

Although incremental technological improvements have made significant reductions in CO_2 emission in order the meet the domestic and international requirements, step changes are needed. The following section describes the potential
technologies, costs and environmental potential of CCS in the cement industry.

3.2   Carbon Capture and Cement Manufacture

3.2.1   Technology

There are three potential technologies that could be used to assist the capture of
carbon dioxide from the waste gas stream of clinker production:

- Pre-combustion capture.
- Oxy-fuel firing.
- Post-combustion capture.

3.2.1.1 Pre-combustion Capture.   Pre-combustion capture involves the pre-
treatment of the primary fuel. In the case of cement manufacture, this would
mean steam treatment or gasification of the coal/petcoke fuel to generate
hydrogen to be used as a kiln fuel and, following separation and the catalytic
reaction of carbon monoxide, a concentrated CO_2 gas stream for capture.

The applicability of pre-combustion capture to the clinker production process strongly depends on the technical possibility of using hydrogen as a main

fuel in the kiln,[29] particularly given its explosive properties. Research[30] has identified that, unlike a power station, a cement plant does not have an existing waste steam cycle that could be utilised for the gasification, further adding to cost and complexity.

However, the most significant limitation is that this method would only capture fuel CO_2 during pre-treatment and before kiln firing, and therefore overlook the 60% calcination emission and thus not exploit the full potential of CO_2 capture in the cement industry. As such, pre-combustion capture is not considered the most viable option for the cement industry.[31]

3.2.1.2 Oxy-Combustion. Oxy-fuel or oxy-combustion means that combustion in the kiln takes place in oxygen instead of ambient air. Nitrogen is removed from the air mix using a cryogenic air separation unit to distil the oxygen, and the oxygen-rich stream is then fed into the kiln or pre-calciner, depending on the configuration.

The result from combustion in the oxygen-rich environment is a high CO_2 concentration exhaust gas. The higher concentration of CO_2 makes its separation for storage more cost effective. The benefit of an oxygen enhancement is that the exhaust gas CO_2 concentration rises above 80%, compared to 14–33% for ambient air combustion.[32]

Two studies[33,34] have looked at the possibility of using oxy-combustion in the manufacture of cement. IEAGHG (2008) used a model in their study that focuses on the oxygen-enhanced combustion in the pre-calciner, although Zeman and Lackner (2006) proposed modifications to the kiln in their model for a 'Zero Emission Kiln'.

However, there are limitations to both approaches; first, there is the potential for air intrusion; second, the cost of air separation; and third, the effects of the oxygen-rich environment on burnability, reaction kinetics and additional stress on the fabric of the kiln from the increased heat. However, it is promoted that a greater flue-gas recycling and enhanced fuel pre-processing are key features of a 'Zero Emission Kiln', despite the potential energy penalty associated with separating oxygen from air.

3.2.1.3 Post-combustion Capture. Post-combustion technology is already in operation in power plants.[xvi] As the name suggests, CO_2 is captured post-combustion, *i.e.* from the exhaust gas. This end-of-pipe technology lends itself to retrofitting, as less process intrusion is required.

Current technologies use an aqueous amine (possibly monoethylamine) which undergoes a reversible reaction with CO_2. However, oxides of sulfur and oxides of nitrogen also react with the amine, and the concentrations of these impurities in the flue gas need to be carefully controlled, as does the flue-gas temperature, primarily to avoid degradation of the amine.

[xvi] Mongstad (Norway) – natural gas CHP, post combustion with separate hydrogen production, providing 1.3 Mt p.a. CO_2 for North Sea-based off-shore aquifer storage in the Johansen formation. http://www.dynamis-hypogen.com/

Under the post-combustion model, equipment additions to a normal kiln system therefore include a flue-gas desulfurisation (FGD) unit, to pre-treat the exhaust gas before amine scrubbing to remove the CO_2. The need for an FGD also increases the power demand, and further fuel use is needed in an additional steam boiler or CHP plant for the regeneration of the amine. Compression also requires additional power, but the use of a CHP plant could mean that the post-combustion-fitted cement plant could be a net exporter of power.

3.2.2 Investment and Operation

The most detailed research to date provides an indication of both the capital and operational costs of a cement plant with carbon capture. In the study,[35] the researchers have assumed a $1\,Mt\,yr^{-1}$ cement output from a modern five-stage pre-calciner kiln based in the UK, with a base-case capital cost of €263M. The capital cost estimates for post-combustion and oxy-combustion fitted plants are €558M and €327M, respectively. The capital and operating cost summary is provided in Table 1. From this summary, it can be seen that for a post-combustion plant there is a significant increase in fuel consumption, mainly due to the need for a CHP plant providing steam for amine absorbent

Table 1 Summary of Cement Plant Costs with and without CO_2 Capture for a $1\,Mt\,yr^{-1}$ Cement Plant. (Source: © IEA Greenhouse Gas R&D Programme, 2008).[54]

	Unit	Base Case (no capture)	Post-combustion Capture	Oxy-combustion Capture
Capital Costs[a]	€M	263	558	327
Operating costs				
Fuel	€M yr^{-1}	6.7	21.5	6.9
Power	€M yr^{-1}	4.0	−1.1	6.4
Other variable operation costs	€M yr^{-1}	6.1	10.6	6.4
Fixed Operating Costs	€M yr^{-1}	19.1	35.3	22.8
Capital charges	€M yr^{-1}	29.7	63.1	36.9
Total Costs	€M yr^{-1}	65.6	129.4	81.6
Cement Production Cost	€t^{-1}	65.6	129.4	81.6
CO_2 abatement costs				
Cost per tonne of cement product	€t^{-1}	-	63.8	16.0
Cost per tonne of CO_2 captured	€t^{-1}	-	59.6	34.3
Cost per tonne of CO_2 emission avoided[b]	€t^{-1}	-	107.4	40.2

[a]Note that the capital costs include miscellaneous owners costs but exclude interest during construction, although this has been taken into account in the calcualtion of the overall production costs.
[b]The costs per tonne of CO_2 emissions avoided take into account the emissions associated with imported and exported power

regeneration and power to drive an FDG (Flue Gas Desulfurisation) unit, although this is partially off-set by the reduction in externally sourced power. The consequence for the post-combustion plant is a doubling of the cement production cost, due to an annual capture cost of €63M, which leads to a cost per tonne of CO_2 avoided of €107.4 and a cement production cost of €129.4 tonne^{-1}. This means that in order to equalise the additional cost of abatement, CO_2 allowance costs, under the EU Emissions Trading Scheme, would need to be around €63.8 t^{-1} CO_2.

The criticality for domestic cement manufacture will be one of competitiveness. Before EU allowance prices reach such a level, it will be much more cost effective to import clinker from non-carbon constrained economies than to invest in post-combustion capture-ready plant. Of course, the importation of the CO_2-intensive intermediate product does nothing to combat climate change, so this in turn creates a dilemma for policy makers: drive hard for clean technology transfer and place domestic manufacture at a cost disadvantage, encouraging imports, or, delay and face failure of domestic climate-change targets.

By comparison, the oxy-combustion model demands much less increase in fuel use, but more than doubles the power demand, mainly due to the need for an Air Separation Unit. The result is a cement production cost of €81.6 t^{-1}, 24% greater than the base case. As a consequence, the lower CO_2 abatement cost of €16 t^{-1} CO_2 which leads to €40.2 t^{-1} CO_2 emissions-avoided cost, or 57% lower than the post-combustion equivalent. Superficially, this makes the oxy-combustion model potentially the preferred approach, but the fundamental redesign of the kiln needs to be considered.

For comparison the estimated[36] cost of CO_2 capture and compression (excluding CO_2 transport and storage) is $27–39 tonne^{-1} of CO_2 emissions avoided for coal-fired plants, and $48–102 tonne^{-1} for natural gas combined-cycle plants.

The environmental integrity of post-combustion is also debatable. In terms of absolute CO_2 emission it can be seen from Table 2 that the base-case plant would normally emit 770.4 kt CO_2 yr^{-1} (overall net) and, due to the additional fuel and power consumption, the post-combustion plant would generate more CO_2 (1244.3 kt CO_2 net). However, with a high capture efficiency (1067.7 kt CO_2) the overall CO_2 emissions avoided could be up to 77%. On the other hand, the oxy-combustion model generates only 9% more CO_2 (overall net) than its base-case equivalent, but only captures 465 kt CO_2, which results in a net emissions avoided (including power imports and exports) figure of 396.8 kt CO_2 (52%).

If the technical constraints on the use of both post-combustion and oxy-combustion were equal (which they are not), then the line of reasoning for a cement company is essentially balanced between two poles. First, the high cost, high energy demand but better emissions-avoided potential of the post-combustion model, compared with the comparatively lower generation of CO_2, lower cost but less potential for emissions avoided for the oxy-combustion approach.

Table 2 Summary of Cement Plant Performance with and without CO_2 Capture for a $1 \, Mt \, yr^{-1}$ Cement Plant. (Source © IEA GHG R&D Programme, 2008).[54]

	Unit	Base Case (no capture)	Post-combustion Capture	Oxy-combustion Capture
Fuel and Power				
Coal Feed	$kt \, yr^{-1}$	63.3	291.6	72.1
Petroleum coke feed	$kt \, yr^{-1}$	32.9	32.9	27.1
Total Fuel Consumption (LHV basis)	MW	96.8	304	97.8
Average Power Consumption	MW	10.2	42.1	22.7
Average on-site power generation	MW	–	45	0.7
Average net power consumption	MW	10.2	−2.9	22
CO_2 emitted and captured				
CO_2 captured	$kt \, yr^{-1}$	–	1067.7	465
CO_2 emitted on-site	$kt \, yr^{-1}$	728.4	188.4	282.9
CO_2 emission avoided at the cement plant[a]	$kt \, yr^{-1}$ %	– –	540 74	445.6 61.0
CO_2 associed with power import/export	$kt \, yr^{-1}$	42	−11.8	90.8
Overall net CO_2 emissions	$kt \, yr^{-1}$	770.4	176.6	373.7
CO_2 emissions avoided including power import and export	$kt \, yr^{-1}$ %	– –	593.8 77	396.8 52

[a]The CO_2 emission avoided are the emissions of the base case plant without carbon capture minus the emission of the plant with CO_2 capture.

3.2.3 Future Work

The work commissioned by IEA GHG programme was based upon the current state-of-the-art technology most likely to be developed into full-scale operational plant. In a fast-growing area such as CCS, new developments are frequently reported, and more efficient post-combustion capture may be possible through the use of "advanced amines" or amine/carbonate mixes, or a chilled ammonia process (CAP).[xvii] Similarly the potential benefits of oxy-combustion are likely to initiate further work to extend this option to existing as well as new plant configurations.

In addition, other capture technology, such as that based upon algae, has been identified as a potential alternative to amines, although this is in an early

---

[xvii] ALSTOM is installing this cutting edge technology in the Pleasant Prairie Power Plant owned and operated by We Energies. The pilot project will capture CO_2 emissions from a slipstream from one of the two boilers operating at the plant. The aim is to demonstrate the technology's capabilities on actual flue gas, gather field operating data and evaluate system energy consumption.

stage of development, and questions about scale remain for high-volume emitters such as cement.

3.3   Removing Barriers to Development

3.3.1   Legislation

The tranche of law associated with carbon capture and storage is a complex mix of international and domestic provisions, and the relatively recent priority assumed by the technology has necessitated the modification of existing legislation, in addition to the introduction of CCS-specific measures.

Addressing the regulation of capture processes has been relatively straightforward since, as an industrial process, no new concepts have had to be addressed and, within the EU, this has been achieved by modification of existing legislation. However, the control of transport, injection and storage has been more complex, particularly in the case of off-shore injection, where transboundary transport of CO_2 and other international measures are involved.

With regard to trans-boundary issues between Member States, the competent authorities are jointly required to meet the requirements of the CCS-specific and all other general Community legislation relating to CO_2 transport, storage sites and storage complexes.

The Commission's Energy Package agreed on 17th December 2008 contained two measures of relevance to CCS – the Directive amending the EU Emissions Trading Scheme, and the Directive on the Geological Storage of Carbon Dioxide.

Whereas the EU Emissions Trading Directive is essentially a driver for the introduction of CCS through carbon cost considerations, the Carbon Capture and Storage Directive covers the geological storage of carbon dioxide. It was introduced to address the issues within Europe associated with storage, and much of its content is concerned with these issues. However, Chapter 7 makes amendments to a number of existing legal instruments in order to make them compatible with the regulation of the capture process for on-shore storage.[37] These are discussed in more detail, below.

3.3.2   Scope of the CCS Directive

The Directive applies within the territory of the Member States, their exclusive economic zones and on their continental shelves (see Figure 2).

3.3.3   Regulation of Capture Operations

Within the EU, the permitting and regulation of capture operations falls within the Integrated Pollution Prevention and Control Directive,[38] to which only a minor change was necessary to bring these into the ambit of the IPPC regime.[39]

A further issue to be addressed to the operator of the capture facility is the composition/purity of the CO_2 stream that is collected. In practical terms this is likely to be academic for post-combustion capture, since current amine technology requires a low level of impurities in the collected gas stream to retain a high degree of capture efficiency. However, CO_2 purity has been identified as one of the concerns of stakeholders concerned that CCS may be used as an alternative disposal route for conventional air pollutants.

Specific provisions relating to CO_2 composition are not included in the modification to the IPPC Directive, on the premise that compositional requirements will be ensured through the application of its BAT provisions.[40]

In addition, a pipeline/storage facility operator may only accept and inject CO_2 streams subject to a satisfactory risk analysis and analysis of the composition of the gas streams, which must include corrosive substances, and demonstrate that the contamination levels satisfy the composition criteria referred to in the CCS Directive.

The Commission has identified a possible need to modify the BAT Reference Documents (BREFs) for a number of manufacturing processes: those falling within the Large Combustion Plant Directive; cement and lime; mineral oil and gas refineries. It has also indicated the possibility of introducing a horizontal BREF for CO_2-capture technologies.

Reference has been made to the modification of the Large Combustion Plants Directive 2001/80/EC[41] in relation to current and future requirements for new plant to be "capture ready".

3.3.4   Waste Issues

Whilst it was envisaged[42] that the revision of the Waste Framework Directive (WFD) would remove captured carbon from its ambit through an addition to the provision relating to "waste regulated elsewhere",[xviii] this was not included in the final version, Directive 2008/98/EC (ref. 43), of 19th November 2008.

The Preamble to the Waste Framework Directive states:

"(21) Disposal operations consisting of release to seas and oceans including sea bed insertion are also regulated by international conventions, in particular the Convention on the Prevention of Marine Pollution by Dumping of Wastes and Other Matter, done at London on 13 November 1972, and the 1996 Protocol thereto as amended in 2006"

and Annex I-D 7 includes "Release to seas/oceans including sea-bed insertion" as a Waste Disposal Operation.

However, through Article 35 of the CCS Directive, the Waste Framework Directive was subsequently amended by modifying of Article 2(1)(a) to become:

"gaseous effluents emitted into the atmosphere and carbon dioxide captured and transported for the purposes of geological storage and geologically stored in accordance with the provisions of Directive 2009/31/EC of the European Parliament and of the Council of 23 April 2009 on the geological storage of carbon

---

[xviii] *"CO_2 streams that are transported for the purpose of storage, injected, or stored in accordance with the provisions of Directive 2008/98/EC are not considered to be waste . . . "*

dioxide, [OJ L140 5.6.2009, p114] or excluded from the scope of that Directive pursuant to its Article 2(2).

It is significant that this modification removes captured CO_2 from the scope of the WFD, rather than identifying it as a "waste regulated elsewhere", as initially envisaged. This is important in relation to on-shore storage and trans-boundary issues, although off-shore storage will be governed by the relevant international provisions.

Similarly, "shipments of CO_2 for the purposes of geological storage in accordance with the provisions of Directive 2009/31/EC of the European Parliament and of the Council" of 23 April 2009 on the geological storage of carbon dioxide, [OJ L140 5.6.2009, p114] are excluded from the provisions of the Regulation on the transboundary shipments of waste, Article 36 of the CCS Directive.

3.3.5 Water

The Water Framework Directive (2000/60/EC) is of relevance to on-shore injection, and through Article 32 has been modified to permit *"– injection of carbon dioxide streams for storage purposes into geological formations which for natural reasons are permanently unsuitable for other purposes, provided that such injection 'is made in accordance with of Directive 2009/31/EC of the European Parliament and of the Council' of 23 April 2009 on the geological storage of carbon dioxide,[OJ L140 5.6.2009, p114] or excluded from the scope of that Directive pursuant to its Article 2(2)."* [xix]

The conditional nature of this modification is interesting and it is difficult to envisage how permanent unsuitability for *any* other purpose will be determined. As with the modification relating to waste, the final version of the CCS Directive contains an *exclusion* rather than an *exemption* from the Directive referred to: *i.e.* not within the ambit of these Directives, as opposed to within their ambit but not subject to their provisions.

3.3.6 Environmental Assessment and Post-Closure Issues

In the EU, major projects cannot proceed unless a satisfactory environmental impact assessment has been undertaken, and a requirement to carry out such an assessment for capture, transport and storage operations was introduced through a modification of the Environmental Impact Assessment Directive (85/337/EEC).[44] However, Member States retain the right to determine the areas within their territory from which storage sites may be selected.

With regard to environmental liability, the *operation* of storage is brought within the controls of the Environmental Liability Directive, Directive 2004/35/EC, under Article 35 of the CCS Directive.

3.3.7 Off-shore Operations

Although the CCS Directive does not address international issues associated with transport and storage, the Preamble notes[45] that legal barriers to the

---

[xix] This modification is inserted after third indent Article 11(3)(j).

geological storage of CO_2 in sub-seabed geological formations have been removed through the adoption of related risk-management frameworks, both under the 1996 London Protocol to the 1972 Convention on the Prevention of Marine Pollution by Dumping of Wastes and Other Matter (1972 London Convention), and under the Convention for the Protection of the Marine Environment of the North-East Atlantic (OSPAR Convention).

Amendments to the 1996 London Protocol to the 1972 London Convention in 2006 were adopted by the Contracting Parties which allow and regulate the storage of CO_2 streams from CO_2 capture processes in sub-seabed geological formations.

The Contracting Parties to the OSPAR Convention in 2007 adopted amendments to the Annexes of the Convention to allow the storage of CO_2 in geological formations under the seabed, a Decision to ensure environmentally safe storage of carbon dioxide streams in geological formations, and OSPAR Guidelines for Risk Assessment and Management of that activity. They also adopted a Decision to prohibit placement of CO_2 into the water-column of the sea and on the seabed, because of the potential negative effects.

3.3.8 Transportation Infrastructure

Pipeline authorisations are specific to the substance conveyed, and the licensing of re-use for existing pipelines is uncertain. Indications are that a CO_2 pipeline transport system is unlikely to use any existing pipes, and the cost of a new, dedicated CO_2 pipeline infrastructure system will be considerable. Research[46] into CO_2 transport suggests that costs depend on the diameter, length and cost of the pipe and how many boosters are needed. Two potential configurations for a CO_2 transport network are suggested:

- *Direct Connect Network*: where dedicated pipes transport the CO_2 to terminal sites from where it is transported to a storage site.
- *Hub & Spoke Network*: where CO_2 is transported to dedicated hubs, where it is aggregated and transported to terminal sites in larger pipes.

An example of the latter is given in Figure 5, and comparison with the legal zones of the sea (see Figure 2) indicates the complexity of the relevant legislation.

There are a number of other potential scenarios that may involve transport to temporary on-shore storage sites, before being transferred to larger off-shore sites. This may be attractive to smaller volume CO_2 producers in the non-power generating sector.

3.3.9 Demonstration Projects

The UK Committee on Climate Change has stated[47] that it is now essential to invest in projects that demonstrate the effectiveness of *various CCS technologies* in large-scale installations, and which identify the feasible timescales and likely costs of extensive deployment.

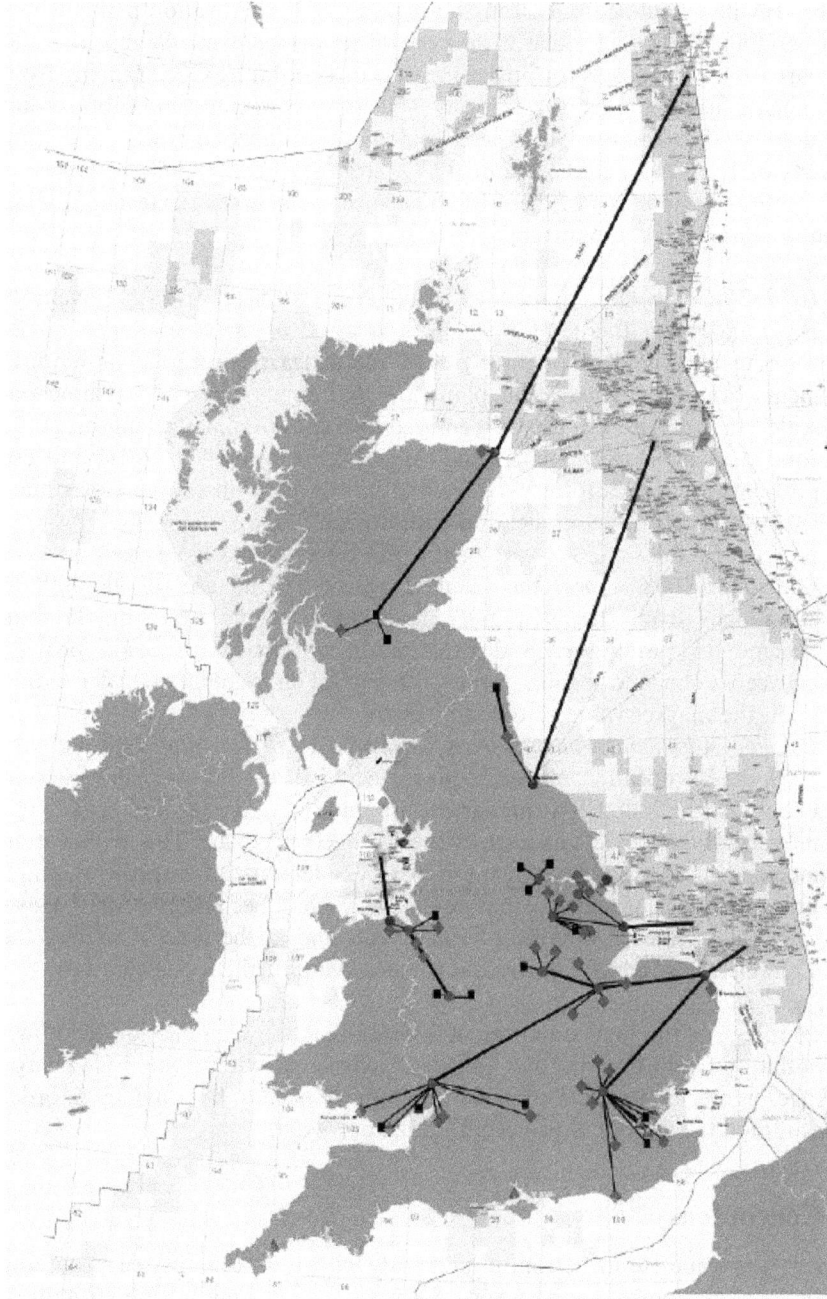

Figure 5 Indicative Transport Network. (Source: Pöyry Energy Consulting).[53]

The IEA has identified only four full-scale CCS demonstration plants which are operating worldwide, none with a coal-fired power plant. Within the UK there are *ca.* ten proposals for power projects incorporating CCS, ranging from technologies using pre-combustion, as well as post-combustion capture and advanced oxy-fuel combustion. These would account for total power generating capacity of 12 500 MW and an annual CO_2 storage of *ca.* 60 million tonnes.

To realise the potential of these 10 UK projects, a source of external funding is required and there are currently three sources:

- *UK CCS Competition.* Launched by the Prime Minister on 19th November 2007; under this the government will cover all of the capital costs of the winning plant and its operating costs for at least 15 years. The requirements of the demonstration plant are that it must be: of "commercial scale"; capture 90% of the CO_2 emitted by a coal-fired power station of the equivalent of 300 MW generating capacity; operational by 2014; and use post-combustion capture.[xx] However, the competition progress was marred by a high profile application withdrawal.[48]
- *EU Energy Package.* As part of the EU Energy Package agreed on 17th December 2008, *supra*, the Council agreed to the use of 300 million allowances from the New Entrant Reserve of the Emission Trading Scheme for funding CCS demonstration projects. This could provide between €6bn and €9bn support for capital investment, *i.e.* depending upon the market value of the allowances.
- *Possible Additional Funding from Unspent EU Budget.* On 28th January 2009, the European Commission presented proposals for the reallocation of €5bn unspent 2008 EU from agriculture funding to energy and broadband infrastructure projects as part of the EU recovery plan. This reallocation included €1.25bn for investment in five CCS demonstration projects, and a provisional list of projects that could benefit had been drawn up.[xxi] However, this created a number of legal issues and at the time of writing the Commission's redrafted proposal was receiving further consideration.

Regardless of the final outcome, it is clear that the Commission acknowledges that, in addition to the financial provisions within the EU Energy Package, significant levels of external funding will need to be provided in order to encourage the first CCS projects to be undertaken.

4 Conclusions

Cement and concrete are essential materials with the capability of providing sustainable solutions to modern society for a range of long-lasting structures:

[xx] However, the definition of post-combustion has been extended to cover oxy-fuel plants that burn pulverised coal in pure oxygen to produce a stream of very pure CO_2.

[xxi] Under the proposal, projects in Germany, the Netherlands, Poland, Spain (with Portugal) and the UK would each benefit from a contribution of £250 million. Potential projects in the UK include: Kingsnorth, Longannet, Tilbury and Hatfield.

domestic, public and commercial buildings; transport infrastructure; essential utilities, as well as their use in renewable energy facilities, and coastal and river protection schemes. Their use is based upon their basic properties of durability, security, flood/fire resistance, service life and thermal mass energy efficiency.

The whole-life evaluation of cement and concrete products demonstrates a positive environmental benefit, but within this it is nevertheless important to reduce the emissions from the cement manufacture component. Not only does this contribute to the government's necessarily ambitious climate change targets, but reduction of manufacturing CO_2 emissions also improve indirectly the carbon footprint of the resulting products.

Carbon capture and storage is one of the few options available to the industry to achieve these goals in the long term, and has the potential to provide significant reductions in the direct emissions from cement manufacture, leading towards a net zero-emission production when combined with the extensive use of biomass and waste-derived fuels.

Oxy-fuel combustion and post-combustion capture have been identified as the two most likely processes that could be applied to cement manufacturing, although to date only desk-top studies have been undertaken. Whilst the oxy-fuel option has the lower abatement cost, current technology is more applicable to new cement kilns, and unlike post-combustion capture, there are more technical uncertainties that need to be resolved at a laboratory scale.

Nevertheless, oxy-fuel plant configurations different from that in the current research[49] are being considered as a potential development of this technology, and more efficient post-combustion capture work may be possible through the use of "advanced amines" or amine/carbonate mixes or a chilled ammonia process (CAP). In addition, other capture technology, such as that based upon algae, is in the early stages of development and questions over its applicability to high volume sources such as cement are yet unanswered.

Cost remains a major barrier both to the development and future application of CCS techniques to cement manufacture but, equally, access to a CO_2 transport infrastructure will be a major consideration, and this issue is yet to be resolved for the UK development project in the power sector due to be on stream in 2015.

The application of CCS to cement manufacture now requires laboratory work to resolve the technical issues relating to flue-gas composition identified in the desk-top studies (*i.e.* importance of SO_x, NO_x and particulate levels). It is likely that this will be followed by work on a pilot-plant scale before the first cement demonstration plant is built.

However, the issues facing CCS as a whole are ones that only a multi-actor approach will resolve:

- Practical experience on the operation of CCS at a commercial scale, which will require substantial government involvement in the establishment of demonstration projects and a collection infrastructure.
- The establishment of a stable carbon price, based upon international agreement on emissions trading.

- Further legislative changes to facilitate the introduction of CCS, particularly in relation to planning, which should be part of a nationally coordinated strategy. Public acceptance to the need for CCS is necessary as a transition to a low carbon economy.

In order to accelerate the deployment of CCS in the cement industry, public funding will be needed. This funding could be usefully sourced from hypothecated revenues from CO_2 allowance auctions carried out under the EU ETS regime, *i.e.* without direct cost to the public finances.

References

1. *2002/358/EC*: Council Decision of 25 April 2002 concerning the approval, on behalf of the European Community, of the Kyoto Protocol to the United Nations Framework Convention on Climate Change and the joint fulfilment of commitments there under, *Official Journal L 130*, 15/05/2002, P. 0001– 0003.
2. D. N. Pocklington, *Eur. Environ. Law Rev.*, 2001, **10**(7), 220.
3. *Framework for the UK Emissions Trading Scheme*, Defra, 2001. See also: *Appraisal of Years 1–4 of the UK Emissions Trading Scheme – A Report by ENVIROS Consulting Limited*, December 2006, Defra.
4. D. N. Pocklington and R. Leese, in *The Finance of Climate Change*, ed. K. Tang, Risk Books, 2005.
5. D. N. Pocklington, *Eur. Environ. Law Rev.*, 2002, **11**(7), 210.
6. N. Stern, *The Economics of Climate Change: The Stern Review*, Cambridge University Press, Cambridge, UK, 2007.
7. IPCC, in *Climate Change 2007: The Physical Science Basis. Contribution of Working Group I to the Fourth Assessment Report of the Intergovernmental Panel on Climate Change*, ed. S. Solomon, D. Qin, M. Manning, Z. Chen, M. Marquis, K. B. Averyt, M. Tignor and H. L. Miller, Cambridge University Press, Cambridge, UK and New York, NY, USA, 2007.
8. Reported by J. Randerson, The Guardian, 7th August 2008.
9. *Directive 2009/31/EC* of the European Parliament and of the Council of 23 April 2009 on the geological storage of carbon dioxide, [OJ L140 5.6.2009, p114,]; and *Directive 2009/29* of the European Parliament and of the Council of 23 April amending *Directive 2003/87/EC* so as to improve and extend the greenhouse gas emission allowance trading scheme of the Community, [OJ L140, 5.6.2009, p140].
10. *Building a Low-Carbon Economy – The UK's Contribution to Tackling Climate Change*, Climate Change Committee, December 2008, 47.
11. *Integrated Pollution Prevention and Control Draft Reference Document on Best Available Techniques in the Cement, Lime and Magnesium Oxide Industries*, European Commission, Joint Research Centre, Institute for Prospective Technologies, European IPPC Bureau, 2009, 47.

12. K. A Baumert. T. Herzog and J. Pershing, *Navigating the Numbers Greenhouse Gas Data and International Climate Policy*, World Resources Institute, 2005, 57.
13. BSI, *BS EN 197-1 Cement — Part 1: Composition, Specifications and Conformity Criteria for Common Cements.*
14. M. T. Taylor, *Fact Sheet 12: Novel cements: Low Energy and Low Carbon Cements*, British Cement Association. http://www.cementindustry.co.uk.
15. C. Rose, *The Dirty Man of Europe*, Simon & Schuster Ltd, March 1991.
16. IPCC, in: *Climate Change 2007: The Physical Science Basis. Contribution of Working Group I to the Fourth Assessment Report of the Intergovernmental Panel on Climate Change*, ed. S. Solomon, D. Qin, M. Manning, Z. Chen, M. Marquis, K. B. Averyt, M. Tignor and H. L. Miller, Cambridge University Press, Cambridge, UK and New York, NY, USA, 2007, 13.
17. T. Kerr, *A Call to Action on CO_2 Capture and Storage*, IEA Day, UN FCCC COP, Poznan, 2008.
18. IEA Greenhouse Gas R&D Programme (IEA GHG), *CO_2 Capture in the Cement Industry*, 2008/3, 2008.
19. N. Müller and J. Harnisch, *A Blueprint for a Climate Friendly Cement Industry - How to Turn Around the Trend of Cement Related Emissions in the Developing World*, a report prepared for the WWF–Lafarge Conservation Partnership, 2008.
20. T. de Saulles, *Thermal Mass Explained*, The Concrete Centre, 2009.
21. European Commission, MEMO/08/798, *Questions and Answers on the Directive on the Geological Storage of Carbon Dioxide*, 17th December 2008.
22. Commission Communication, *Sustainable Power Generation from Fossil Fuels: Aiming for Near-Zero Emissions after 2020*, COM(06) 843, 10th January 2007.
23. Article 33, (reference 9).
24. *Green Light for Three New Power Stations to Power 4 Million Homes*, DECC Press Release, 2009.
25. L. Smith, *The Times*, 25th September 2008.
26. *Towards a Sustainable Cement Industry – Study 8: Climate Change*, WBCSD CSI, March 2002.
27. D. N. Pocklington and R. Leese, in *The Finance of Climate Change*, Risk Books, ed. K Tang, July 2005.
28. S. Holloway, C. J. Vincent and K. L. Kirk, in the *UK Report No. COAL R308*, DTI, 2006.
29. *Carbon Capture Technology – Options and Potentials for the Cement Industry Technical Report*, TR 044/2007, European Cement Research Academy (ECRA) GmbH, 2007. http://www.ecra-online.org/ecra/. Accessed 3.9.2009.
30. IEA Greenhouse Gas R&D Programme (IEA GHG), Reference 18.
31. D. J. Barker, S. A. Turner, P. A. Napier-Moore, M. Clark and J. E. Davison, in *Energy Procedia 1*, Elsevier, 2009, pp. 87–95.

32. ECRA, 2007, p27. Reference 29.
33. IEA Greenhouse Gas R&D Programme (IEA GHG), Reference 18.
34. F. S. Zeman and K. S. Lackner, *Int. Cement Rev.*, 2006, May, 55–58.
35. IEA Greenhouse Gas R&D Programme (IEA GHG), Reference 18.
36. J. Davison, *Energy*, 2007, **32**(7), 1163–1176.
37. *Directive 85/337/EEC*, (Environmental Impact Assessment); *Directive 2000/60/EC*, (Water Framework); *Directive 2001/80/EC*, (Large Combustion Plant); *Directive 2004/35/EC*, (Environmental Liability); *Directive 2006/12/EC*, (Waste Framework); *Regulation (EC) 1013/2006*, (Transfrontier Shipment of Waste); *Directive 2008/1/EC*, Integrated Pollution Prevention and Control).
38. *Integrated Pollution Prevention and Control Directive*, 2008/1/EC[0].
39. Under Article 37 of the CCS Directive, Annex I to Directive 96/61/EC has been modified through the addition of point 6.9: "*6.9 Capture of CO_2 streams from installations covered by this Directive for the purposes of geological storage pursuant to Directive 2009/.../EC of the European Parliament and of the Council ... [on the geological storage of carbon dioxide]*".
40. See Recitals 16, 27 and 45 of the CCS Directive, (reference 9).
41. Through Article 33 of the CCS Directive. See also Recital 47.
42. S. Brockett, *Regulatory and Legal Issues: Development of an Enabling Framework for Carbon Capture and Storage in the EU*, Stakeholder Consultation Meeting, 8th May 2007.
43. Directive 2008/98/EC of the European Parliament and Council on waste and repealing certain Directives, 19th November 2008, Official Journal 22nd November 2008, L312/3.
44. Council Directive 85/337/EEC of 27 June 1985 on the assessment of the effects of certain public and private projects on the environment was modified through Article 31 of the CCS Directive, (reference 9).
45. Recitals 12 to 14 inclusive of the CCS Directive, (reference 9).
46. *Analysis of Carbon Capture and Storage Cost-Supply Curves for the UK*, DTI, Pöyry Energy Consulting, Oxford, January 2007.
47. Reference 10, Executive Summary, page xvi.
48. R. Van Noorden, *RSC Chemistry World*, July 2007. www.rsc.org. Accessed 9.3.2009.
49. IEA GHG R&D Programme, Reference 18.
50. D. Aeron-Thomas, *The UK Cement Industry - Benefit and Cost Analysis*, 2005.
51. *Legal Aspects of Storing CO_2*, OECD/IEA, 2005, 22.
52. IEA, Reference 17.
53. Pöyry Energy Consulting, Reference 46.
54. IEA GHG R&D Programme, Reference 18.

Geological Storage of Carbon Dioxide

NICK RILEY

1 Introduction

Geological storage of carbon dioxide (CO_2) captured from large stationary industrial sources comprises injecting it into porous rocks deep in the Earth's crust so as to isolate this gas from the atmosphere. The whole industrial process chain involves CO_2 capture, transport and storage, commonly referred to as CCS. This chapter is concerned only with CO_2 storage.

It is difficult to see how the world can reduce greenhouse gas (GHG) emissions at the rate required (50% plus by 2050) without widespread deployment of this technology, especially since it appears that fossil fuels will dominate primary world energy supply until at least mid-century, and possibly beyond (International Energy Agency). It is the only technology that has the potential, on human timescales, to permanently avoid CO_2 accumulations in the atmosphere from fossil fuel use at significant scale. Deployment of non-fossil-based energy technologies, or more efficient use of fossil fuels, although essential in reducing or even reversing global emission growth, cannot guarantee that all available fossil fuel resources will not be burnt with the resultant cumulative build up of CO_2 in the atmosphere and ocean.

Injection of gases (including CO_2) into the sub-surface is routine in the oil and gas industry, where these techniques are used to enhance oil and gas production (see Section 3). Gases such as natural gas (mainly methane) and hydrogen are already stored underground in many parts of the world, including the UK[1] and often close to population centres. Without underground natural gas storage, many countries would not be able to maintain strategic or operational security of gas supply, especially during prolonged periods of cold

Issues in Environmental Science and Technology, 29
Carbon Capture: Sequestration and Storage
Edited by R.E. Hester and R.M. Harrison
© Royal Society of Chemistry 2010
Published by the Royal Society of Chemistry, www.rsc.org

weather. Despite the maturity of underground gas storage technology, large-scale storage of CO_2 globally still poses many technical and social challenges. These are largely due to the scale and permanence required for CO_2 storage to be an effective and safe GHG-mitigation technology and uncertainties about the performance of saline aquifers (also known as saline formations), which are the most ubiquitous and volumetrically the largest geological units in which CO_2 might be stored. The fact that large underground accumulations of CO_2 occur in nature[2] bears testament to the fact that CO_2 can be safely stored underground, in the right geological conditions, for millions of years.

There are numerous CCS research and demonstration projects and networks worldwide that have been developed over the last two decades. A regularly updated database is published by the International Energy Agency Greenhouse Gas Programme.[3] In recent years the CCS profile has increased in international and regional emission-control policy, with Europe (through an EC Directive)[4,5] being the first large region in the world to provide comprehensive legislation for CCS projects.

2 Geology and CO_2 Storage

2.1 Rock Characteristics

Rocks are classified by geologists into three basic categories: sedimentary, igneous and metamorphic. *Sedimentary* rocks are formed from sediments that accumulate and are buried. Sediments (*e.g.* sands, clays) may be sourced by erosion from other rocks, through bio-geochemical deposition (*e.g.* coals, limestones, salt) or a combination of processes (*e.g.* beach sands). *Igneous* rocks are formed from molten rock as it cools (*e.g.* granite, basalt, lava, ash). *Metamorphic* rocks (*e.g.* slates and schists) are formed by modification of sedimentary or igneous rocks by intense pressure and heat. All rocks have open pores within them which can be filled with gas, water or oil. This is especially so for sedimentary rocks; hence, they host most of the world's oil and gas fields and underground water supplies (aquifers). Sedimentary rocks are, therefore, the most promising hosts for storing CO_2. Thick accumulations occur where the Earth's crust has subsided to form sedimentary basins. Geologists use the term "porosity" to describe the volume of pores within a rock. Rock minerals include the original mineral grains and crystals, as well as new minerals that form during burial through a process termed diagenesis. In sedimentary rocks, diagenetic minerals cement the sedimentary grains together and provide mechanical strength. These "cements" can also line and/or progressively occlude the original rock pores (primary porosity) as they crystallise out of solutions that are trapped within, or pass through the rock. As a general rule, porosity diminishes with depth as heat and pressure redistribute minerals and compact the rock. Diagenesis can also result in new pores being formed as some minerals dissolve during burial or deep weathering (secondary porosity). Diagensis can also cause a volume change through chemical reactions (*e.g.* calcite conversion to dolomite), or by dewatering (*e.g.* some limestones), or a

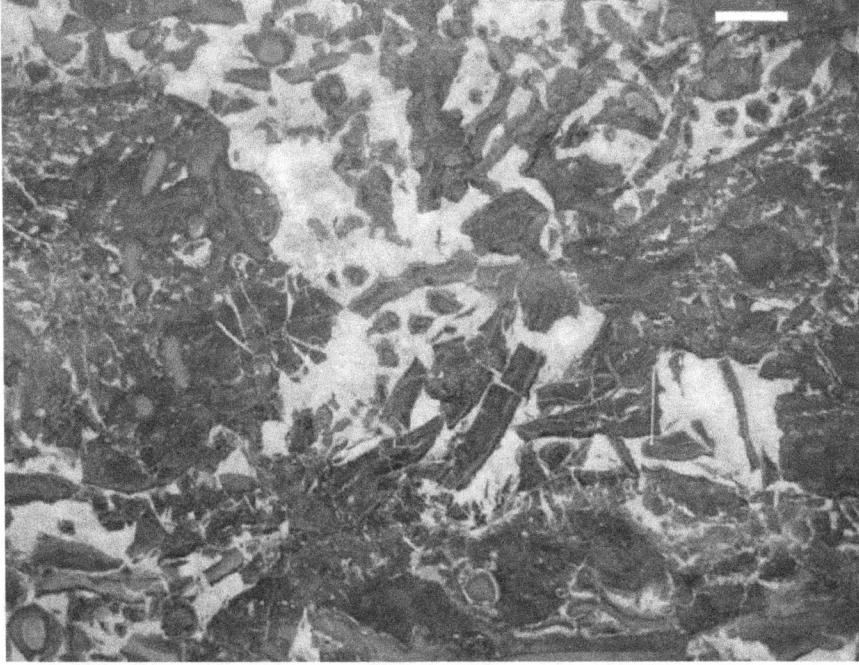

Figure 1   Polished slab of Carboniferous Limestone from Belgium (white bar scale is
approx. 10 cm) showing secondary porosity developed by fractures and
dissolution, later in-filled with calcite burial cements (white crystalline
mineral). The refittable edges of fractures are clearly visible, as are small
vesicular-shaped patches where the original limestone was dissolved to form
a mouldic porosity.

mixture of both processes (*e.g.* conversion of peat to bituminous coal). Such
volume changes can cause the rock to fracture and form open cracks (fracture
porosity; see Figure 1). Fracture porosity can also be caused by external stress,
when rock is flexed through folding or faulting (tectonic stress). As with primary porosity, secondary porosity can be occluded by burial cements. In
extreme cases, rocks can be dissolved by volcanic process (*e.g.* hydrothermal
vents) or by deep weathering (dissolution of limestone by groundwater to form
cave systems) forming large voids and pathways to the surface.

"Permeability" describes the amount of interconnectivity between pores and
how easily fluids (gas, water, oil) can migrate through the rock. With very
narrow pore connections, surface tension, viscosity, molecular binding and
friction are significant factors that can inhibit or prevent fluid movement. Good
pore connectivity results in high or even infinite permeability. A good reservoir
rock is one where the permeability is open enough for fluids to migrate easily.
For successful CO_2 storage (as with oil, gas and water production), permeability is the most crucial reservoir characteristic. A rock which has very low
permeability cannot easily be injected with CO_2. Usually, reservoir rocks are

filled with water (formation water), but they can be gas- or oil-filled (*e.g.* hydrocarbon fields). On or near land, the water at shallow depths is commonly fresh (potable water aquifer), but with depth it becomes increasingly salty (saline aquifer). Beneath the sea, formation waters are nearly always brines.

A "cap rock" is one where the permeability is so low that fluids cannot pass through easily, or at all. An effective cap rock (also known as a seal) that lies on the top surface of the reservoir is a vital feature in CO_2-storage containment. Similarly oil and gas accumulations, just like CO_2, require a reservoir rock to be overlain by a cap rock; this is because they are buoyant fluids that rise through the reservoir, the cap rock preventing upward escape to overlying rocks, or even to the surface. Such situations are called "traps". Traps usually require rocks to be folded or faulted, so as to prevent the buoyant fluids from moving so far horizontally along the reservoir cap rock interface that they eventually find a pathway through the cap rock (*e.g.* through a fracture), or escape around its edge. In some circumstances, lateral movement can be limited by "hydrodynamic trapping", where water pressure holds the buoyant accumulation in place beneath a horizontal cap. This situation only occurs in large regional saline aquifers. It is akin to a bubble of air being trapped and immobilised beneath a sheet of ice on a frozen lake or river. Sometimes a reservoir rock may be lens- or ribbon-shaped, completely sealed on all sides by cap rock. These traps are known as "stratigraphic traps" and are usually associated with lenticular, ribbon, or wedge-shaped deposits, such as buried river channels or debris fans.

No reservoir rock has the same porosity or permeability throughout. Reservoirs composed of rocks which have a wide variation in permeability are termed "heterogeneous reservoirs". Heterogeneity may be expressed horizontally or vertically. The degree of reservoir heterogeneity is a major factor in predicting whether a reservoir is suitable for CO_2 storage and how the resultant buoyant CO_2 plume will behave as it migrates from the injection point to the base of the reservoir seal (see Figure 2).

2.2 CO_2 *Properties and Geological Storage*

Carbon dioxide has several physico-chemical properties that affect its behaviour underground. Effective storage is achieved by using these properties and their interaction with the geology to best advantage.

2.2.1   *Thermodynamic Properties*

Most important is the phase behaviour of CO_2, especially where the temperature and pressure cause CO_2 to become a dense-phase gas. In this state, one tonne of CO_2 that would occupy a volume of $509\,m^3$ at STP, will occupy only $1\,m^3$ in its dense phase. Injecting underground to the depth and pressure where the CO_2 will remain in its dense phase, maximises use of available storage volume in the pore spaces within the reservoir (saturation). On land, such conditions are achieved on average at about 700 m below the surface, dependent on the local geothermal gradient and the hydrostatic head (essentially the

Figure 2 Outcrop of Sherwood Sandstone Formation at Nottingham Castle, UK. This rock face, comprising of thick, stacked, channel sandstones and thin silt beds, was laid down by an ancient river system during the Triassic Period. The outcrop shown is about 3 m high and displays vertical (thin, low permeability, fine-grained beds separating thicker, more permeable, coarse-grained, pebbly beds) and horizontal heterogeneity (cross-bedding in the coarser-grained beds) in bed forms. Rocks of this type and age extend across much of NW Europe and are traditionally referred to as "Bunter Sandstone". The Bunter Sandstone hosts important aquifers and hydrocarbon-bearing reservoirs and is an interval of major potential for CO_2 storage in many parts of Europe, including under the North Sea. About 800 m beneath Berlin, natural gas is stored in similar rocks in order to maintain the city's gas supply.

pressure exerted by a column of water equivalent to the depth of the rock stratum). Beneath the sea bed, the additional water pressure of the sea water column may allow shallower storage (in terms of geological depth) in this phase. Because dense-phase CO_2 is still a gas, it is further compressible with depth, but the benefits of the slightly increased density are offset by the increased energy (and therefore cost) required to inject deeper. The optimum depth for storage is therefore between 1 and 3 km depth.

2.2.2 Buoyancy Trapping and Plume Behaviour

Buoyancy trapping is the dominant mechanism for storage during the injection and early post-injection phase of a storage operation. Since CO_2 is buoyant relative to brine, injection is best done near the base of the reservoir, or in the down-dip part of a dipping storage reservoir. The CO_2 thus rises through the reservoir, along its longest trajectory, to form a plume. Observations from Sliepner,[6–8] the world's first large-scale saline aquifer CO_2-storage project, show that plume shape is controlled by subtle vertical and lateral changes in porosity and permeability, with thin clay or silt layers within the reservoir acting as horizontal baffles (vertical heterogeneity) to vertical CO_2 movement. These baffles divert the CO_2 laterally until it finds a pathway through into the next highly permeable layer above. Once reaching the reservoir top, the CO_2 spreads out laterally along the cap rock base. This top-most layer of the plume defines the areal extent of the plume footprint. Lateral and vertical heterogeneity of this type has the advantage that it prevents the plume rising too quickly through the reservoir, and spreads the plume out laterally within the reservoir in successive layers. This has two beneficial effects: that of filling a broad vertical column of the reservoir with CO_2, thus improving storage efficiency, and increasing the surface area of contact of CO_2 gas with the formation waters, thus enhancing CO_2 immobilisation through dissolution and consequent mineral reactions. If a reservoir is strongly heterogeneous in one dimension (anisotropic), the CO_2 by-passes much of the reservoir, focussing its migration along the most permeable zones (called "channelling" or "fingering"). In such situations the plume contact with the cap rock may not be established or, if made, can be significantly off-set from the injection point. If heterogeneity is so strong as to compartmentalise the reservoir, plumes will be restricted to each injected compartment and may, or may not, meet the base of the cap rock.

The lower boundary of buoyant plumes in structural traps will closely follow the depth contours of the structure, as the brine–CO_2 gas contact migrates down the structure as it is filled with CO_2. It is important that care is taken not to fill the structure with so much CO_2 that the plume base will spill out of the lower part of the structure through any spill points (unless intended). It should also be noted that the base of a plume may not be horizontal, especially if the underlying formation waters are flowing along a regional hydrostatic pressure gradient. Pressure is highest around the injection point. This can cause the CO_2 to move in a radial fashion, as gravity-driven migration is over-ridden by injection pressure. This effect can distort the plume base downwards, close to the injection point (see Figure 3).

2.2.3 Residual Gas Trapping

When CO_2 gas migrates through the reservoir, either through buoyancy drive or by the injection pressure, tiny bubbles of gas get trapped in the pore spaces and are immobilised by capillary forces, even though water itself can still flow. This process is termed "residual gas trapping". It is a very important process in the early immobilisation of CO_2 and in the attenuation of CO_2 in the cover rocks (overburden) should it migrate outside the storage reservoir.

2.2.4 Solubility Trapping

Carbon dioxide, unlike hydrocarbon gases, readily dissolves in water. This has several consequences that are advantageous to storage. The resultant CO_2 brine solution is heavier than the native brine, thus causing the dissolved CO_2 to sink by advection. This characteristic is extremely important in preventing the CO_2

from finding a pathway to the surface over the long term. Where the buoyant gas cap is relatively small compared to the overall saline aquifer volume, all the gaseous CO_2 will dissolve over time. The amount of CO_2 that can be dissolved is dependent on the salinity, pressure, temperature and volume of available brine that can come into contact with the CO_2 (see Figure 4).

2.2.5 Mineral Trapping

In the presence of water, CO_2 gas becomes reactive and can form new minerals, causing a fall in pH (hence it is called an "acid gas" by the oil and gas industry). This too acts to lock up the CO_2 in a solid mineral or a dissolved bicarbonate phase, preventing migration to the surface. Typical mineral reactions that have been predicted[9] include the following:

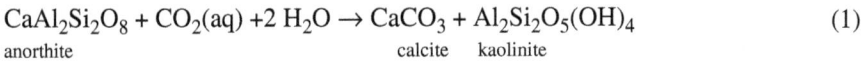

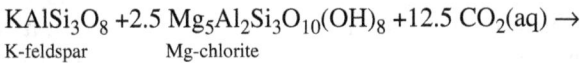

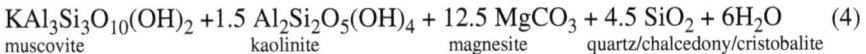

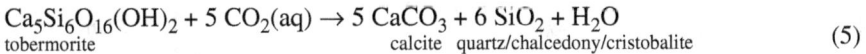

$$CaAl_2Si_2O_8 + CO_2(aq) + 2\ H_2O \rightarrow CaCO_3 + Al_2Si_2O_5(OH)_4 \tag{1}$$
anorthite $\qquad\qquad\qquad\qquad\qquad\qquad$ calcite $\quad$ kaolinite

$$KAlSi_3O_8 + Na^+ + CO_2(aq) + H_2O \rightarrow NaAlCO_3(OH)_2 + 3\ SiO_2 + K^+ \tag{2}$$
K-feldspar $\qquad\qquad\qquad\qquad\qquad$ dawsonite $\qquad$ quartz/chalcedony/cristobalite

$$CaCO_3 + CO_2(aq) + H_2O \rightleftharpoons Ca^{2+} + 2\ HCO_3^- \tag{3}$$
calcite

$$KAlSi_3O_8 + 2.5\ Mg_5Al_2Si_3O_{10}(OH)_8 + 12.5\ CO_2(aq) \rightarrow$$
K-feldspar $\qquad\quad$ Mg-chlorite

$$KAl_3Si_3O_{10}(OH)_2 + 1.5\ Al_2Si_2O_5(OH)_4 + 12.5\ MgCO_3 + 4.5\ SiO_2 + 6H_2O \tag{4}$$
muscovite $\qquad\qquad\quad$ kaolinite $\qquad\qquad\quad$ magnesite $\quad$ quartz/chalcedony/cristobalite

$$Ca_5Si_6O_{16}(OH)_2 + 5\ CO_2(aq) \rightarrow 5\ CaCO_3 + 6\ SiO_2 + H_2O$$
tobermorite $\qquad\qquad\qquad\qquad\qquad$ calcite $\quad$ quartz/chalcedony/cristobalite $\qquad$ (5)

Prediction of CO_2-reacted mineral phases is based on observations of natural CO_2 systems and analysis of produced fluids and rocks obtained from CO_2 injection operations in enhanced oil recovery and in laboratory experiments. The latter involve flooding cores, or rock samples, containing CO_2 gas or CO_2 in solution, with a "synthetic" formation fluid that mimics the formation waters found in the "*in situ*" reservoir or cap rock. Core-flood experiments can be carried out at reservoir temperature and pressure in specially designed pressure vessels. Such experiments may take from many months to years to complete. Reaction fluids are periodically drawn off and analysed. Observed changes in the reaction fluids can then be related to mineral changes in the rock in response to the CO_2–rock and formation fluid reactions. At the end of the experiment the core can be analysed microscopically to visually confirm the reactions. Some mineral phases (*e.g.* dawsonite, $NaAlCO_3(OH)_2$) are only stable at reservoir temperatures and pressures. They cannot be directly observed at STP and therefore can only be inferred as being present using the reaction-fluid analysis results. Batch experiments require the rock to be ground down to a fine powder and reacted with CO_2. This provides a large surface area between the rock minerals and the CO_2, thereby accelerating reactions. Trying to predict the reacted mineral species, and the time in which they will form, requires the construction of modelling codes derived from the inferred chemistry of all these observations. Although reactions can start immediately injection starts, it takes from many centuries to millennia for CO_2–rock reactions to fully exhaust themselves at field scale. That exhaustion limit can be reached before all potential minerals are reacted, because the new minerals produced can form a barrier to further reactions. Some reactions improve porosity and permeability (*e.g.* dissolution of calcite), whereas others can cause permeability deterioration (*e.g.* production of silica). Clearly, over the long term, these reaction processes may have a positive or negative effect on the storage capacity of the reservoir and on the integrity of the seal, dependent on the original mineralogy.

2.3 Pressure

We will first consider pristine saline aquifers in this context and then hydro-carbon fields. Pressure is a fundamental consideration in CO_2 storage. It is highest around the injection point (in order to push the CO_2 into the reservoir) and diminishes radially away from it. Another region of raised pressure is found where the plume reaches to cap rock. This pressure is a combination of the radial pressure effect of the injection, plus the pressure effect of the buoyant column of CO_2 pressing against the base of the reservoir seal. In storage, it is vital that reservoir pressure does not exceed the strength of the cap rock

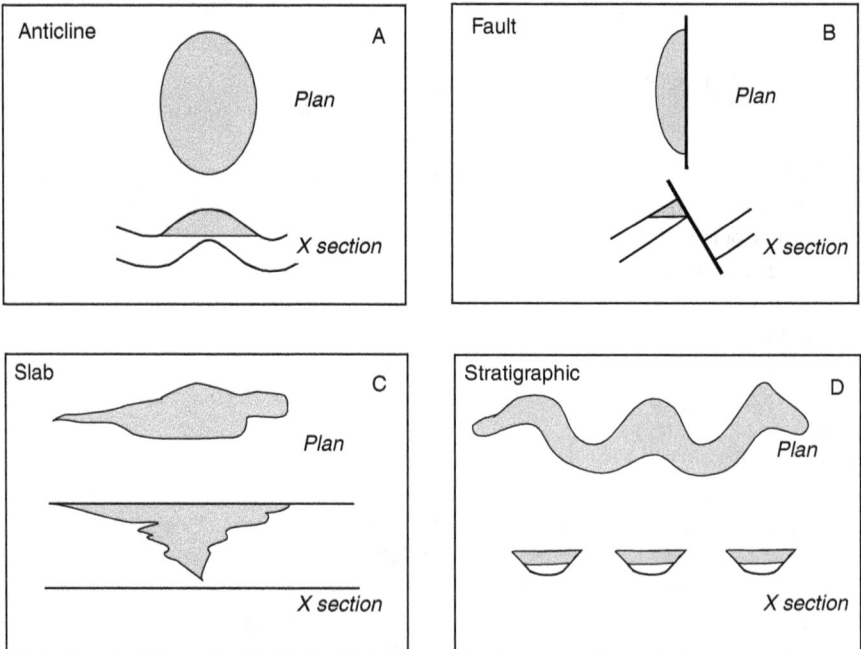

(capillary entry pressure, or mechanical strength), otherwise the cap rock will fail and CO_2 will ingress into it. Pressure is primarily accommodated in the reservoir by displacement of the formation waters and elasticity of the crust. Over time, pressure equalises across the reservoir. The highest pressures are therefore reached during injection and dissipate relatively quickly during the post-injection phase. The maximum pressure reached is dependent on reservoir size, the injection rate/duration and the permeability. The rate at which pressure equalises mainly depends on: the volume of the reservoir; its hydrostatic connectivity within the storage structure and the surrounding geology; the mechanical responses of the crust; and whether any CO_2 gas cap is retained,

---

Figure 3   Schematic diagrams of CO_2 plume outlines (shaded), in plan and cross-section, controlled by buoyancy trapping. Scenarios (A) and (B) are structural traps, where the rocks have been folded to form an anticline (A), or faulted (B). Note the plan view of the plume in (A) is elliptical, denoted by the structure contour on the fold at the depth of the base of the buoyant part of the CO_2 plume. In (B), the ellipse in plan view is truncated by the fault. Note also, that in (A) if the base of the buoyant plume extends below the edge of the fold structure it will spill out into an adjacent structure. If the intention is to contain all the injected CO_2 in anticlinal structure, then cumulative injection needs to cease before the base of the buoyant plume can encroach on this spill point. However, a storage project may deliberately plan to fill the primary storage structure to its spill point so that the CO_2 spills into adjacent (secondary) storage structures. This increases storage efficiency by maximising the volume of the primary storage structure accessed by CO_2, and by causing the CO_2 to take a long trajectory through the reservoir rock into the secondary storage structure. A long trajectory is also achieved within a primary storage structure if the CO_2 injection point is placed towards the reservoir base or down dip on a dipping reservoir, such as in (B). In (B), not only is the cap rock overlying the reservoir acting as the main seal preventing buoyant CO_2 escaping from the storage site, but also the fault zone itself is acting as a seal. This brings out the important principle that faults need not always be pathways for leakage, but can act as effective seals. In (C), which is a horizontal slab of reservoir rock representative of storage sites such as Sliepner, the vertical movement of the CO_2 as it rises from the injection point is impeded by several thin low-permeability horizontal baffles within the reservoir. Each time the plume reaches the base of a baffle, it is forced to move horizontally until it finds a way through the baffle into a more permeable layer above. A series of such baffles have caused multiple stacked layers of buoyant CO_2 to be trapped between the injection point, close to the reservoir base, and the cap rock seal at the top, giving rise to an anvil shape in the cross-sectional aspect of the buoyant plume. Once reaching the top of the reservoir, the buoyant plume can only migrate laterally and, because the reservoir is horizontal, the plume shape is irregular, reflecting subtle vertical undulations on the cap rock/ reservoir rock contact surface. In (D), the storage reservoir, in this case a meandering channel of sand, is completely enclosed within an impermeable unit. This is a stratigraphic trap. The plan and cross-sectional shape of the buoyant CO_2 plume reflects the channel geometry. As the reservoir is completely sealed on all sides this is the most difficult storage site to inject into, as formation water cannot transmit pressure (through displacement) out of the storage unit and the amount of formation water into which the CO_2 can dissolve is limited.

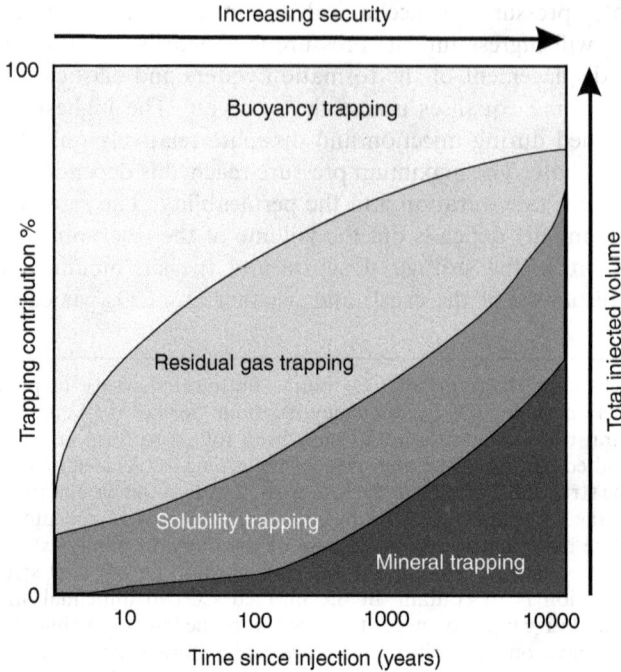

Figure 4 Summary of main trapping mechanisms for CO_2 in a CO_2 storage reservoir, modified from the *IPCC Special Report on Carbon Dioxide Capture & Storage* (2005). Note that solubility trapping and residual gas trapping are the first trapping mechanisms to act on the CO_2 during injection, respectively caused by CO_2 dissolving in the formation water and some gas becoming immobilised in pore spaces. Buoyancy trapping increasingly comes into play as the rising CO_2 plume is impeded vertically by impermeable zones (if present) within the storage reservoir, and ultimately accumulates under the base of the cap rock (or seal) at the top of the storage reservoir. Mineral trapping, caused by the CO_2 reacting with the reservoir rock and formation brines to form new minerals, is the slowest process, but becomes cumulatively more significant with time as the CO_2 plume slowly dissolves into the formation brines which then sink into deeper parts of the storage complex. If the final CO_2 injected volume is very large relative to the storage reservoir volume, especially in structural or stratigraphic traps with limited access by the CO_2 to the formation brines, some buoyancy trapping may persist as a permanent feature of the storage site. The combination of trapping mechanisms cumulatively immobilising the CO_2, together with the dissipation of injection pressure relatively quickly after CO_2 injection ceases, means that the storage site becomes increasingly secure and even less likely to leak over time.

thereby maintaining buoyancy pressure against the cap rock. Small geological traps are more pressure-sensitive than large ones, especially if they have no hydrostatic connection to the regional geology (*e.g.* some stratigraphic traps). Low-permeability reservoirs require higher pressure injection because it is harder to displace the formation brines with CO_2. The pressure gradient away from the injection point will therefore be steeper. The least pressure-sensitive storage sites are large horizontal slab aquifers with good permeability, such as Sleipner.[6–8] These are so vast that injection is easily accommodated and hydrostatic connectivity is effectively unhindered.

Oil and gas fields are proven traps with known original pressure. The records of oil and gas production also give accurate information on the porosity and relative permeability. In depleted/depleting oil and gas fields, storage pressure is lower than the original formation pressure due to the production of hydrocarbons. When storing in gas fields this can be a major issue, especially if the field has been "blown down" close to atmospheric pressure in order to maximise gas production before field closure. In such cases, management of the CO_2 injection during the early injection phase of the CO_2 storage operation has to prevent an immediate pressure drop across the injection well/rock interface, from dense phase to ambient reservoir pressure, otherwise there is a risk of well and formation damage and permeability loss through gas hydrate formation. This latter effect is due to the latent heat absorbed by the CO_2 as it changes phase to a low-density gas. Managing injection in this scenario is an active area of research.[10] Another issue to consider with depleted gas fields is water ingress. Gas fields often have much lower permeability than oil fields and therefore, if water has ingressed into the pore spaces vacated by natural gas during gas production, it may not be possible to overcome the capillary forces of the ingressed water so as to displace it when CO_2 is injected. Conversely, gas fields that have retained their gas cap after production will readily receive CO_2, because there will be no capillary forces that need to be overcome in order for the CO_2 to migrate through the reservoir from the injection point.

3 CO_2 Storage through Enhanced Hydrocarbon Recovery

The use of CO_2 in enhanced hydrocarbon recovery is a way in which CCS infrastructure can be deployed and operated when there is an insufficient value chain for CCS dedicated solely to carbon abatement to be economic. This technique can also aid energy security.

3.1 Enhanced Oil Recovery (EOR)

Carbon dioxide injection has become routine practice in North American oilfields, spurred on by the world oil-price shocks of the 1970s, as a means for improving domestic oil production from depleting on-shore fields (the Carter Administration's "Oil Windfall Tax"). A considerable CO_2 pipeline infrastructure has grown since then, supplying over 70 fields. The main objective in these operations is to produce as much oil as possible, by using the minimum amount of CO_2 (as this has to be purchased from a supplier). Storage is therefore a passive by-product and is equivalent to all the CO_2 that remains in the field after oil production, by trapping processes, such as residual gas and buoyancy trapping, dissolution in formation waters and mineralisation.

Carbon dioxide in its dense and liquid phase is an excellent solvent for hydrocarbons. This makes it an ideal gas for use in accessing oil that cannot be produced under natural pressure drive or from pumping. When in contact with CO_2 at reservoir conditions, oil swells and becomes less viscous. The CO_2 also selectively dissolves the lighter oil fraction (and hydrocarbon gases). Carbon dioxide EOR is therefore most effective for lighter oils, but it can also improve production from heavy oils when combined with thermal techniques. When dissolved in water, CO_2 can also improve porosity/permeability through the dissolution of carbonate (if present), further improving migration of hydrocarbons to the production wells. The key in a successful EOR operation is to achieve maximum contact between the oil and the CO_2. This is done by injecting the CO_2 into the reservoir so that the CO_2 is at miscible or near-miscible ("sub-miscible") pressure with respect to the oil.

After primary production, oil fields are at reduced pressure compared to their original pristine state. Water may then be injected to push oil out to the production wells. This "water flood" phase is known as secondary recovery. Potentially recoverable oil that still remains after the water flood is then targeted by CO_2. This is known as tertiary recovery. The technique used is "water alternating gas" (WAG), which involves alternating injections of dense-phase CO_2 and water in order to produce the remaining oil as quickly as possible. Depending on the field characteristics, this tertiary phase can produce an extra 5–15% of the original oil in place and extend field life by several decades. The injected CO_2 is guided through the parts of the field where recoverable oil still remains. This is done by injecting water, along lines of boreholes positioned either side of the reservoir area to be swept by the CO_2 so as to produce a corridor of diminishing pressure gradient, focussed towards the production well. These well layouts are known as "panels". The CO_2 and water are removed from the produced oil and gas, and reinjected. In WAG, the final injection of a panel is by water, so as to flush out all the recoverable oil and CO_2. Another potential method of CO_2 EOR which has yet to be attempted commercially is by using gravity-stable gas injection (GSGI). This involves injecting CO_2 in the region of the original oil–water contact (oil floats above water) in the oil-field flank. Over a long period of time the field is re-pressurised. The rising CO_2–oil front sweeps the oil to the production well on the crest of the field structure. The extra oil produced using this method is significantly greater than for WAG, but it takes many years before production is stimulated; hence it is less attractive commercially over the short-to-medium term. The volume of CO_2 used (and therefore passively stored) is much greater than with WAG.

3.2   Enhanced Gas Recovery (EGS)

In enhanced gas recovery the intention is to inject CO_2 using GSGI into a depleting gas field. In an ideal case, the strategy is to inject CO_2 into the base of

.the natural gas cap. Since CO_2 is denser, it displaces the methane (CH_4) upwards toward the production well located on the crest of the trap.

3.3 Enhanced Coal Bed Methane Recovery (ECBM)

Coal Bed Methane (CBM) recovery is a commercial process by which methane contained in coal is released and collected. In gassy coals that are mined underground it is an essential health and safety technique used to reduce explosion risk caused by CH_4 ("firedamp") released at the coal face. Horizontal boreholes are drilled into the coal face prior to coal cutting. The mechanical shock caused by the drilling improves the coal fracture permeability (cleat) and encourages desorption of the weakly bound CH_4, which drains out through the coal face as a result of the pressure gradient ("pressure swing") between the virgin coal and the coal face (the latter is near atmospheric pressure). Modern mines use this methane for power generation, rather than venting the resultant air–methane mix ("gob gas") to the atmosphere *via* the mine ventilation. This practice is a valuable GHG-mitigation strategy (as methane's GHG effect is equivalent to 23 times that of CO_2, molecule for molecule).

In un-mineable coals, the intention is to produce CH_4 only. Wells are drilled at intervals from the surface into the coal seams. Water is usually then injected at high pressure to fracture ("hydrofrac") and mechanically "shock" the coal matrix to stimulate CH_4 release. Sand, or other solid sand-sized particles ("propants"), may be injected with the water to hold the resultant fractures open so that permeability is maintained. Production wells are then pumped to reduce pressure (pressure swing) in the fractures, thus desorbing CH_4 from the coal matrix and drawing the produced gas to the well *via* the cleat fractures.

In enhanced coal bed methane (ECBM) production[11,12] additional CH_4 is displaced from the micropores and fractures ("coal cleat") in black coals by injecting nitrogen gas. This is a well-established commercial technology. The concept of CO_2 ECBM is to use CO_2 instead of N_2. Methane gas production is stimulated because of chemical bonding of the CO_2 with the coal, and a preferential sorption of CO_2 onto the coal as compared with methane. A major downside of using CO_2 in ECBM, unlike N_2, is that the coal swells in the presence of CO_2 and becomes more "plastic", thus reducing permeability and thereby inhibiting migration of the displaced CH_4 to the production well. Further injection of CO_2 is also impaired. For these reasons (as well as the lower cost of N_2 supply) CO_2 ECBM has yet to be proven commercially. Carbon dioxide storage in coal has been piloted at various test sites.[13]

3.4 Shale Gas

Shale gas is a new technology which has proven very successful in North America over the last ten years. As with coal, methane is bound in the organic component of black shales (oil and gas source rocks) and can reside as a free gas in fractures. Hydrofrac of black shales releases the bound CH_4 and enables its migration to a production well. Tests have shown that CO_2 can be adsorbed onto the organic matrix of black shales, displacing CH_4 in a similar way to the process in coal.[14] There is, therefore, future potential to utilise CO_2 in enhanced shale-gas production, as well as to store CO_2. Since black shales are much thicker stratigraphic units than coals, it should be easier to ensure that all the injected CO_2 remains in the shale. It is not yet known what the volumes of black shale are world-wide that may be suitable for storage, but clearly this needs to be assessed as a potential energy and CO_2-storage resource.

4   Storage Options

As noted in the previous sections, saline aquifers and hydrocarbon fields are already being used in CO_2 storage operations. These are the main options for CO_2 storage. The following sub-sections describe other options.

4.1   CO₂ Storage in Salt Caverns

Salt cavern storage is a mature technology used for underground gas storage, such as natural gas, hydrogen, compressed air, and also for liquid hydro-carbons. Carefully designed sub-spherical caverns are produced by deliberate dissolution of salt through injection of water. The produced brine can then be used by the chemical industry (salt caverns are expensive to produce, especially if no market for the brine is available) or discharged into the sea. On decadal scales, salt caverns are ideal underground storage containers, but over time they gradually lose volume. This is due to a process known as "salt creep". For this reason, salt caverns have to be purged of stored contents when decommissioned. As CO_2 has to be stored for thousands of years, it cannot be guaranteed that over such a long period cavern stability can be maintained; hence, it is unlikely that such sites will be suitable for long-term CO_2 storage. However, salt caverns could be used as a temporary buffer store (as with natural gas) within a CO_2 pipeline grid network.

4.2   Underground Coal Gasification Cavities

Underground Coal Gasification (UCG; see also Chapter 4 in this book) is a technology where a well is drilled horizontally along a coal seam to meet a vertical well. An oxygen–nitrogen–steam mix is then injected along the horizontal injection well and ignited. The coal burn at the injection well tip is moderated by varying the relative proportion of oxygen to other gases, and the resultant gases are produced through the vertical well. Of these, carbon monoxide and hydrogen can be used for electricity generation or synthetic natural gas production. The horizontal well tip is gradually retracted along the seam leaving a burn cavity in its wake, as more coal is gasified. It has been postulated that CO_2 could be captured from the produced syngas and then re-injected into the burn cavities for storage. If this technology is to be viable it would have to be proved that CO_2 storage would be secure, despite the thermal and mechanical effects of the UCG process on the surrounding rocks, and that the cavities would not collapse over time (creep).

4.3 CO_2 Storage as CO_2 Hydrates

This theoretical technique[15-17] suggests that CO_2 could be injected into sediments at high latitudes, where temperatures and pressures are suitable for gas hydrates to form (see also Chapter 10). Hydrates are ice crystals in which gas is held in the ice-crystal lattice. At high latitudes, hydrates are associated with permafrost and marine sediments. At lower latitudes they are found in the high-pressure, cold conditions of the deep ocean. There are several issues to be overcome if this technique is to be realised, the main one being injection, as new hydrate formation around the injection point might occlude permeability. There is also a risk that existing methane hydrates might be perturbed, leading to CH_4 release to the atmosphere, exacerbating greenhouse gas release. As with coals, CO_2 is preferentially held in hydrates as compared with CH_4. This characteristic may be an opportunity to recover CH_4 from hydrates for energy use and to store CO_2.

In deeper geological CO_2-storage scenarios at high latitudes, it has also been posed that hydrates could act as a seal, preventing CO_2 leakage to atmosphere.

4.4 CO_2 Storage in Igneous/Metamorphic Rocks

This storage strategy[18-21] proposes that CO_2 could be injected into igneous and metamorphic rocks (*e.g.* basalts, serpentinites, ophiolites) which contain minerals reactive to CO_2 (*e.g.* olivine). Several issues need to be overcome if this is to succeed. Such rock types have poor average permeability, but can have fracture systems through which the CO_2 will preferentially flow, thus by-passing much of the rock (and hence surface area) with which the CO_2 could react. Such fractures could also channel the CO_2 to the surface. Mineral reaction times are very slow; hence, this limits the rate at which CO_2 can be injected, or requires that the storage reservoir has an effective seal so that the CO_2 is held long enough to react completely (*i.e.* for thousands of years).

5 Storage Capacity

5.1 The Resource Pyramid

As with any natural resource, calculation of the usable volume of a storage reservoir is based on various levels of uncertainty.[22] In short, these can be summarised as "theoretical" (assuming the most optimistic scenarios), "possible" (discounting an estimate of suspected negative factors that might diminish the volume of theoretical resource that can be exploited), "probable" (which uses known technical data to characterise the estimate of the resource),

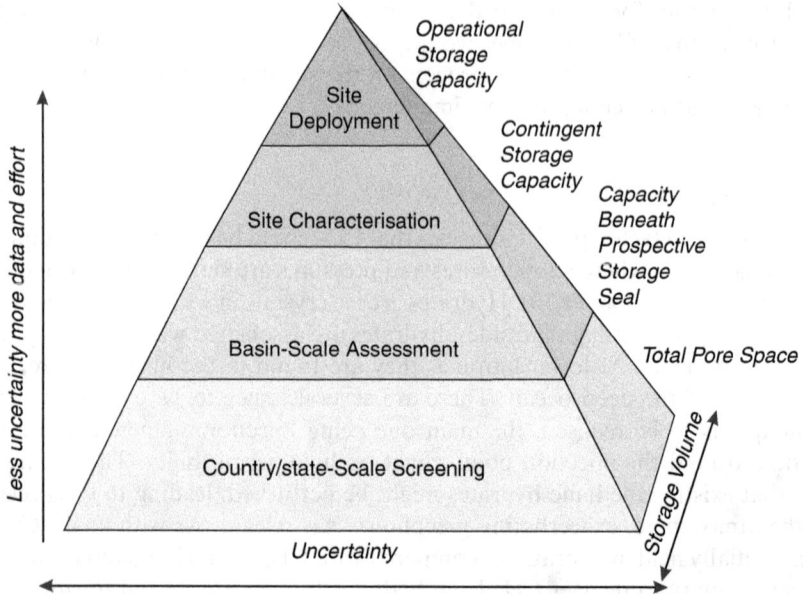

Figure 5 Resource pyramid (based on various sources: ref. 22–24) illustrating the confidence of CO_2 storage capacity estimates at various geographical scales. At the base of the pyramid is the most optimistic estimate of potential capacity for a region or country, and at the apex is the capacity accurately constrained at an operational CO_2 storage site. If CO_2 storage is to be deployed on a large scale, then many sedimentary basins will require significant geological exploration, just as has happened with oil and gas, in order to decrease uncertainty and identify storage sites where the geological characterisation is at a level of confidence that sites can operate.

and "realistic" (which defines the resource under present and immediately foreseen economic and practical conditions). Critical paths for utilisation of the resource may be: access to infrastructure; build costs; spatially related factors, such as ecosystem risk; socio-economic factors, such as public acceptance; legality of the project and policy incentives; as well as technical constraints. Another way to illustrate these concepts is to layer them into a "resource pyramid", with the theoretical resource at the base and the economic resource at the apex (see Figure 5).

5.2 Estimating Storage Capacity

Storage capacity is derived from estimating the volume of pore space that can be occupied by CO_2 at reservoir conditions.[22–24] Even in a highly permeable reservoir with good average porosity, many geological factors will prevent all the pore space being occupied by CO_2. The most accurate estimates can be made from depleted hydrocarbon fields, because the reservoir conditions and behaviour are well known, as is the volume of produced oil or gas. For oil fields lacking a gas cap, CO_2 storage capacity has been calculated[23] using the following parameters:

$$M_{CO_2} = (V_{GAS}(stp).B_o) \cdot \rho CO_2$$
(equation for oil field lacking a gas cap)
\hfill (1)

$$M_{CO_2} = (V_{GAS}(stp).B_g) \cdot \rho CO_2 \quad \text{(equation for a gas field)} \hfill (2)$$

$$M_{CO_2} = V_{PORE} \cdot \rho CO_2 \quad \text{(equation for a saline aquifer)} \hfill (3)$$

where: $M_{CO_2} = CO_2$ storage capacity (10^6 tonnes)
stp = standard temperature and pressure
V_{GAS} (stp) = volume of ultimately recoverable gas at stp ($10^9\,m^3$)
V_{PORE} = pore volume within the storage site
B_o = oil formation volume factor (the ratio between a volume of oil and the dissolved gas that it contains at reservoir temperature and pressure and the volume of the oil alone at stp)
B_g = gas expansion factor (from reservoir conditions to stp)
ρCO_2 = density of CO_2 at reservoir conditions ($kg\,m^{-3}$)

The resultant storage values are then discounted based on whether the depleted hydrocarbon field has been ingressed by water, or not. With respect to water ingress, some studies[23] have assumed that only 65% of the pore space could be refilled with CO_2 (due to capillary effects) compared to 90% for a reservoir with natural gas still remaining. In saline aquifers, the discount factors are derived from average permeability measurements (taken from core testing/analysis or inferred from borehole logs) or by using data acquired from oil and gas operations conducted elsewhere in the same reservoir. In the North Sea, it has been estimated[23] that up to 40% of the pore space could be occupied by CO_2 in saline aquifers within the Bunter Sandstone Formation.

6 Storage Site Operation

6.1 Geological Characterisation

Once a potential site has been selected for a CO_2-storage operation it is necessary to gain as much knowledge as is feasible about the geology. Standard oil and gas industry techniques are used to do this, based on borehole, seismic, outcrop and laboratory test data. The paramount objective is to be able to gain the geological knowledge needed to demonstrate that the expected CO_2 storage capacity can be realised within the time-frame of the injection period, and that the CO_2 will not migrate out of the site.[25] For this, it is necessary to build geological models that portray the three-dimensional geology within and around the site (including up to the surface) that might be affected by the storage operation. In the European Directive on Carbon Capture and

Storage[4,5] this rock volume is termed the "storage complex". Particular attention needs to be paid to the location of features along which CO_2 might escape from the storage reservoir, such as faults, existing wells, mines and, in particular, the thickness, quality and distribution of the cap and reservoir rock. The location of "spill points" on the flanks of the geological storage structure is also crucial. Reservoir simulations[26] using industry-standard software are required to predict the behaviour of the CO_2 plume, especially its expected maximum areal footprint against the cap rock. The design, positioning and management of injection boreholes needs to be informed from the model and plume simulations.

6.2   Risk Assessment

A risk assessment is required to inform the safe design, operation and monitoring plan of a storage site. One way of doing this is to identify all the relevant features, events and processes (FEPs) that exist or might occur within or impact upon the storage complex. Relevant FEPs external to the storage complex also need to be evaluated (*e.g.* overall effect on global climate in the event of leakage, or interruptions in CO_2 supply due to power plant failure). Scenarios are selected (*e.g.* a well failure, or hydrodynamic interference between other storage operations in the same region) and models run, from which qualitative and quantitative consequence analysis can be derived. This process informs the design, operational and intervention strategies needed to mitigate critical risks or remediate any unintended consequences of the storage operation, as well as meet any regulatory requirements (*e.g.* Environmental Impact Assessment). A publicly available and regularly up-dated on-line database tool for FEP methodology is provided by the IEA GHG.[27]

6.3   Measurement, Monitoring and Verification (MMV)

Site characterisation and an Environmental Impact Assessment (EIA) should establish baseline conditions prior to CO_2 injection. Subsequent changes caused by the CO_2 storage operation are evaluated by measuring change using monitoring tools and techniques. Verification that the CO_2 storage operation is going according to plan is crucial. MMV activities should be robust enough to identify any significant departure from the expected site performance that suggests an increase in the risk of leakage or indirect damage/disturbance to other resources (*e.g.* groundwater), infrastructure or people.[28–31] In such cases, operational changes such as temporary CO_2-injection suspension may be required. If these are not successful, site abandonment and/or remediation may be required, or imposed by a regulatory authority. As MMV information accumulates over time it can be used iteratively to fine-tune and ground-truth reservoir and risk models, thus enhancing future prediction of site performance.

Deployment of particular monitoring technologies is site-dependent. A monitoring deployment selection tool, which suggests and explains MMV technologies appropriate to various storage-site scenarios, is publicly available in an on-line tool at the IEA GHG website.[32] This web tool is regularly updated as existing technologies are refined and new technologies emerge.

MMV technologies fall into the following main deployment categories:

- *Borehole Deployment*: *e.g.* fluid and gas sampling, temperature, pressure, electrical conductivity/resistivity, stress, corrosion, acoustics, cross-well seismic, tracer injection.
- *Shallow Sub-Surface and Surface Deployment*: *e.g.* seismic, soil gas probes, electrical conductivity/resistivity/induction, tilt meters, gravimeters, ground-water measurement and analysis, side-scan sonar, ecosystem monitoring.
- *Atmosphere*: *e.g.* open-path infrared laser, direct atmospheric gas sampling and analysis.
- *Remote-Sensing Platforms (Airborne and Satellite)*: *e.g.* InSAR, LIDAR, infrared thermal imaging.

6.4 Leakage

Leakage is the main concern for all stakeholders (operators, regulators and public).[33–37] Under the EU CCS Directive, a site should be designed not to leak in order to obtain its operational licence. However, it is also a requirement of the Directive to conduct risk and Environmental Impact Assessments (EIA) to predict what might happen if the site leaked. Since pressure is highest during and just after injection, this is the time when leakage is most likely to occur. Post-injection, the leakage risk diminishes as the CO_2 is increasingly immobilised by residual gas trapping, dissolution and mineral trapping.

In the sub-surface, potential damage to natural resources[38] and buried infrastructure is the main concern (*e.g.* associated with mineral deposits, other oil and gas operations, potable and agricultural water supplies). This may be caused by the CO_2 itself, or indirectly through fluids and substances that may mobilised or displaced (*e.g.* brines entering useable water resources).

Leakage to the surface poses risks to ecosystems and people. In aquatic systems, studies on natural sub-marine CO_2 seeps[38] have shown that benthonic calcifying organisms are the most vulnerable to CO_2 exposure (see also Chapter 9 in this book). Nektonic organisms are able to move away. Plant responses to elevated CO_2 levels are well known from natural terrestrial seeps[39] and further evidence is mounting about plant and microbial responses.[40–42]

If CO_2 is released in the atmosphere it normally disperses readily, but it can accumulate in topographic lows at ground level[43] or in restricted areas of ventilation (*e.g.* building basements). Populations who live where geological CO_2 is emanating naturally through the ground (*e.g.* at Ciampino, near Rome, Italy), have their basements ram ventilated (similar to basement ventilation of radon). Local by-laws prevent sleeping on the ground-floor of such buildings. Meteorological stations monitor surface atmospheric conditions and are used to warn the community when ground surface wind speeds are low.

The International Energy Agency has published a report[44] on natural releases of CO_2, including the tragic Lake Nyos and Lake Monoun incidents

(Cameroon) where there were fatalities. The relevance of these tragic incidents to underground industrial CO_2 storage risks is controversial, because the geological mechanisms and context are quite different from those associated with CO_2 storage sites.

7 Public Awareness of CO_2 Storage

Public awareness of CO_2 capture and storage is emerging, but is still at a relatively low level compared to other carbon-management technologies (*e.g.* renewables and nuclear energy). Recent studies[45,46] show that public awareness and attitudes to CO_2 storage vary, depending on age, culture, regional geography, nationality, education and gender. As a general rule, people are more positive after technical information on CCS is explained and understood, compared to when they first hear about the concept. Public opinion can be swayed by a minority of opinion formers (*e.g.* politicians, media, NGOs). Scientists and engineers have an important role to play in communicating impartial, accurate information and to give opinion supported by the evidence-base.[47] Strongest public engagement occurs at the local level when communities perceive that they are exposed to unacceptable risk from a particular CO_2 storage project, or see no local benefit great enough to outweigh their risk aversion (of which safety concerns are most prominent). Without substantial local public support or acceptance, it can be very difficult, or even impossible, for a project to proceed (*e.g.* the proposed Barendrecht 10 Mt CO_2 storage project in a depleted gas field in the Netherlands). Even if accurate information/opinion is given and communicated well, it will not necessarily be accepted or trusted. All of us form opinions and make decisions influenced by a variety of factors, aspirations, experiences and perceptions,[48] and it is important to understand these with regard to public engagement about CCS (see Figure 6).

8 Conclusions

Fossil fuels are likely to remain the world's main source of primary energy for the next 50 years, and possibly beyond. They are the root cause of the emissions problem that humanity faces. Geological storage of CO_2 captured from large stationary point sources is therefore an essential part of the mix of technologies urgently needed to be deployed, at large scale, in order to stabilise and reduce global emissions to the required level (50% plus) by 2050. Depleted oil and gas fields offer the quickest route to rolling out significant storage deployment but, unless storage into deep saline aquifers is also developed in parallel, CCS will be limited in capacity and geographic application. Existing oil and gas technologies and techniques can be used to conduct CO_2 storage, but there is still much to learn, particularly in improving predictive performance of storage sites.[49] Gaining public confidence that CCS can be safely deployed at the scale required is a significant factor. Especially for people who may live in proximity to

Figure 6 The surface expression of infrastructure at a CO_2 injection site is extremely small and unobtrusive. This CO_2 injection well at the BP-led InSalah CO_2 storage project, at Krechba in Algeria, is capable of injecting around 300 Mt of CO_2 a year and yet its pad area is only approx. $20\,m^2$. It is also virtually silent. The site injects into formation brines, hosted in marine sands of Carboniferous age, on the flank of an anticline capped by a producing gas field.

storage sites. In terms of being an effective GHG mitigation technology, it is clear that whilst fossil fuel use continues "business as usual", CO_2 will continue to "leak" into the environment, unless CO_2 capture and storage is routinely and successfully deployed.

Acknowledgements

N. Riley publishes by permission of the Executive Director, British Geological Survey (NERC)

References

1. D. J. Evans and J. M. West, *An Appraisal of Underground Gas Storage Technologies and Incidents for the Development of Risk Assessment Methodology*, UK Health and Safety Executive, Research Report RR605, 2008. http://www.hse.gov.uk/research/rrpdf/rr605.pdf.

2. Nascent Project, *Natural Analogues for the Storage of CO$_2$ in the Geological Environment*, European Commission Framework 6 Contract no: ENK5-CT-2000-303. http://www.bgs.ac.uk/nascent/home.html.

3. International Energy Agency Greenhouse Gas Programme. http://www.co2captureandstorage.info/.

4. European Commission, *Directive of the European Parliament and of the Council on the Geological Storage of Carbon Dioxide* and amending *Council Directive 85/337/EEC, Directives 2000/60/EC, 2001/80/EC, 2004/35/EC, 2006/12/EC, 2008/1/EC* and *Regulation (EC) No.1013/2006*, 2009.

5. S. Brockett, *Energy Procedia*, 2009, **1**, 4433–4441.

6. R. A. Chadwick, R. Arts, C. Bernstone, F. May, S. Thibeau and P. Zweigel, *Best CO$_2$ Practice for the Storage of CO$_2$ in Saline Aquifers*, British Geological Survey Occasional Publication, 2008, 14.

7. C. Hermanrud, T. Andresen, O. Eiken, H. Hansen, A. Janbu, J. Lippard, H. N. Bolås, T. H. Simmenes, G. M. Teige and S. Østmo, *Energy Procedia*, 2009, **1**, 1997–2004.

8. R. A. Chadwick, D. Noy, R. Arts and O. Eiken, *Energy Procedia*, 2009, **1**, 2103–2110.

9. C. A. Rochelle, I. Chernichowski-Lauriol and A. E. Milodowski, in *Geological Storage of Carbon Dioxide*, ed. S. J. Baines and R. H. Worden, Geological Society Special Publication, 2004, **233**, 87–106.

10. D. S. Hughes, *Energy Procedia*, 2009, **1**, 3007–3014.

11. A. Korre, J. Shi, C. Imrie and S. Durucan, *Energy Procedia*, 2009, **1**, 2525–2532.

12. S. J. Friedmann, R. Upadhye and F. Kong, *Energy Procedia*, 2009, **1**, 4551–4557.

13. S. Reeves, R. Gonzalez, S. Harpalani and K. Gasem, *Energy Procedia*, 2009, **1**, 1719–1726.

14. B. C. Nutall, C. F. Eble, J. A. Drahovzal and R. M. Bustin, *Analysis of Devonian Black Shales in Kentucky for Potential Carbon Dioxide Sequestration and Enhanced Natural Gas Production*, University of British Columbia and Kentucky Geological Survey, 2005. http://www.uky.edu/KGS/emsweb/devsh/final_report.pdf

15. G. Ersland, J. Husebø, A. Graue and B. Kvamme, *Energy Procedia*, 2009, **1**, 3477–3484.

16. J. Husebø, G. Ersland, A. Graue and B. Kvamme, *Energy Procedia*, 2009, **1**, 3731–3738.

17. J. S. Levine, J. M. Matter, D. Goldberg and K. S. Lackner, *Energy Procedia*, 2009, **1**, 3647–3654.

18. J. M. Matter, W. S. Broecker, M. Stute, S. R. Gislason, E. H. Oelkers, A. Stefánsson, D. Wolff-Boenisch, E. Gunnlaugsson, G. Axelsson and G. Björnsson, *Energy Procedia*, 2009, **1**, 3641–3436.

19. D. Goldberg and A. L. Slagle, *Energy Procedia*, 2009, **1**, 3675–3682.

20. V. Prigiobbe, M. Hänchen, G. Costa, R. Baciocchi and M. Mazzotti, *Energy Procedia*, 2009, **1**, 4881–4884.

21. S. C. Krevor, C. R. Graves, B. S. Van Gosen and A. E. McCafferty, *Energy Procedia*, 2009, **1**, 4915–4920.
22. S. Bachu, D. Bonijoly, J. Bradshaw, R. Burruss, S. Holloway, N. P. Christensen and O. M. Mathiassen, *Int. J. Greenhouse Gas Control*, 2007, **1**(4), 430–443.
23. S. Holloway, C. J. Vincent and K. Kirk, *Industrial Carbon Dioxide Emissions and Carbon Dioxide Storage Potential of the UK*, Report No. Coal 308, DTI, 2006.
24. Scottish Government, *Opportunities for CO2 Storage around Scotland-an Integrated Strategic Research Study*, 2009. www.scotland.gov.uk/Resource/Doc/270737/0080597.pdf
25. International Energy Agency, *Geologic Storage of Carbon Dioxide-Staying Safely Underground*, IEA Working Party on Fossil Fuels and IEA Greenhouse Gas R&D Programme, 2008. http://www.cslforum.org/publications/documents/geostoragesafe.pdf accessed 10 June, 2009.
26. C. Green, J. Ennis-King and K. Pruess, *Energy Procedia*, 2009, **1**, 1823–1830.
27. International Energy Agency Green House Gas Programme, *Risk Assessment*, CO2 FEP database. http://www.quintessa.org/co2fepdb/PHP/frames.php
28. J. P. Morris, R. L. Detwiler, S. J. Friedmann, O. Y. Vorobiev and Y. Hao, *Energy Procedia*, 2009, **1**, 1831–1837.
29. A. Mathieson, I. Wright, D. Roberts and P. Ringrose, *Energy Procedia*, 2009, **1**, 2201–2209.
30. L. Zheng, J. A. Apps, Y. Zhang, T. Xu and J. T. Birkholzer, *Energy Procedia*, 2009, **1**, 1887–1894.
31. J. A. Apps, Y. Zhang, L. Zheng, T. Xu and J. T. Birkholzer, *Energy Procedia*, 2009, **1**, 1917–1924.
32. Intenational Energy Agency Green House Gas Programme, *CO2 storage monitoring selection tool*. http://www.co2captureandstorage.info/co2tool_v2.2.1/index.php
33. A. Grimstad, S. Georgescu, E. Lindeberg and J. Vuillaume, *Energy Procedia*, 2009, **1**, 2511–2518.
34. M. Stenhouse, R. Arthur and W. Zhou, *Energy Procedia*, 2009, **1**, 1895–1902.
35. S. Bachu and T. L. Watson, *Energy Procedia*, 2009, **1**, 3531–3537.
36. W. Crow, D. B. Williams, J. W. Carey, M. Celia and S. Gasda, *Energy Procedia*, 2009, **1**, 3561–3569.
37. V. Meyer, E. Houdu, O. Poupard and J. Legouevec, *Energy Procedia*, 2009, **1**, 3595–3602.
38. J. M. Hall-Spencer, R. Rodolfo-Metalpa, S. Martin, E. Ransome, M. Fine, S. M. Turner, S. J. Rowley, D. Tedesco and M. Buia, *Nature*, 2008, **454**(3), 96–99.
39. A. Raschi, F. Miglietta, R. Tognetti and P. van Gardingen, *Plant Responses to Elevated Carbon Dioxide-Evidence from Natural Springs*, Cambridge University Press, 1997.

40. J. M. West, J. M. Pearce, P. Coombs, J. R. Ford, C. Scheib, J. J. Colls, K. L. Smith and M. D. Steven, *Energy Procedia*, 2009, **1**, 1863–1870.
41. P. Maul, S. E. Beaubien, A. Bond, L. Limer, S. Lombardi, J. M. Pearce, M. Thorne and J. M. West, *Energy Procedia*, 2009, **1**, 1879–1885.
42. M. Krüger, J. West, J. Frerichs, B. Oppermann, M. Dictor, C. Jouliand, D. Jones, P. Coombs, K. Green, J. Pearce, F. May and I. Möller, *Energy Procedia*, 2009, **1**, 1933–1939.
43. F. K. Chow, P. W. Granvold and C. M. Oldenburg, *Energy Procedia*, 2009, **1**, 1925–1932.
44. International Energy Agency, *Greenhouse Gas Programme, Natural Releases of CO₂*, 2008 http://www.ieagreen.org.uk/glossies/naturalreleases.pdf
45. J. Anderson, J. Chiavari, H. de Coninck, S. Shackley, G. Sigurthorsson, T. Flach, D. Reiner, P. Upham, P. Richardson and P. Curnow, *Energy Procedia*, 2009, **1**, 4649–4653.
46. F. Johnsson, D. Reiner, K. Itaoka and H. Herzog, *Energy Procedia*, 2009, **1**, 4819–4826.
47. R. Arts, S. Beaubian, T. Benedictus, I. Czernichowski-Lauriol, H. Fabriol, M. Gastine, O. Gundogen, G. Kirby, S. Lomabardi, F. May, J. M. Pearce, S. Persoglia, G. Remments, N. J. Riley, M. Sohrabi, R. Stead, S. Vercelli and O. Vizika-Kavvadias, *What does Geological Storage Really Mean?* Co2GeoNet, 2008.
48. D. Ariely, *Predictably Irrational: The Hidden Forces that Shape our Decisions,* Harper Collins, 2008.
49. K. Michael, M. Arnot, P. Cook, J. Ennis-King, R. Funnell, J. Kaldi, D. Kirste and L. Paterson, *Energy Procedia*, 2009, **1**, 3197–3204.

Carbon Sequestration in Soils

STEPHEN J. CHAPMAN

1 Introduction to the Carbon Cycle in Soil

Carbon sequestration in soils is the process whereby atmospheric carbon dioxide can be fixed into soil such that it is held there in a relatively permanent form, *i.e.* the term 'sequestration' implies a combination of both capture and storage. This, of course, will require that the carbon dioxide is converted to some other chemical form and this will usually be organic rather than inorganic. An understanding of how this might be promoted first requires an understanding of the carbon cycle in soil.

There is an inorganic carbon cycle in soil, whereby carbon dioxide dissolved in rainwater forms carbonic acid which then reacts with basic cations to form secondary carbonates, or with calcium–magnesium silicate minerals during the weathering process to release basic cations that then precipitate as carbonates.[1] However, such processes are extremely slow and are only likely to be of importance in the saline and sodic (alkaline) soils found in arid and semi-arid zones.[2] Hence, the inorganic carbon cycle is not of consequence for most UK and European soils.

Of far greater significance is the organic carbon cycle, whereby atmospheric carbon dioxide is fixed by photosynthesis into plants by forming organic compounds, the bulk of which are cellulose, hemicellulose and lignin, though with additional protein, lipids and other complex compounds. As plants die, these compounds enter the soil and are broken down by the action of soil microorganisms which then release the carbon dioxide back into the atmosphere (see Figure 1). Of course, an important sub-cycle occurs where plants are consumed by animals; part of the carbon is respired, but animal excreta (and the animals themselves as they die) ultimately finds its way into the soil only to be decomposed along with the plant remains.

Issues in Environmental Science and Technology, 29
Carbon Capture: Sequestration and Storage
Edited by R.E. Hester and R.M. Harrison
© Royal Society of Chemistry 2010
Published by the Royal Society of Chemistry, www.rsc.org

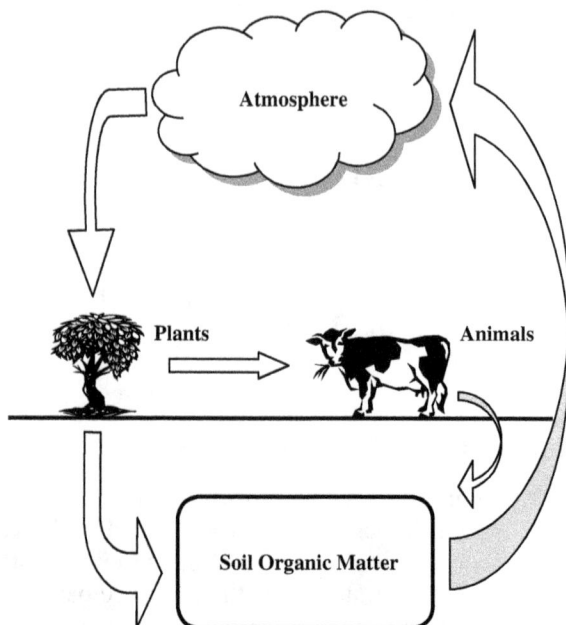

Figure 1   The Carbon Cycle.

1.1   *Plant Production*

Plant productivity, measured as the annual (or seasonal) input of carbon (C) to the whole plant, including both shoots and roots, varies greatly across natural ecosystems, going from deserts to tropical rainforests, with mean values of less than $0.5\,t\,C\,ha^{-1}\,a^{-1}$ to over $10\,t\,C\,ha^{-1}\,a^{-1}$, respectively (see Table 1). Some intensive cropping systems can have greater productivity, but usually with high inputs of fertiliser, pesticides and irrigation. On a global scale, the fixation of carbon amounts to $120\,Pg\,a^{-1}$ ($1\,Pg = 10^{15}\,g = 1\,Gt$; $t = tonne$; $a = annum$, or year; $ha = hectare$), but half of this immediately returns to the atmosphere in shoot and root respiration (see Figure 2). Thus $60\,Pg\,C\,a^{-1}$ is available to enter the soil.

An important aspect of plant production is that what is seen above ground is only part of the story. Up to 40% of the carbon captured by photosynthesis is directed towards the roots. Part of this forms what is known as rhizoexudates: soluble carbon compounds released by the roots into the soil, root cells 'sloughed off' into the soil and dead roots, which are part of the ongoing root turnover that most plants exhibit.

1.2   *Decomposition*

Decomposition is the process of carbon mineralisation, whereby organic carbon is converted back to carbon dioxide which is then released back into the atmosphere. This is also referred to as soil respiration. Often the initial step is

Table 1 Mean Net Primary Production (NPP) for some major vegetation zones.[45]

Ecosystem	NPP $(t\,C\,ha^{-1}\,a^{-1})$
Desert	0.5
Tundra	1
Needle-leaf forest	3
Grasslands	4
Summer-green broad-leaf forest	5.5
Sub-humid woodlands	6.5
Ever-green broad-leaf forest	8
Tropical rain forest	10

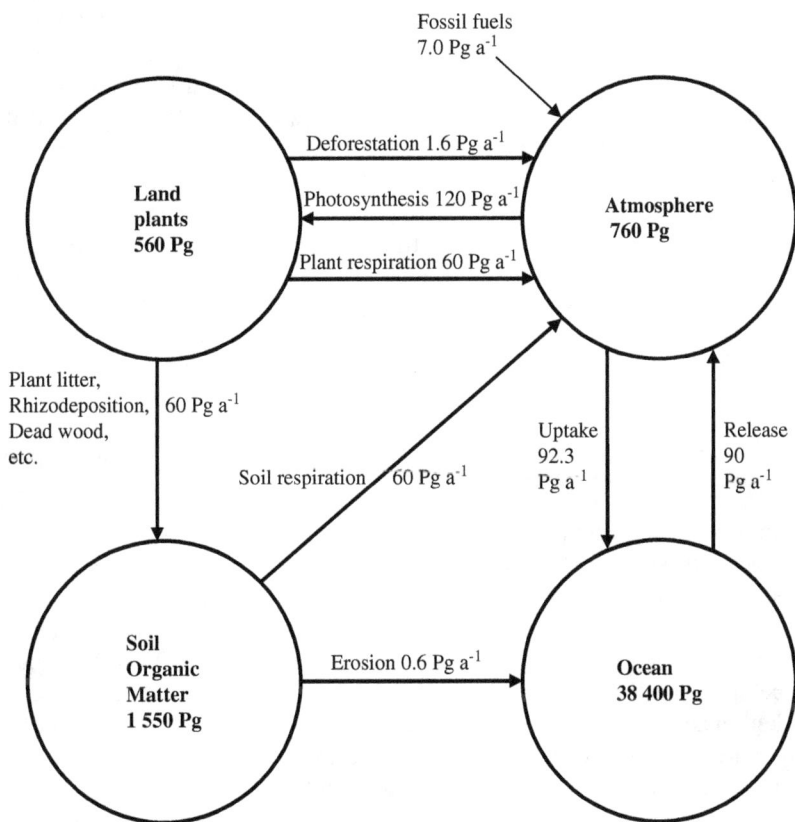

Figure 2 Global carbon pools and fluxes. Modified from Lal (2008).[35]

the consumption of plant debris by soil animals, ranging from a host of larger invertebrates (the 'macrofauna') like woodlice, centipedes and earthworms; to smaller animals (the 'mesofauna') like mites, springtails and enchytraeid worms; and to the smallest animals (the 'microfauna') such as nematodes and protozoa.[3] These groups mainly act by communition, breaking down the plant material into smaller pieces. Ultimately, however, the component plant structural compounds are broken down by fungi and bacteria. It is these groups that possess the necessary enzyme complexes that can convert cellulose and lignin into soluble compounds, that can then be assimilated and metabolised.

Often the decomposition of above-ground plant material will begin on the soil surface and only with time, through the activity of soil animals (bioturbation), the action of weather, or through mechanical means such as ploughing, will it become incorporated into the soil. Roots, of course, are already within the soil environment. In some soils, typically peats, the dead plant material remains essentially at the surface and only becomes buried as more dead plant material is added on top of it.

During the decomposition process, carbon is channelled into the soil bacteria and fungi, known collectively as the microbial biomass, and used for both energy and for building more microbial biomass. That used for energy is converted to carbon dioxide – soil respiration. That converted into new microbial biomass eventually dies, or is eaten, and enters a second cycle of decomposition as the dead 'biomass' is broken down by other soil bacteria and fungi. This process continues, with the carbon going through many cycles of decay, each time getting smaller in quantity, until the original carbon eventually disappears. On a global scale, the carbon entering the soil and the carbon mineralised in soil respiration are approximately in balance, such that soil respiration puts 60 Pg $C a^{-1}$ back into the atmosphere (Figure 2). Even on a biome scale there is a close correlation between Net Primary Production (NPP) and soil respiration.[4]

1.3   Soil Organic Matter

Soil organic matter (SOM) is not a single chemical entity but a complex range of compounds, of which the precise nature of many is unknown. Part of the SOM will consist of newly added plant material and in many environments this will be under seasonal control. As this plant material undergoes decomposition, the more readily available and simple constituents, sugars, amino acids, nucleic acids, proteins, *etc.*, are broken down first. The structural polymers, pectin, hemicellulose and cellulose are then more slowly degraded. Finally, lignin is attacked once most other constituents have been exhausted. However, in much plant material, these chemical entities are not present singly, but rather in varying degrees of physical and chemical complexity. Hence, cellulose is often complexed (intimately mixed, probably with some covalent bonding) with lignin, and the lignin component then offers some resistance to the cellulose against decay. Some proteinaceous material may also survive longer when closely associated with lignin. Other complex plant components, such as tannins and cutins (waxes), can also offer some protection to otherwise rapidly decomposable substrates.

The microbial biomass itself forms a vital part (literally!) of the SOM. It is generally found that the microbial biomass makes up 1–3% of the total SOM.[5]

The chemical structures of bacteria and fungi differ from those of plant (and animal) material, and so add another layer of complexity to the organic chemistry of the soil. Of particular note are the melanins produced by certain fungi, that persist in soil and confer a darkening pigmentation to the SOM.

As decomposition proceeds, the original structure of the plant material gradually becomes unrecognisable and intimately associated with compounds of microbial origin. It is considered that not only is there breakdown of polymers, but also a random re-synthesis of chemical bonds by various condensation reactions outside of living cells. This is the process of humification, leading to the formation of humic substances which, because of their random assembly, are highly resistant to further enzymic degradation.[6] As decomposition proceeds, the carbon content of the SOM increases from the *ca.* 42% found in fresh plant material to *ca.* 58%. A further important step in many soils is the association (binding or complexation) of SOM, particularly the humified fraction, with mineral particles. The finer particles, such as the clay fraction, are most effective in this. This physical, and possibly chemical, association gives some protection of the SOM against further decomposition.

1.4 Characteristics and Age of Soil Carbon

Globally there is a balance between plant inputs and soil respiration and this is often true for specific soils. However, this does not mean that all the carbon added in a particular year is mineralised that year. Rather, a fraction will decay this year, less next year, even less the year after, and so on. Thus, the carbon dioxide respired during one year will have come from organic material added that year and over all the preceding years, in ever decreasing amounts as the cycles of decay progress (see Section 1.2). This gives an age structure to the SOM, which is often expressed in terms of turnover time, with fresh plant material having a turnover time ranging from months to a few years. The precise turnover time will vary with soil temperature, soil moisture, the season, soil disturbance, soil nutrient status and other factors. SOM associated with mineral fractions will have a turnover time of decades while the most highly humified material may have a turnover time beyond a thousand years.[7,8] This latter fraction is considered to be passive or virtually inert, and is often needed in models of SOM turnover for them to describe C cycling in soil adequately. Such turnover times have been confirmed by the ^{14}C dating of SOM fractions. In practice, there will be a continuum of ages, though the age distribution may be uneven.

1.5 Losses to Water

An additional component of the soil carbon cycle is the loss to water, as precipitation passes through the soil and enters streams, rivers, and ultimately the oceans. Part of this will be dissolved organic matter but some may be particulate organic matter. The latter becomes more significant where soil erosion is taking place. Also waters may carry inorganic carbon as dissolved carbon dioxide and bicarbonate. Globally this accounts for 0.6 Pg C being transferred to the ocean (Figure 2). The fate of the organic carbon lost in this way is unclear. Some, possibly half, will be respired as carbon dioxide, while the remainder may become locked away in ocean sediments. Losses to water may be particularly significant for peatland soils.[9]

2   Factors Influencing Carbon Accumulation

What controls the level of organic matter in soils? This is a question to which we do not have a full answer. However, if we can begin to understand some of the controlling factors then we can begin to suggest ways of increasing the sequestration of carbon in soils. The rate of change of SOM in a particular soil is the difference between the rate of addition of plant (and animal) material and the rate of decomposition. However, the overall decomposition rate is a first-order reaction, *i.e.* it is directly proportional to the amount of material that is there. Hence, over time, the system becomes 'self-regulating' until the decomposition rate balances the addition rate of plant material. We then say that the soil is in 'equilibrium'. At this point the SOM content ceases to change over time but has arrived at a fixed level. Basically, there two ways of increasing the C sequestration in soils: either we increase the rate of C input or we decrease its rate of decay.

What happens when conditions change? This is illustrated in Figure 3. If we consider a soil that has attained an equilibrium level of 100 t C ha^{-1} and at point A experiences a change where C input decreases or decomposition rate increases, then we will see an exponential decrease in the carbon stock until a new equilibrium level is reached at point B. The changes are initially rapid, but

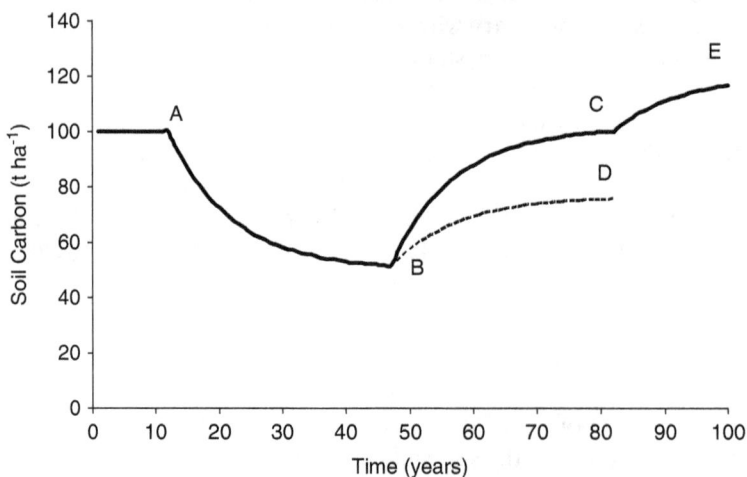

Figure 3   Time-course of change in soil carbon following different scenarios.

then slow down and the new equilibrium level is reached asymptotically. Such changes occur when natural ecosystems are converted to agriculture and historically have occurred wherever man has cultivated the soil. The more recent and well-documented decreases in soil carbon have been those observed in North America with the ploughing up of the prairies for continuous cereal production. Over 50–100 years, 50% of the soil C can be lost in a temperate climate, but losses can be 50–75% over 10–20 years in the tropics.[1] If conditions are reversed, *e.g.* ploughed land is allowed to revert to a semi-natural ecosystem, then it is possible, at least in theory, to return to the initial level (point C in Figure 3). In practice, one might arrive at a lesser carbon density at point D. Alternatively, if the new system is managed in a way that increases carbon inputs (or decreases outputs) over and above the original ecosystem, then it is possible to arrive at point E.

One consequence of having a SOM fraction that is inert or passive is that we have a pool that may be entered but never left. Hence, unless it was formed when the soil was formed, it must continue to be added to as the soil develops. For this reason, some soil scientists believe that a true equilibrium state is never reached but that soil C is always slowly accumulating. Even if the 'inert' fraction is not completely inert but is lost at a very slow rate, the time to equilibrium is extremely long and would exceed the age of many northern temperate soils that have only formed since the last ice sheet retreat, about 10 000 years ago. This means that many soils may have a greater capacity to sequester carbon than current SOM levels would indicate.[10]

2.1 Climate

Even a cursory glance at a map of soil carbon content across the globe will indicate that climate plays a major role in determining its level. This is not due to inputs; in fact, these tend to decrease from the equator to the poles (Table 1). Rather, temperature and moisture are major determinants of the rate of decomposition since they directly affect the activity of the microbial biomass. Rates increase with increasing temperature, but can also decrease under very wet conditions as soils become anaerobic. Decomposition rates are then limited by the availability of oxygen. Such a process accounts for the accumulation of large stocks of carbon in the peats and organic soils of the cool temperate and boreal zones. That soil moisture can have an over-riding impact is seen in the accumulation of carbon in tropical peatlands, where water-logging assumes greater importance than temperature. There are also certain arid environments, either cold or hot, where the absence of soil moisture also inhibits decomposition and organic matter can accumulate, although primary productivity also tends to be very low.

2.2 Plant Inputs

Generally speaking, the soil carbon stock is proportional to the annual input of plant material. Hence any way of increasing this total will have the effect of increasing the soil carbon. In arable systems, plant inputs can be increased by using cover crops, *i.e.* reducing the amount of time that the soil lies bare. This particularly applies to rotations, that include a bare fallow for improving water conservation. Changes in the plant species or strains of plant species grown can also increase overall yield and hence the proportion that gets returned to the soil. Crop rotations that include grass leys also benefit as grasses tend to generate more plant biomass, especially below ground.[11] For example, a change to deeper rooting grass species increased the SOM level by up to $70\,t\,C\,ha^{-1}$ in some savanna soils of Colombia.[7]

Increasing crop residue returns is also beneficial. This means not removing straw or other plant residues off site, but leaving them *in situ* or at least returning them after processing or composting. For example, the return of rice hulls to the soil is particularly advantageous as the high silica content renders them slow to decompose. In a similar way, residues with a high lignin or phenolic content, such as composted wood bark, as long as they are not phyto-toxic, are also helpful. While straw burning has effectively ceased in the UK, the burning of crop residues is still practised in many parts of the world following harvest. For example, hand harvesting of sugar cane demanded that most of the non-cane trash was burnt off prior to cutting the cane. However, mechanised cutting (even with some fossil fuel input) means that such residues can be returned to the soil.[12]

Improving overall fertility will increase crop yields and hence the proportion that is incorporated into the soil. A meta-analysis on 137 sites[13] where the response of soil carbon had been measured, showed a positive response to the nitrogen (N) fertiliser applied, as well as a positive response to rainfall, and what was termed the 'cropping index', which was the number of crops per year. There was a negative response to temperature and to an index of soil texture, where greater carbon increases were seen in more coarsely textured (sandy) soils. This is probably because sandy soils tend to start off at a lower level of SOM than finely textured (clay) soils. The negative response to temperature meant that there was a net benefit to N fertilisation in temperate climates, but not in tropical systems. This is because, even though there is a response to nitrogen, it is not enough to offset the carbon cost of producing and applying the nitrogen fertiliser. Every kg of N fertiliser costs *ca.* 1–2 kg C through energy needs for production, distribution and application.[12]

2.3   Other Organic Inputs

Increasing the overall input of organic material other than that grown *in situ* will also increase carbon sequestration. Hence, the return of composts and manures to land should be encouraged. While this would be normal practice in the UK, this is not always the case elsewhere; in many areas, manures are dried and used as fuel for domestic cooking. Other sources of organic matter such as composted municipal ('yard') waste and sewage sludge may also be applied, though in the case of the latter a careful eye has to be kept on the levels of heavy metals or other pollutants that may also be added. There is a further subtlety in that the returns should be targeted to those soils that would benefit the most. For example, it is better to return such residues to arable soils rather than to grassland soils.[1,14]

2.4 Tillage

Tillage is the general term applied to all sorts of ploughing or cultivating the soil. As mentioned above, ploughing was responsible for much of the C loss from agricultural soils in comparison to the grassland or forest soils from which they were originally derived. Disturbance of the soil exposes SOM that may have been physically protected, particularly by the soil mineral components, to further decomposition.[11] Greater aeration of the soil and changes in temperature and moisture may also contribute. There is a general stimulation of the microbial biomass and increased soil respiration. Soil carbon can decrease even after one tillage event, with up to 11% being lost.[15] Greater losses occur as tillage intensity increases. Reduced till (or minimum till) refers to some cultivation by shallow or deep tine (ripping) or by discs, which is much less disruptive than ploughing; the former gave 6% loss in carbon compared to 27% with ploughing.[15] No-till or zero-till refers to the absence of any tillage operation. A related term is conservation tillage, which is where plant residues are left on the soil surface to conserve water and reduce soil erosion. A meta-analysis on the impact of reducing tillage at 161 sites[13] showed a mean increase in soil C of $2.1\,t\,ha^{-1}$. Indications were that over time, this might increase to *ca.* $12\,t\,ha^{-1}$. While zero or minimum tillage has seen some uptake in the United States, its implementation in the UK has been rather slow, covering no more than 3% of arable land at the most.[16]

2.5 Grazing

While grasslands tend to have much greater soil C levels than arable soils, correct management of grazing levels can improve the quality of the grass sward, increase overall plant returns and may raise soil C levels by up to $1.3\,t\,C\,ha^{-1}\,a^{-1}$ (ref. 2), though the results are not always clear-cut.[12] On the contrary, over-grazing will lead to sward impoverishment, bare areas, erosion and C loss.

2.6 Drainage/Irrigation

The impacts of water status on soil carbon are quite complex. While drainage can improve plant productivity, decomposition is inhibited greatly in water-logged soils. Hence drainage usually leads to carbon loss, especially in soils where organic matter has accumulated under wet conditions, such as in fens and bogs. Within the UK, this has been most dramatic in East Anglia where drainage has caused the level of peat to fall nearly four meters.[17] Some of this is due to shrinkage of the peat on drying but a large proportion reflects loss as carbon dioxide. Reducing the loss is best implemented by raising water tables.[18]

In contrast to drainage, irrigation is often beneficial since it relieves limitations due to water stress, and increases plant yield, residue inputs to the soil and hence soil carbon.[12,19]

2.7   Erosion

Soil erosion, whether by water or wind, inevitably removes SOM from the site of origin. If such carbon is merely displaced then it will not result in any net loss. However, the disruptive forces of erosion are similar to those of tillage, and will probably promote increased decomposition.[20] An alternative view (held by some sedimentologists), is that the deposited, and probably buried, carbon may be protected from decomposition.[20] This may be partly true where eroded SOM eventually finds its way to ocean sediments. However, this is an area of great uncertainty and we really do not know what happens to much of the carbon displaced by erosion. Of course, erosion is also harmful in that plant yields will be decreased in heavily impacted areas so that returns to the soil are also decreased.

2.8   Fire Cycles

In many ecosystems, fire is part of the natural cycles of events. Its occurrence may be almost annual to perhaps once every century. It applies to forest, peatland, moorland, shrubland and grassland alike. The cause may be natural (lightening strikes), accidental or arson. The impact of lightening strikes on carbon accumulation has been observed in islands off northern Sweden, where much greater carbon sequestration is seen in the smaller islands; the risk of strike is proportional to the area, and hence fire incidence is low.[21]

Either annual or periodical burning may be used as a management technique to control vegetation: the removal of old biomass will stimulate new growth, such as in 'muirburn' where moorland is burnt to stimulate new heather growth. This reduction in above-ground biomass inevitably leads to less input to the soil and potentially a loss of soil carbon.[22] However, this may be compensated for by an increase in plant productivity. Additionally, there will be some generation of 'charcoal' or 'black carbon' which will add to the passive soil C fraction (see below). Nevertheless, the greatest danger is where a 'controlled' burn becomes uncontrolled and sets fire to the underlying soil organic matter. Even if the risk of this is very low, once it occurs, it undoes centuries of slow carbon sequestration.

The most serious consequences arise where deep peat soils are involved and the fire moves from consuming above-ground biomass to the peat itself. Such was the situation in Indonesia in 1997, when slash-and-burn practices set fire to the underlying peat that burnt for months, releasing an estimated 0.81–2.57 Pg C (ref. 23). Unfortunately, this was not an isolated incident and fires continued to burn in subsequent years which, with the carbon dioxide released from drainage of these tropical peats, released an estimated $2\,Pg\,C\,a^{-1}$ (ref. 24). This is almost 8% of global emissions from fossil-fuel burning, so clearly any effort to halt this destruction will have a major impact on efforts to conserve soil carbon. Within the UK, fire has been a regular visitor to the North York Moors[25,26] and has been implicated in the initiation of widespread peat erosion in the Southern Pennines.[27]

3 Land-Cover Classes and their Carbon-Sequestration Characteristics

Each major land-cover type has characteristic soil-carbon levels. These will depend upon the combination of factors outlined in the previous section. In the following, we consider the UK environment, but the values will be similar within the other cool temperate regions of Europe. Across most land-cover classes the carbon stocks within UK soils tend to be greater than the global mean (see Table 2). This is a reflection of the relatively cool climate. Even within the UK there is a marked increase in soil carbon as one moves northwards.

3.1 Arable

Arable soils will generally have the lowest soil C values since they are inevitably cultivated to varying degrees. Even those now in zero or minimum till will have been subjected to ploughing in the past. An exception will be cultivated peat-lands, but such soils are always in transition and losing carbon at a rapid rate. Typical values are in the range $120–150\,t\,C\,ha^{-1}$ (Table 2).

Table 2 Soil carbon stocks (0–100 cm) within different ecosystems. The global values show two independent estimates of the mean, while the UK values show the range of values in the different regions (England, Scotland, Wales, Northern Ireland).

Ecosystem	Carbon density (Global)[46] $(t\,C\,ha^{-1})$	Carbon density (UK)[47] $(t\,C\,ha^{-1})$
Tropical forest	122–123	
Temperate forest	96–147	170–370
Boreal forest	247–344	
Tropical savanna and grassland	90–117	
Temperate grassland and shrubland	99–236	130–230
Deserts and semi-deserts	42–57	
Tundra	127–206	
Cropland (arable)	80–122	120–150
Wetland	643	230–390[a]

[a]values are for all "semi-natural", not just wetland.

3.2   Grassland

Grasslands will nearly always have greater soil carbon levels than arable soils. The values in Table 2 refer to more permanent grasslands. Where grassland is in rotation with arable crops (a ley-arable rotation) the values will be intermediate between the two, depending upon the length of time in grass.

3.3   Forest/Woodland

Forest or woodland soils tend to have carbon stocks similar to, or slightly greater than, grassland. However, the coniferous forests, typical of the north and west of Britain, have at least twice the soil C stock of the deciduous/broadleaf forests, typical of the south and west of Britain.[28]

3.4   Semi-Natural

Semi-natural land, which includes wetland, usually has the greatest carbon stock. This particularly applies to peatland where the soil carbon density in some areas of deep peat can exceed $1000\,\mathrm{t\,ha^{-1}}$ (ref. 29).

3.5   Land-Use Change

Given the typical soil C stocks found within the various broad land-cover classes (Table 2), it is clear that a land-use change from one class to another will initiate a change in soil carbon, leading eventually to a new equilibrium level. This will follow the changes as illustrated in Figure 3 with either an exponential loss or gain of carbon. The conversion of grassland, woodland or semi-natural vegetation to arable will lead to the loss of soil carbon.[30] The conversion of semi-natural vegetation and, in most cases, woodland to grassland will also lead to carbon loss. The conversion of semi-natural vegetation to woodland may have variable effects, depending upon the soil carbon status of the semi-natural vegetation. The afforestation of deep peats almost certainly leads to a loss of carbon as the peat dries, compacts and oxidises.[30] This is exacerbated by the necessary peatland drainage prior to forest planting. Forests planted on shallow organic soils may not cause any net change in soil carbon, though this assumes no significant soil disturbance (*e.g.* ploughing) prior to planting and limited drying out of the organic soil horizon. Any losses may be compensated for by the formation of a litter layer, particularly under coniferous forest. Forests planted on mineral soils should show a net increase in soil carbon. Reversing any of the above land-use change scenarios should have the reverse impact on soil carbon. In particular, taking arable soils out of cultivation should initiate a substantial increase in soil carbon. Such a change has been seen in the past in the UK and elsewhere with the generation of 'set-aside' land.

These land-use changes are the basis of the estimation of the AFOLU (Agriculture, Forestry and Land Use Change) category during the calculation of national greenhouse-gas inventories, which is a requirement of parties to the

UNFCCC (United Nations Framework Convention on Climate Change). Using a land-use change matrix and the estimated changes in soil carbon with each land-use change, it is possible to estimate the losses (or gains) of carbon due to the land-use change part of this category.[30]

4 Climatic Zones other than Cool Temperate

While the major focus of this chapter is on soils of the cool temperate zone, such as are found within the UK and northern Europe, it may be useful to briefly consider the soils of warmer climates.

4.1 Warm Temperate

Soil carbon levels in southern Europe tend to be much lower than those in northern Europe, and consequently the potential for soil carbon sequestration is correspondingly lower. For example, estimates of the 1990 carbon stock in forest soils of central, north-western and northern Europe was three to five times that in soils of southern Europe, and the potential carbon sink was up to ten times greater.[31]

4.2 Tropical

The potential for soil carbon sequestration in tropical soils is also much less than for temperate soils. While for temperate soils the carbon stock in the soil may at least equal that in the vegetation but often may be many times greater, for tropical soils the reverse is true; most of the ecosystem carbon is held by the vegetation. The exception to this would be the tropical peatlands. Many areas are arid or semi-arid and water supply limits productivity. Such areas are often subject to periodic, if not annual, fire. Where the soil is cultivated, soil carbon losses are rapid at the high soil temperatures. Vegetation inputs are often limited due to other uses such as fodder and fuel. Estimated annual sequestration rates are only about half of those for temperate soils, though similar gains can be made when applying agroforestry, improved grazing and soil restoration.[1] Some benefit can also be had through the use of zero-till, cover crops and green manures.[11]

5 The Quantification of Carbon-Sequestration Strategies

A number of questions may be asked. What is the potential for carbon sequestration in soil? How much carbon can be stored in this way and how does it compare with current carbon emissions? What are the best strategies to achieve this? How quickly can it be done? And where is it best carried out? Not all the answers to these and further questions[32] are fully known and in some areas further trials and experiments are required. However, over the past eighty years, soil scientists, agronomists, foresters and ecologists have amassed enough information to give strong indicators of the best options. More recently, many models of carbon turnover in soil have been developed that can be used to predict the future soil carbon content given a range of management scenarios.

Precise quantification of the carbon sequestration potential is difficult. Usually this involves up-scaling from data gathered at specific sites and on specific soils. Such data are not uniformly distributed. Hence there tends to be many more studies based in North America and Europe than elsewhere. The underlying assumptions can be quite variable, *e.g.* how much arable land might be converted to other land uses. Some data include carbon in above-ground biomass as well as soil. Also there is the question as to whether values are 'theoretical', 'realistic' or 'conservative' (see Section 5.1).[33]

5.1   Worldwide Soil Carbon Sequestration Potential

Several authors have attempted to estimate the total amount of carbon that might be sequestered in soil. Global averages for the conversion of arable to forest and arable to grassland are 0.34 and $0.33 \, t \, C \, ha^{-1} \, a^{-1}$, respectively,[8] but these values will vary considerably with the climate, soil and vegetation. One starting point is that since soils have lost *ca.* $66 \pm 12 \, Pg$ since man began removing forests and cultivating the ground, then a similar amount could be restored to the soil. In Figure 3, having gone from A to B, we would then go to C. Restoring 50 Pg over 50 years would give an annual rate of $1 \, Pg \, a^{-1}$ (ref. 32). The rates for some major countries summarised in Table 3 total 0.33–$0.60 \, Pg \, a^{-1}$ (ref. 34) but obviously this doesn't cover everywhere. Another review gave $0.4 \, Pg \, a^{-1}$ for restoring degraded soils in the tropics, 0.4–$0.7 \, Pg \, a^{-1}$ for desertification control of soils in arid and semi-arid regions and 0.4–$1.2 \, Pg \, a^{-1}$ for the implementation of zero-till agriculture, giving a total of 1.2–$2.3 \, Pg \, a^{-1}$ (ref. 35). Others have cited a 'theoretical' 2–$4 \, Pg \, a^{-1}$ (ref. 33) and $0.9 \pm 0.3 \, Pg \, a^{-1}$ (ref. 36). This latter figure would be 13% of the global fossil fuel emissions (see Figure 2).

Wetlands, and particularly peatlands, present a rather special case in that they naturally sequester carbon without man's intervention. This has been estimated as being historically *ca.* $0.1 \, Pg \, a^{-1}$ (ref. 35). However, drainage, erosion, atmospheric pollution, over-grazing, peat cutting, afforestation and cultivation have all conspired to reduce the area of active carbon sequestration.

Table 3   Worldwide estimates of the soil carbon sequestration potential.[34]

Country	Sequestration potential $(Tg \, C \, a^{-1})$
United States	75–208
Canada	24
European Union	90–120
China	105–198
India	39–49

Hence restorative efforts are required to reinstate this sink function where at all possible.

5.2 Soil Carbon Sequestration Potential for Europe

Carbon sequestration rates on a per-hectare basis from one authority[1] (see Table 4) are of a similar order to the global values given above, and very similar to values cited for Canadian agriculture.[34] The greatest rates are where arable soils that have depleted SOM levels are put back into woodland, grassland or semi-natural land (usually, in time, this will also succeed to woodland). Second to this is improving cropland by employing zero- or minimum-till; returning straw and other organic residues such as animal manure, sewage sludge and various composts to the land; using cover crops and improved crop rotations that include grass leys or legumes. More modest returns are achieved by improving pastures and permanent crops (*e.g.* fruit trees), by optimising the fertilisation regime and plant strains to increase overall plant productivity and particularly root biomass residues. Judicious fertilisation will also give limited increase in soil carbon under woodlands and forests though, of course, the greatest benefit here is in increased above-ground yield. A sequestration value for north-west European forest soils of $0.4\,t\,C\,ha^{-1}\,a^{-1}$ has been given using a modelling approach.[31] However, in this case the value reflects the maturing of forests throughout Europe as many were planted in the second half of the last century and have yet to reach full productivity, both in the tree biomass and in the development of the forest soil.

Table 5 gives carbon sequestration potentials from a second authority.[37] This listing focuses primarily on arable soils and gives more of a breakdown on the various possible management options. Lower rates are given for set-aside but higher rates for some of the management alternatives for improving cropland, particularly the high estimates for farm-yard manure (FYM) and slurry additions. A lot of these differences reflect uncertainty in the underlying data, as well as variation in the starting conditions and time over which the sequestration has been observed. Greatest increases are likely in the initial years following an intervention, which then decreases as time progresses and some upper limit is approached.

Table 4 Estimates of European soil carbon sequestration potential (1).[1]

Strategy	Sequestration rate $(t\,C\,ha^{-1}\,a^{-1})$	Sequestration potential $(Tg\,C\,a^{-1})$
Restoration of degraded soils as set-aside	0.2–0.6	18.9–56.6
Improved management practices on cropland	0.1–0.4	19.7–78.7
Improved pastures	0.1–0.2	18.3–36.6
Improved permanent crops	0.1–0.2	1.7–3.4
Improved forests/woodlands	0.05–0.1	7.9–15.8
Total		66.5–191.1

Table 5 Estimates of European soil carbon sequestration potential (2).[37]

Strategy	Sequestration rate[a] $(t\,C\,ha^{-1}\,a^{-1})$
Zero-tillage	0–0.4
Reduced tillage	0–0.2
Set-aside	0–0.2
Convert to permanent crops and perennial grasses	0–0.6
Deep-rooting crops	0–0.6
Solid animal manure (farmyard manure, FYM)	0.2–1.5
Slurry	0.2–1.5
Crop residues	0.1–0.7
Sewage sludge	0.1–0.3
Composting	0.2–1.5
Improved rotations	0.17–0.76
N fertilisation (inorganic)	0.1–0.3
Irrigation	0.05–0.1
Bioenergy crops (soil)	0–0.6
Extensification/de-intensification	0–0.5
Organic farming (arable)	0–0.5
Convert arable to woodland	0.3–0.5
Convert arable to grassland	0.3–1.9

[a] Low and high estimates.

The total soil carbon sequestration potential for Europe of 70–190 Tg a^{-1} (Table 4; ref. 1) is compatible with that given in Table 3 (90–120 Tg a^{-1}; ref. 34) as the latter value is only for the smaller area of the European Union. Another estimate for the EU of 165–210 Tg a^{-1} (ref. 38) is actually similar since only half is soil carbon sequestration; the rest is forest biomass, fossil carbon offsets from biofuels and fuel savings from reduced tillage. A 'theoretical' estimate for Europe (the EU15) arrived at 200–500 Tg a^{-1} (ref. 33). A separate modelling exercise of sequestration in just the forest soils of the EU gave 26 Tg a^{-1} for 1990, expanding to 43 Tg a^{-1} in projections to 2040 (ref. 31). An optimal carbon sequestration potential will probably be realised through a combination of strategies, bearing in mind that some may be applied to the same area of land while others are mutually exclusive. Additionally, some strategies may not be practically applicable everywhere, even though they are theoretically possible. In particular, land given to biofuel production, a strategy with strong mitigation potential since it can both increase soil carbon and offset fossil-fuel consumption, needs to be within reach of processing facilities. Using a combined and optimised approach, 103 Tg a^{-1} was possible,[39] of which 28 Tg a^{-1} was from biofuel offset. A central plank of this strategy was that 10% of arable land was surplus. Half would be used for biofuel and half converted to woodland. On the remaining arable land, maximum rates of organic residues would be added and zero-till used throughout. This level of carbon sequestration would be able to offset 8% of the EU's 1990 CO_2-C emissions, which is the EU's reduction target for the first commitment period (2008–2012).

Table 6 Estimates of soil carbon sequestration potential for England.[14]

Strategy	Sequestration rate $(t\,C\,ha^{-1}\,a^{-1})$
Zero-tillage	0.15–0.24
Reduced tillage	0.04
Set-aside field margins on arable land	0.49–0.73
Manure to arable rather than grassland	0.05–0.21
Straw residues	0.53–0.72
Sewage sludge	0.61
Bioenergy crop – *Miscanthus*	0.49–0.73
Bioenergy crop – willow	0.55–0.83
Extensification – outdoor pigs on grass	0.48
Extensification – break-crops to grass	0.48
Organic farming (arable)	0.48
Organic farming (with livestock)	0.48
Convert arable to woodland	0.55–0.83

5.3 Soil Carbon Sequestration Potential for the UK

As may be expected, soil carbon sequestration potentials for the UK are not too dissimilar from those for Europe in general. A list of sequestration rates compiled for England,[14] which we may take to be representative of the UK as a whole, are shown in Table 6. Generally, the rates are at the higher end of the European rates given in Table 5. The total soil carbon sequestration potentials for the UK have been variously estimated as a 'theoretical' value of 30–70 $Tg\,a^{-1}$ (ref. 33), 9 $Tg\,a^{-1}$ (England only)[14] and 10.4 $Tg\,a^{-1}$ (includes 1.4 $Tg\,a^{-1}$ from biofuel offset).[16] The last figure was obtained using a combined and optimised approach, as was used above for Europe as a whole.[39] The most significant departure from the European scene was that, within the UK, it was recognised that zero-till would not be applicable to all soil types and so could not be applied across all arable areas. The value of 10.4 $Tg\,a^{-1}$ would be able to offset 6.6% of the UK's 1990 CO_2-C emissions.

5.4 Biochar Additions

As has been mentioned earlier in this chapter, the passive fraction of SOM can consist partly of 'black carbon' or charcoal as a result of fire cycles. It has been estimated from a number of studies that about 1–5% of the biomass carbon that is burnt during a fire may be converted into black carbon. Globally, it has been estimated that this may sequester 0.05–0.27 $Pg\,C\,a^{-1}$ (ref. 40). While much of this remains in the soil, some is known to move *via* air or water to the ocean where it remains locked in deep sediments.

Black carbon or pyrogenic carbon is not a single compound but a continuum ranging from partially charred biomass, through charcoal to soot, and eventually pure graphite, though this is only formed under high heat and pressure. Chemically, this continuum is characterised by decreasing hydrogen and oxygen contents. For example, charcoal has an atomic H : C ratio of 0.6 and an O : C ratio of 0.4, while charred biomass has an H : C ratio of 1 and an O : C ratio of 0.6 (ref. 41). Hence, black carbon is not pure carbon, but has a central core of fused aromatic sheets with varying degrees of additional side-chain functional groups containing oxygen and hydrogen. This structure renders black carbon relatively inert to enzymic attack or breakdown by soil microorganisms. There may be some limited degradation of the side chains and further oxidation over time. However, in soil these functional groups can increase the cation exchange capacity of the soil, *i.e.* the ability of the soil to retain nutrients like calcium, magnesium and potassium. This then has the potential to increase soil fertility.

The natural level of black carbon in soil is variable, ranging from a few percent to 40% or more in soils subject to regular fire. Dating has shown that much of this carbon can be many thousands of years old and the result of slow accumulation. However, there are many soils in which the contribution to the soil carbon by black carbon is unknown.[41] Certain soils of the Amazon basin have been found to contain extraordinarily high levels of black carbon, to the extent that they are called 'black earths' or '*terra preta*' in Portuguese. It appears that these soils are the result of the deliberate addition of charred biomass by the Amerindian population who cultivated the region over many centuries before the European invasion. The quantity of soil carbon in the top 100 cm was increased from *ca.* 100 up to $250\,t\,C\,ha^{-1}$ (ref. 42). Such increased levels of soil carbon were not the result of 'slash and burn' agriculture, but rather the addition of charcoal from wood or other biomass brought in from outside these black-earth patches, which then increased the fertility and productivity of these soils.

These observations have prompted an interest in replicating the addition of black carbon to soil, not so much as a way to improve fertility but as a means of sequestering carbon. However, any improvement in fertility can be used as an added incentive to this practice. The term 'biochar' has been applied to charred biomass that is generated specifically for soil carbon storage. Besides the benefits of fertility and carbon sequestration, biochar production can also produce a certain amount of biofuel. During the pyrolysis or gasification processes some 'bio-oil' or a mixture of hydrogen and carbon monoxide is generated, which can then be used as an energy source. Hence the biochar process can be considered to be a 'win-win-win' situation. The advantage from the carbon sequestration point of view is that, although half of the carbon in the original biomass is lost, the remainder that is added to the soil is extremely resilient, with a turnover time of centuries or longer. If the same biomass is added directly to the soil, while 100% of the carbon is there initially, within one year 60–70% will have been lost and by five years we will be fortunate if 10% remains.[42]

What materials can be used? Almost any biomass that has been adequately dried is possible, though greater yields of biochar come from biomass rich in lignin. There are obvious advantages in using 'waste' materials such as certain agricultural wastes, forestry residues, municipal wastes, waste wood, *etc*. Nevertheless, for biochar to be a serious contender as a carbon sequestration methodology, one would have to consider growing biomass specifically to be converted to biochar. In the UK we would be considering plantation wood-land, short-rotation coppicing of willow or something like *Miscanthus* (elephant grass). Greater options are available in tropical regions with wastes like bagasse, pulp waste from oil palm, rice husks and various nut shells, as well as biomass that might otherwise be used as bio-fuel. It should be remembered that using biomass, grown by the fixation of carbon dioxide through photo-synthesis, as a biofuel is carbon neutral (no net loss or gain of CO_2); using biomass to generate biochar is carbon negative (a net uptake of CO_2), though obviously less energy is produced.

Are there any downsides to biochar production? First, we have to be sure that adding biochar to soil does not lead to any long-term toxicity. The experience of the *terra preta* soils would suggest few problems, though we do not know the exact conditions under which the biochar was produced or from what. Contemporary biochar loadings of $50\,t\,C\,ha^{-1}$ have shown benefit and loadings of $140\,t\,C\,ha^{-1}$ have been tolerated by most plants, though there have been limited studies on this.[42] Secondly, there is some concern that redirecting 'wastes' away from direct application to the land, or following composting, may deplete the soil of essential organic inputs that maintain the microbial biomass, soil fertility and soil structure, leaving it more open to erosion and degradation.[43] We would have to be assured that the timely reapplication of biochar would make up for any deficiencies.

How much carbon may be sequestered using biochar? In one sense there is no upper limit. While the addition of organic materials to soil will eventually lead to a new equilibrium level, the annual addition of biochar will give a continual linear increase in carbon. Any final levels where negative effects may begin to show will almost certainly exceed those possible by the addition of organic residues. Using biochar as a means to generate renewable energy in place of current direct bio-mass combustion would lead to the production of $0.18\,Pg\,C\,a^{-1}$ (ref. 42). This is a relatively small, but not insignificant flux when compared to those on the global scale (Figure 2), although it has the potential to be much larger with greater commitment to producing more energy from biomass production.

5.5   Other Greenhouse Gases and Carbon Equivalents

While this chapter has concentrated on sequestering atmospheric carbon dioxide into the soil mainly as SOM, any strategies should also take into consideration the possible impact on the emission of other greenhouse gases (GHGs).[12] Natural wetland (peatland) soils and paddy (rice cultivation) soils are major sources of methane, while a third major source of methane is livestock. At the same time, a great many soils have varying capacities to take up small quantities of methane and remove it from the atmosphere. The use of nitrogen fertiliser is a major source of nitrous oxide. The worst offender is the direct use of nitrate, but ammonium fertiliser may be slowly oxidised to nitrate and even organic nitrogen will be mineralised to ammonium and then converted to nitrate. Smaller quantities of methane and nitrous oxide may be released from manures and during biomass burning. Both methane and nitrous oxide are produced when oxygen in the soil is limited and so soil wetness is a major factor. While the absolute quantities of these two gases that may be released from soil is much less than the carbon dioxide released in soil respiration, because they are many times more effective as GHGs they still play a significant role in global warming effects. Often their impact is recorded as 'C equivalents', *i.e.* the quantity of carbon dioxide that they are equal to in terms of the greenhouse effect. Methane is *ca.* 20 times more effective, and nitrous oxide *ca.* 300 times more effective, in global warming terms, than carbon dioxide.[12]

It is possible that changes in land use or management practices that aim to sequester carbon may have reduced efficacy, either complete or in part, if other GHGs are emitted at the same time. For example, conversion to wetland or efforts to sequester soil carbon by reducing or eliminating drainage may lead to greater methane, and possibly nitrous oxide, emissions. The use of extra nitrogen fertiliser to promote plant biomass production may result in further nitrous oxide emissions. The use of minimum till may increase the moisture content of surface soils such that nitrous oxide production ensues. The conversion to grassland may imply greater animal stocking with the associated methane emissions.

5.6   Whole Cycle Analysis

During the assessment of the benefits of soil carbon sequestration strategies, it is important that a whole life-cycle approach is taken. This means including every step along the chain of events that lead to sequestration. We have already discussed the possibility of other GHGs being evolved. Another critical factor is the use of fossil fuels. These may be required for producing fertilisers, transportation of fertilisers, crop residues and other wastes to and from the land, and for the cultivation processes used. Conversely, the production of biofuel crops will be designed to substitute for fossil fuel use. However, where food production is displaced to produce biofuels, there is potentially an associated set of emissions from land use change in a new location. The whole cycle approach would examine all of these, both for the existing land-use/management system and compare it with any changed land-use/management. Additionally, any indirect or knock-on effects should also be considered. For example, a reduction in the total arable area may mean that food has to be imported from elsewhere and this may have a greater cost in carbon terms. Using a whole-cycle way of thinking adds to the complexity of the system with outcomes that may not be predicted at the outset or that may be the opposite of what was first envisaged.[44]

6   Limitations and Challenges

6.1   Realistic Goals

During the consideration of using soils to sequester carbon from the atmosphere, it is tempting to think that soils can be used to remove all the fossil fuel carbon emitted during the industrial era, or can recover all the carbon that has been lost from them as a result of deforestation and cultivation, or, at the very least, can mitigate current fossil fuel emissions and be used to hold atmospheric carbon dioxide at its current concentration. Unfortunately, none of these are true and it is important to seek realistic goals so that the contribution that soils can make can be put into perspective with all the other options for sequestering carbon. In discussing soil carbon sequestration potential, what might be attained has been usefully put into the categories of what is 'theoretical', what is 'realistic' and what is 'conservatively achievable'.[33] The 'theoretical' value has few or no practical constraints, *i.e.* it would be restoring all soils to their original, natural capacity, or even enhancing it by fertilisation, irrigation or other inputs. The 'realistic' value takes account of most constraints, but remains optimistic about the social, political and economic will to pursue this potential. The 'conservatively achievable' value has few optimistic assumptions, is based on current trends and is a cautious, pragmatic scenario. The range of these values are summarised for the world, for the EU15 and for the UK in Figure 4. Clearly there is an almost ten-fold decrease in magnitude in going from worldwide, to the EU15 and then to the UK scenario. Another significant difference between the regions is the extent to which the realistic and achievable sequestration rates are less than the theoretical. On a worldwide basis the

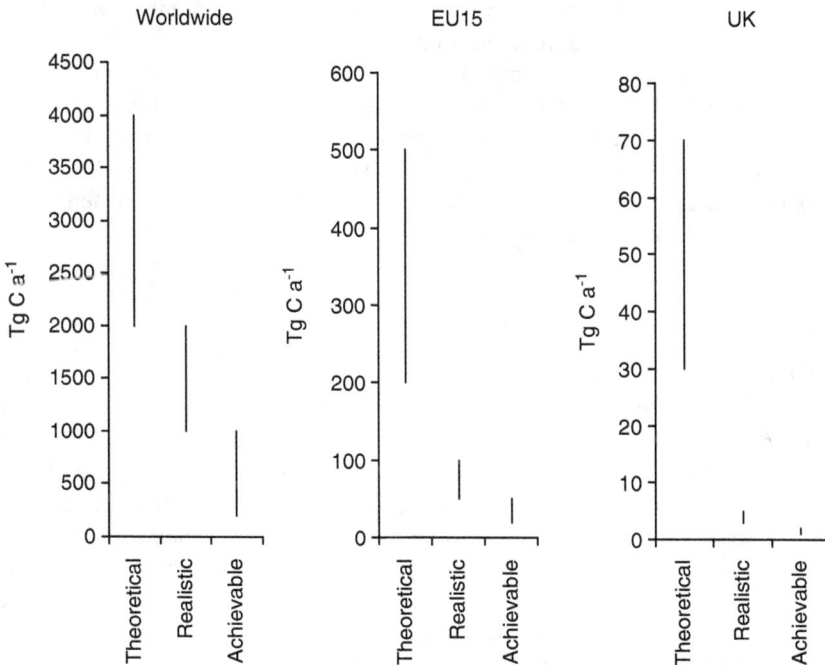

Figure 4   Carbon sequestration potentials estimated on a theoretical, realistic and what might be conservatively achievable basis, worldwide, for the European Union (EU15) and the UK.[33]

ranges almost overlap, whereas for Europe, and to a greater extent for the UK, the range of the realistic rate and then for the achievable rate is very much less than the theoretical. This is partly a reflection of the greater population densities in Europe and the UK which then impose more constraints on the available land and its utilisation.

6.2   Upper Limits and Timescales

As well as having realistic goals in terms of annual soil carbon sequestration rates, it is important to bear in mind that there are upper limits on the total quantity of carbon that can be put away. Once the 'carrying capacity' of a particular soil has been attained then it will not sequester further carbon; the accumulated stock will decompose at a rate at which carbon is added. In other words, it will have reached its new equilibrium value. This quantity will approximate what has been lost from the soil over recent history unless novel ways are found of increasing the stability of the sequestered carbon. The use of biochar is one approach that may be able to raise this upper limit. Given this restriction, it is unlikely that soil carbon sequestration can be actively pursued for much longer than 50–100 years. This obviously depends upon the rate of sequestration and faster rates will reach the limit sooner. Also, as illustrated in Figure 3, there is a gradual slowing down of the carbon addition as the upper limit is approached. Hence, the strategies that we implement now will have ever-decreasing benefit as time goes on.

A further consideration is that most of the carbon sequestered by the various strategies can be lost again if such strategies are relaxed; if we take our foot off the pedal then we can start to go backwards. Again biochar may be an exception to this as once this is in the soil it should be there to stay. Also, where land is devoted to biofuel production, there will an ongoing benefit in the substitution of fossil fuels, which does not have a time limit.

6.3   Competing Processes

Using the soil as a means of sequestering carbon also has implications for the way in which the land is used. Basically, arable land is used for "food, fuel and fibre" but there is a problem obtaining all three at the same time. A choice has to be made, which will be dictated not by the need to sequester carbon so much as by the need to supply the commodity most in demand. In economic terms, that will be the one that gives the best price and this may be complicated by the various subsidies that impact farming in many countries. Hence, there is an ongoing controversy about the place of biofuels and their impact on food supply and food prices. In addition, there is the dimension that carbon trading, the concept of receiving payment for sequestering a certain quantity of carbon, might add to the debate. When we come to woodlands, grasslands and semi-natural land there are additional societal needs – areas for sport, recreation, relaxation, aesthetic value, conservation and biodiversity. All these added pressures need to be taken into account when we are advocating certain pathways to sequester carbon. The need is great and while carbon sequestration in soil in reality can only give modest returns, it is clear that there is no 'silver bullet' available to tackle climate change. All avenues will need to be explored and carbon sequestration in soil needs to be there within a whole basket of strategies that will aim to reduce the current GHG burden of the atmosphere.

References

1. R. Lal, *Nutrient Cycling in Agroecosystems*, 2008, **81**, 113.
2. R. Lal, R. F. Follett and J. M. Kimble, *Soil Sci.*, 2003, **168**, 827.
3. J. Anderson, *New Sci.*, 1983, **6–10**, 29.
4. W. H. Schlesinger and J. A. Andrews, *Biogeochemistry*, 2000, **48**, 7.
5. D. A. Wardle, *Biol. Rev. Cambridge Phil. Soc.*, 1992, **67**, 321.
6. M. H. B. Hayes and C. E. Clapp, *Soil Sci.*, 2001, **166**, 723.
7. R. Lal and J. M. Kimble, *Nutrient Cycling in Agroecosystems*, 1997, **49**, 243.
8. W. M. Post and K. C. Kwon, *Global Change Biol.*, 2000, **6**, 317.
9. D. Hope, M. F. Billett, R. Milne and T. A. W. Brown, *Hydrol. Process.*, 1997, **11**, 325.
10. T. Wutzler and M. Reichstein, *Biogeosciences*, 2007, **4**, 125.
11. K. Paustian, O. Andren, H. H. Janzen, R. Lal, P. Smith, G. Tian, H. Tiessen, M. Van Noordwijk and P. L. Woomer, *Soil Use Manage.*, 1997, **13**, 230.
12. J. Bellarby, B. Foereid, A. Hastings and P. Smith, *Cool Farming: Climate Impacts Agric. Mitigat. Potential*, 2007, **44**.
13. R. Alvarez, *Soil Use Manage.*, 2005, **21**, 38.
14. J. A. King, R. I. Bradley, R. Harrison and A. D. Carter, *Soil Use Manage.*, 2004, **20**, 394.
15. R. T. Conant, M. Easter, K. Paustian, A. Swan and S. Williams, *Soil Tillage Res.*, 2007, **95**, 1.
16. P. Smith, D. S. Powlson, J. U. Smith, P. Falloon and K. Coleman, *Soil Use Manage.*, 2000, **16**, 1.
17. J. N. Hutchinson, *J. Ecol.*, 1980, **68**, 229.
18. A. L. Heathwaite, *Geograph. J.*, 1993, **159**, 203.
19. R. Lal, *Science*, 2004, **304**, 1623.
20. R. Lal, *Soil Tillage Res.*, 2005, **81**, 137.
21. D. A. Wardle, *J. Biogeogr.*, 2002, **29**, 583.
22. J. Holden, L. Shotbolt, A. Bonn, T. P. Burt, P. J. Chapman, A. J. Dougill, E. D. G. Fraser, K. Hubacek, B. Irvine, M. J. Kirkby, M. S. Reed, C. Prell, S. Stagl, L. C. Stringer, A. Turner and F. Worrall, *Earth-Sci. Rev.*, 2007, **82**, 75.
23. S. E. Page, F. Siegert, J. O. Rieley, H. D. V. Boehm, A. Jaya and S. Limin, *Nature*, 2002, **420**, 61.

24. J. O. Rieley, R. A. J. Wüst, J. Jauhiainen, S. E. Page, H. Wösten, A. Hooijer, F. Siegert, S. H. LImin, H. Vasander and M. Stahlhut, in *Peatlands and Climate Change*, ed. M. Strack, International Peat Society, Jyväskylä, 2008, **6**, 148.

25. J. B. Innes and I. G. Simmons, *Palaeogeograph. Palaeoclimatol. Palaeoecol.*, 2000, **164**, 151.

26. E. Maltby, C. J. Legg and M. C. F. Proctor, *J. Ecol.*, 1990, **78**, 490.

27. D. E. Yeloff, J. C. Labadz and C. O. Hunt, *Mires and Peat*, 2006, 1, Art. 4.

28. A. F. Harrison, P. J. A. Howard, D. M. Howard, D. C. Howard and M. Hornung, *Forestry*, 1995, **68**, 335.

29. S. J. Chapman, J. Bell, D. Donnelly and A. Lilly, *Soil Use Manage.*, 2009, **25**, 105.

30. M. G. R. Cannell, R. Milne, K. J. Hargreaves, T. A. W. Brown, M. M. Cruickshank, R. I. Bradley, T. Spencer, D. Hope, M. F. Billett, W. N. Adger and S. Subak, *Climatic Change*, 1999, **42**, 505.

31. J. Liski, D. Perruchoud and T. Karjalainen, *Forest Ecol. Manage.*, 2002, **169**, 159.

32. R. Lal, *Soil Tillage Res.*, 2007, **96**, 1.

33. M. G. R. Cannell, *Biomass and Bioenergy*, 2003, **24**, 97.

34. J. J. Hutchinson, C. A. Campbell and R. L. Desjardins, *Agric. Forest Meteorol.*, 2007, **142**, 288.

35. R. Lal, *Philos.Trans. R. Soc. London, Ser. B: Biological Sci.*, 2008, **363**, 815.

36. P. Smith, *Soil Use Manage.*, 2004, **20**, 212.

37. P. Smith, O. Andren, T. Karlsson, P. Perala, K. Regina, M. Rounsevell and B. van Wesemael, *Global Change Biol.*, 2005, **11**, 2153.

38. I. J. Beverland, J. B. Moncrieff, D. H. Oneill, K. J. Hargreaves and R. Milne, *Q. J. R. Meteorol. Soc.*, 1996, **122**, 819.

39. P. Smith, D. S. Powlson, J. U. Smith, P. Falloon and K. Coleman, *Global Change Biol.*, 2000, **6**, 525.

40. M. S. Forbes, R. J. Raison and J. O. Skjemstad, *Sci. Total Environ.*, 2006, **370**, 190.

41. C. M. Preston and M. W. I. Schmidt, *Biogeosciences*, 2006, **3**, 397.

42. J. Lehmann, J. Gaunt and M. Rondon, *Mitigat. Adaptat. Strategies Global Change*, 2006, **11**, 403.

43. R. Lal, *Soil Tillage Res.*, 2008, **99**, 1.

44. H. Kim, S. Kim and B. E. Dale, *Environ. Sci. Technol.*, 2009, **43**, 961.

45. G. A. Alexandrov and T. Matsunaga, *Carbon Balance Manag*, 2008, **3**, 8.

46. IPCC, *Climate Change: The Scientific Basis*, Cambridge University Press, Cambridge, UK, 2001.

47. R. I. Bradley, R. Milne, J. Bell, A. Lilly, C. Jordan and A. Higgins, *Soil Use Manage.*, 2005, **21**, 363.

Carbon Capture and Storage in Forests

MARIA NIJNIK

1 Introduction: The Role of Forestry in Climate Change Mitigation

The Kyoto Protocol to the United Nations Framework Convention on Climate Change[1] was adopted in 1997 and became legally binding (on its 128 Parties) in 2005. The Parties have committed themselves to actions directed towards stabilising atmospheric Greenhouse Gas (GHG) concentrations, including those of carbon dioxide (CO_2). This is to be achieved by reducing emissions (reduction of sources) and removing GHG from the atmosphere (enhancement of sinks), including by carbon capture and storage (CCS) in forests. (Note: the term 'carbon sequestration' used elsewhere in this chapter may be considered as equivalent to both capture and storage of carbon, but, for simplicity, 'sequestration' usually is used as a substitute for 'storage'). Article 3.3 of the Kyoto Protocol (KP) states that biological carbon sinks enhanced through afforestation, reforestation and the decreasing of deforestation rates since 1990, should be utilised for meeting the commitments of the countries during the stipulated period. *Afforestation* is an expansion of forest on land which more than 50 years ago contained forests but has since been converted to other use. *Reforestation* is a restoration of degraded or recently (20–50 years ago) deforested land.[8] In this chapter, we do not make any distinction between these terms. Since the Conference of the Parties,[2] afforestation, reforestation, forest management and soil carbon have become eligible climate policy measures.

In Europe, aid for woodland development is provided by the programmes of Member States and by the EU initiative that focuses mainly on marginal agricultural land, 1 Mha of which was afforested in 1994–1999 (ref. 3). The total area of EU forests (113 Mha) has expanded by 3%, with 1 Mha having been afforested between 1994 and 1999 (ref. 4). If this trend continues, the

Issues in Environmental Science and Technology, 29
Carbon Capture: Sequestration and Storage
Edited by R.E. Hester and R.M. Harrison
© Royal Society of Chemistry 2010
Published by the Royal Society of Chemistry, www.rsc.org

carbon sequestration potential of $3.84\,Mt\,C\,yr^{-1}$ will be achievable during the first commitment period. Taking into account 25 Member States, a technical sequestration potential of $34\,Mt\,C\,yr^{-1}$ could be reached in the long-run.[5] However, only a fraction of this amount could be accounted for, under the current rules, because most of this carbon uptake is not "additional" with reference to the 1990 baseline. Carbon sequestration forestry activities are supported by afforestation schemes and rural development regulations. The principal forest policy initiatives that increase CCS are recognised as maintaining and enlarging carbon pools by improving existing forests through their protection and sustainable management; expanding the forest area through afforestation (largely with species adapted to local conditions); replacing fossil fuels with fuel wood from sustainable managed forests; and replacing high-energy products with industrial wood products.[6]

Along with afforestation schemes, the European Commission (EC) adopted the White Paper that identifies the need to raise energy production from renewable sources from 6% in 1998 to 12% by 2010 (ref. 7), including the increasing use of woody biomass in energy production. Successful implementation of this policy contributes to carbon sinks provided by standing forests, and adds to the reduction of CO_2 concentrations in the atmosphere from using wood instead of fossil fuels, after timber is harvested. In Europe, the potential to sequester carbon from short rotation timber plantations (SRTP) and from substitution of biomass for fossil fuel, is in the range of 4.5–$9\,Mt\,C$ per year.[5] Depending on SRTP development, even higher carbon savings could be achieved in the future, though this would require proper incentives and links between the Common Agricultural Policy (CAP) and climate policy measures, particularly concerning changes on set-aside and marginal lands.[9]

The last few years have seen an upsurge in the number of scientific papers and reports addressing different aspects of carbon sequestration through forestry-based activities, both internationally and in the UK. Various aspects of using forests to mitigate climate change have been discussed in the literature for Canada, Finland, the Netherlands, Russia, the UK, USA, and other countries.[10–18] Terrestrial carbon sink is on the agenda of international conferences, including the Conferences of Parties.

Carbon sequestration through afforestation is commonly considered as: cheap (cost efficient); clean (it may concurrently provide other ecosystem services); proven (many countries have the legacy of tree-growing); effective in the short-term, providing almost immediate effect after the tree-planting; and as a less resource and energy consuming climate policy measure. It can be incorporated in multi-functional forest use to simultaneously enlarge timber production and bring a variety of other benefits, and can provide economic incentives for sustainable forest management.[19]

The Stern Review[20] examined the socio-economic impacts of climate change. The Carbon Trust[21] published recommendations on how best to deliver carbon emission reductions and different bio-energy options for the UK. A report by AEA Technology[22] analysed the potential of biomass for renewable heat. More specific documents have assessed the relevance of biomass options in regional contexts,[23,24] the importance of climate policies in setting business strategies, and manifold implications of policy decisions and their effects on the way that businesses operate.[25] The most recent publications highlight that social and spatial issues are important in determining the range of land types which are likely to become available for new woodland development; that the main difficulties associated with the use of wood for energy have been policy-oriented and socio-economic, and technological, rather than fuel-related; and that comparative indicators of the cost-effectiveness of alternative climate change mitigation strategies in forestry are needed.[26]

Surprisingly, despite the size of their forests and large areas of marginal agricultural land, there remains only limited room for forest sector policies to sequester carbon in the major wood-producing countries, such as Canada, Finland, Sweden or Russia.[27] In Canada, for example, there is a limit to the amount of carbon offset credits that can be claimed on existing forestland (largely, publicly owned), and the focus is now shifting to afforestation of agricultural land, where the role of private landowners is important and the potential of afforestation is around 1 Mha.[31]

The analysis of the role and place of forestry to mitigate climate change is more relevant to countries which have a substantial potential for forestry development.[28,71] Therefore, carbon inventory and monitoring, cost-effectiveness of afforestation and forest management, social acceptability of various carbon sequestration options, existing challenges and opportunities of woodland development on high carbon soils, using wood in renewable energy projects, and in wood products, are highly relevant topics, for instance in Scotland. The reports by the Sustainable Development Commission[29] and the Fraser of Allander Institute[30] have provided information on the potential of the wood fuel market and on competitiveness of different wood fuel scenarios. A review addressing biomass production and consumption in Scotland has been published by SEERAD.[26]

The reports provide a broad picture concerning technological aspects, GHG life-cycle emissions, air pollution impacts of biomass production and consumption. However, an overall assessment of the role of forests in climatic and atmospheric changes is needed to develop a better understanding and, where appropriate, to improve, simplify and extend the manner in which this role is taken into account. Through the analysis of biogeochemical processes involved, and by assessing the opportunities for forestry to sequester and store carbon, it becomes possible to suggest climate policies and measures at various spatial levels and to advise on their proper sequencing in time. Institutional and economic aspects of CCS in forests are areas that merit special attention.

It is anticipated that forestry-based activities could help reduce CO_2 concentrations in the atmosphere by increasing biotic carbon storage, decreasing emissions and producing biomass as a substitute for fossil fuels. Reducing rates of deforestation, increasing forest regeneration, agroforestry, improving forest and land-use management, and growing energy crops are activities that are supposed to assist countries in coping with the changing climate.[32] In practice, however, existing opportunities are only partially used, as this chapter will discuss further.

The chapter presents the results of analysis of opportunities for terrestrial CCS in forests to mitigate climate change. It is supported by official documentation, including that of DEFRA[33-35] and the Forestry Commission,[36,37] and by other literature available on this topic.[18,38-40,71] The chapter first presents carbon pools and flows in forests. The ecological perspectives of an increased CCS in forests are then discussed, and the carbon sequestration rates and the potential of carbon sequestration in forests are analysed across several European countries. A general overview of the situation in densely wooded regions of the world is given, along with the analysis of the opportunities and challenges of CCS in forests. The focus then shifts towards the social and economic considerations of terrestrial carbon sink. The importance of proper institutions in the development of conditions for creating carbon offsets from forestry is highlighted. The chapter concludes by offering some insights into the feasibility of carbon sequestration in forests and the level of institutional development, as well as by providing some ideas for future research.

2 Carbon Pools and Flows in Forests

Forests cover about 4 Gha of the Earth, or over 30% of the land area, and store around 120 Gt of carbon, more than the total amount in the atmosphere.[37] Forests contain 77% of carbon stored in land vegetation and, of this total, approximately 60% of carbon is stored in tropical forests, 17% in temperate and 23% in boreal forests.[41] Forests account for 90% of the annual exchange of carbon between the atmosphere and the land, and their growth is one of the few ways of taking CO_2 out of the atmosphere.[41] The role of forests in relation to climatic changes is observed in the carbon cycle. Forests are also involved in the cycles of water and GHG, and play a role through their reflectance characteristics (albedo). In return, internal and external drivers, including the changing climate, are affecting forest ecosystems and therefore the carbon cycle.

Trees absorb CO_2 from the atmosphere through photosynthesis and use light energy to run enzyme-catalysed reactions. Much of the carbon eventually goes for production of cellulose, but some is released to the air through respiration. The absorbed carbon goes to form the above-ground biomass (stem wood, branches and leaves), as well as roots. Carbon accumulated in leaves comes back to the atmosphere after a relatively short period of time, when the fallen leaves decompose. Carbon in wood is stored for years. The time depends on tree species, tree-growing conditions and forest management, and on various uncertain occurrences, such as forest fires or diseases. The dry wood is 50% formed from carbon. A widely held assumption is that forests approach carbon saturation at maturity, and when trees reach it they stop sequestering carbon, but with a continuous cover forests could act as long-term storage of carbon. When trees die some of the carbon remains in the forest, being stored in the soil.

Afforestation affects the climate in more ways than through CCS, but the effect is very much case-specific, depending, for instance, on whether the trees are planted on mineral or highly organic soils. Also, some models, for example, predict that older forests could become net emitters of CO_2 as the relative rate of decomposing wood to new growth becomes unfavourable.[44] There is also some evidence to argue that warming as a result of planting trees in the boreal zone might overcome the cooling effect due to carbon sequestration by new forests. Moreover, as forest canopies reflect sunlight differently than open spaces (covered with snow in winter), in the zone of distribution of boreal forests, tree-planting might not be helping to alleviate global warming.[45]

An example of carbon exchange associated with an oak forest in England of general yield class $6 \, m^3 \, ha^{-1} \, yr^{-1}$ and gross primary productivity of *ca.* 14.0 t $ha^{-1} \, yr^{-1}$ is shown in Figure 1. The figure explains the exchange of carbon between the atmosphere and all carbon pools in the forest, *i.e.* above ground

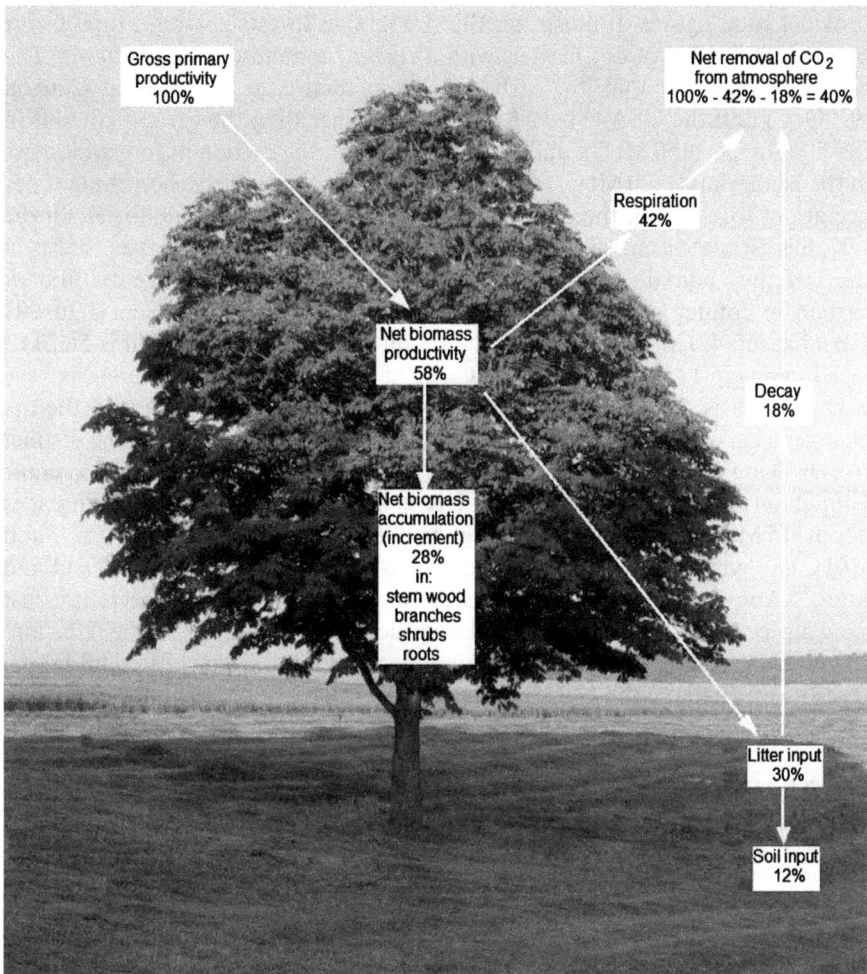

Figure 1 Representation of carbon exchange associated with forest components. (Source: adapted from the summary given in ref. 16).

and roots, and forest soil and litter. Accumulation of carbon proceeds until equilibrium is reached. Afterwards, if the forest is not maintained, carbon is released through wood decay or burning. The rate, dynamics and patterns of carbon sequestration in the forest (*i.e.* when carbon is "locked" into a more stable carbon stock), depends on tree species, their characteristics, and naturally on tree-growing conditions, particularly on temperature, CO_2 concentration and forest management. Over the lifetime of the forest more carbon is captured than is released, and there is a net sink of carbon.

About 48 Gt C is exchanged globally between forest ecosystems and the atmosphere each year.[42] A generally accepted estimate of carbon content in a forest derives from forest biomass. As an example, the biomass content (dry matter) and carbon content of forests in Ukraine is presented in Table 1. The table shows the estimates for coniferous, hard and soft deciduous species and provides total figures. It illustrates that Ukrainian forests possess a total forest biomass of 956 Mt of dry matter, with a carbon content of about 474 Mt. The estimates are lower than for carbon pools in such large countries as Canada (9.3 Gt C) and the USA (11–12.6 Gt C), but higher than the carbon content in the UK forests (150 Mt C), and much higher than the carbon in forest biomass in the Netherlands (20 Mt C; ref. 37,46). With respect to the carbon content per hectare of forest land, the estimates are comparable between countries. In the UK, this estimate ranges from 30 to 60 t C per ha (mature broadleaved forests in the UK may contain up to 250 t C per ha; the long-term average amount of carbon in conifer plantations with rotation length of *ca.* 50 years is 70–90 t C per ha; ref 47); in Canada, it is 38.3 t C on average; in the USA, it is 56.2 t C; in Ukraine, 55.1 t C; and it is 59.7 t C per ha of forest in the Netherlands.[47]

On the basis of the estimates of carbon content in forest biomass, the net change in carbon pools can be assessed. For example, largely due to the distinct tree-growing conditions and the different areas covered by forests, the average annual net uptake by forests in the UK is about 3 Mt C, and in Ukraine it is about 15 Mt C. The estimates roughly correspond to $1.1\,t\,C\,ha^{-1}\,yr^{-1}$ and $1.6\,t\,C\,ha^{-1}\,yr^{-1}$, respectively, excluding the carbon sink in the forest soil and litter.[46] Annual carbon uptake in excess of $4\,t\,C\,ha^{-1}$ can be expected in fast growing conifer stands, *e.g.* of Sitka spruce.[42] The figures for the UK and Ukraine are comparable with the estimates provided for temperate and boreal forests in Canada,[48] the Netherlands,[11] and Finland.[49] Average net uptake by

Table 1 Forest biomass and carbon content in Ukrainian forests, Mt.[43]

Phytomass Components	Coniferous	Hard wood	Soft wood deciduous	Total
Foliage	31.65	7.81	2.66	42.10
Crown wood	45.25	53.44	9.65	108.30
Stemwood	289.35	290.05	58.42	637.80
Stump and roots	55.03	45.78	18.38	119.20
Understory	14.63	24.80	9.78	49.2
Total	435.93	421.88	98.90	956.70
Total Mt carbon	215.65	209.31	48.83	473.80

forests in Canada is $0.64\,t\,C\,ha^{-1}\,yr^{-1}$; in the USA, it is $1.6\,C\,ha^{-1}\,yr^{-1}$; and in the Netherlands, it is $2.0\,t\,C\,ha^{-1}\,yr^{-1}$ (ref. 47).

The IPCC[32] identified the following measures to be implemented to increase the forestry potential in terms of carbon sequestration:

- The afforestation of abandoned and marginal agricultural land.
- Forest management to increase carbon density at the stand and landscape levels, *e.g.* maintaining forest cover, minimising forest carbon soil losses, increasing rotation lengths, increasing growth and managing drainage.
- Increasing off-site carbon stocks in wood products.
- Enhancing product and fuel substitution.

Afforestation is the most straightforward policy measure to enlarge the carbon sink in forests and it has been widely analysed in the literature. For the reason that carbon sequestration positively correlates with the growth rates of trees, it is advocated to plant the most fast-growing tree species, *e.g.* hybrid poplar or Sitka spruce, where appropriate,[31,50–52] or to establish SRTP for the purpose of carbon uptake. The choice of species depends on the location and tree-growing conditions, on the purpose of tree- growing, and other factors.

The carbon sequestration estimates per hectare of forest expansion in the USA and Ukraine, for example, come close and are higher than annual net uptake per hectare of forest development in Canada,[46] because of the presence of vast Canadian boreal forests with lower rates of growth. However, in the Carpathian Mountains, for instance, spruce stands grow rather slowly, and so carbon sequestration in spruce forests proceeds gradually. But when the trees reach maturity, they accumulate a volume of stem of $500\,m^3$ and higher, and so it is their carbon storage function, rather than that of carbon uptake, that is important.

Estimated[46] cumulative carbon uptake by fast-growing tree species across the main forestry regions in Ukraine, shown in Figure 2, describes carbon in the stem volume, plus carbon taken from the atmosphere and stored in branches, leaves and roots of the trees. Carbon in understory, forest soils and litter was not taken into account.

Forests add to the reduction of CO_2 from the atmosphere as long as there is a net growth. In the figure, a 40-year time horizon is considered, because, in the observed conditions, the growth rates of trees chosen for planting increases until the trees reach 40 years of age.[46] Consequently, the highest rates of carbon uptake are observed within this time horizon.

When trees are cut, the above-ground biomass minus the commercial part of the bole that constitutes a log enters the litter account. Later on, when a new generation of trees comes up, there is a re-growth of the non-bole biomass and a re-growth of the volume of stem wood. The process is assumed to continue indefinitely with new generations of trees coming in place of old ones. This observation allows us to model at once all the above-ground biomass of the trees.[28,50] In addition to the above-ground biomass, the root component of forest plays a role in the carbon budget. For example, the carbon sequestered by the root pool of poplar stands in British Colombia, Canada was estimated[48]

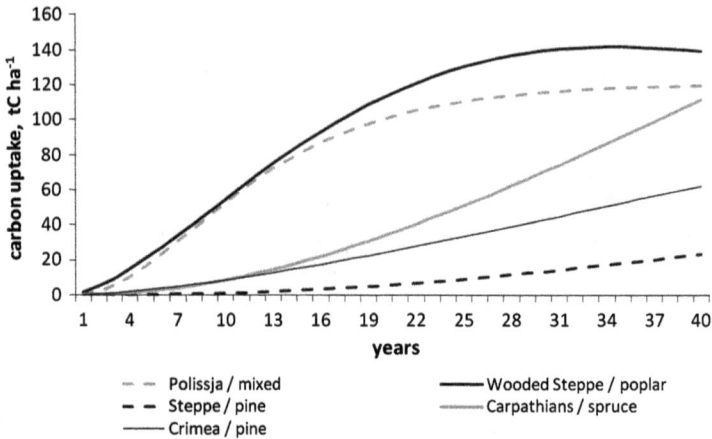

Figure 2 Cumulative carbon uptake by fast-growing tree species across regions in Ukraine, 0% discount rate.[46]

according to the following relationship:

$$R(G) = 1.4319\,G^{0.639}$$

where: R is root biomass (m^3) and G is above ground biomass (m^3).

This relationship might not hold for other species and for different conditions of tree-growing. Also, in some regions, *e.g.* in Scotland, soil carbon is important and should be included in models. When trees reach the culmination of their annual increment, they stop sequestering carbon, because its removal due to the growth of trees comes into a balance with the loss due to the decay of trees, and the forest no longer acts as a carbon sink. Carbon uptake, under a storage policy option (of carbon fixation), has a one-time effect, and eventually, through the decay of wood, all the above-ground carbon is released back into the atmosphere. However, usually, the trees are cut after they reach their mean annual increment (see Figure 3).

The harvested timber enlarges the supply of wood for industry, whereas carbon stored in wood products is an addition to the total carbon sink. Wood received from forest plantations can also be used as a substitute for fossil fuels, or timber used in wood products can later get burned. These policy solutions contribute to carbon sinks provided by forests during their growth, as well as to carbon storage in wood products, and to reduction of CO_2 concentrations in the atmosphere when wood is burned instead of fossil fuels.

If the energy required for the harvesting and processing of wood is not taken into account, the use of timber as a substitute for fossil fuel is a carbon neutral process. The net gain here is the amount of CO_2 that would have been released by burning fossil fuel if not replacing it with wood. The effects for the avoidance of carbon release to the atmosphere through a continual regeneration of

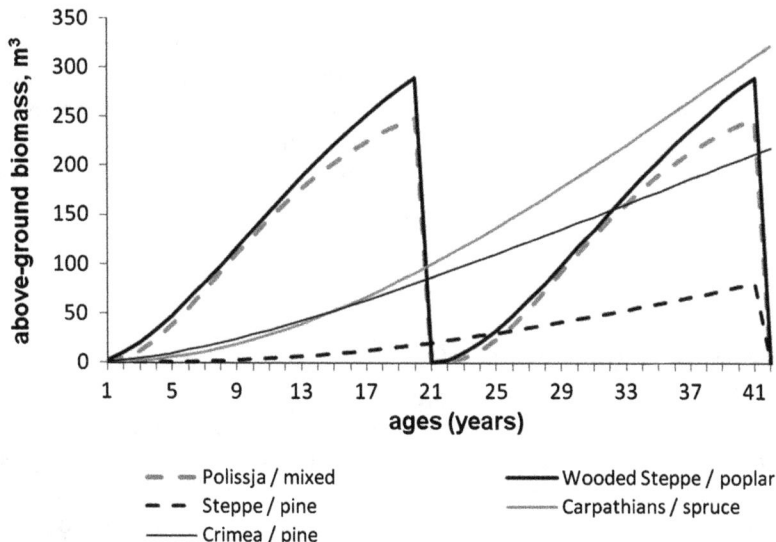

Figure 3 Graphical representation of growth functions of fast growing trees across the regions in Ukraine if the trees are harvested when their groth decelerates.[46]

the forest after harvesting and by the replacement of non-timber materials with wood (especially of carbon-intensive materials and fossil fuels) are repeatable. Therefore, the social benefits (of CCS and substitution effects) under wood product and bioenergy scenarios in the long-run are expected to be considerably higher than under the strategy of carbon fixation. This strategy presumes one-time tree planting, for example for a period of 40 years, without considering future use of wood and land after timber harvesting. Note the assumption can only come along with the assumption that by harvesting the trees, using the revenues to cover future costs of establishing new forests and storing carbon by some means, both the gains and losses in physical and monetary values are relatively balanced.[48] Wood product and bioenergy scenarios, *i.e.* of using wood instead of fossil fuel, are beyond the scope of this chapter. They have been analysed elsewhere,[31] however, for example, for forests in Canada, Ukraine,[50] Slovakia[28] and the UK.[52]

3 Carbon Sink and Storage in Forests: Several Implications from Europe

The potential for CCS in European forests has been analysed, and four examples from countries which differ according to their natural, socio-economic and institutional characteristics, are considered in this section. The UK plays a leading role in global efforts to tackle climate change and shows a sound example of carbon sequestration implications. The selection of Slovakia (EU member state) and Ukraine is based on their high CCS potential and on the similarity between the development paths of these countries, which include the transition to a market economy and the setting up of new institutional frameworks, including those that concern forestry and climate policy. The Netherlands makes a sound comparison with other countries, as in the addressing of Kyoto it can hardly rely on forestry. It invests heavily in energy efficiency and successfully implements the KP Joint Implementation (JI) and Clean Development mechanisms (CDM).

3.1   A Focus on the United Kingdom

The UK has one of the best records in the world in reducing direct GHG emissions within its territorial boundaries. In 2005, UK GHG emissions were reported to be 15.3% below base-year levels, with CO_2 emissions having fallen by 6.4% (ref. 35). The commitment of the UK under the EU burden-sharing target is 12.5% GHG emissions reduction for 2008–2012, relative to the base year. Further, a domestic goal of a 20% reduction of CO_2 emissions by 2010 has been introduced.[9] A series of targets have been set out – including making the UK's targets for a 60% reduction of CO_2 emissions by 2050, and a 26–32% reduction by 2020 (ref. 53). The goal is to achieve the majority of reductions nationally, including through further expansion of forest cover and development of short-rotation forestry, in combination with further using of wood as a construction material and a renewable energy source.[5]

The UK is on track to meet its targets. Its emissions in 2010 are predicted to be 23.6% below base-year levels, and 11.1% lower than required.[53] The Stern Review[20] examined the economics of climate change and explored stabilising GHG in the atmosphere. The UK Climate Change Act[54] – the first of its kind in any country – set out a framework for moving to a low-carbon economy with a target of 80% emissions reduction by 2050. The documents demonstrate the UK's desire to deal effectively with climate change, including by terrestrial carbon capture and storage.[55] The UK Climate Change Programme[34] set out policies and priorities for action. It targets are linked across policy objectives, including those of carbon sequestration forestry projects.

The UK has one of the lowest percentages of forest land in Europe (11.8% compared with the EU average of 38%), but it has significantly expanded its wooded cover in the last hundred years (currently 2.85 Mha; ref. 56). The dynamics in sources and sinks from forestry and land-use changes (LUC) in the UK is summarised in Table 2.

The policy of woodland development is supported by financial instruments which vary across the territory of the UK. In England, the Forestry Commission (FC) administers the English Woodland Grant Scheme (EWGS).[37] Payments differ depending on localities and land categories, tree species (£1800 per ha for broadleaves, £1200 per ha for conifers) and the likelihood of social benefits. Woodland creation grants also encourage farmers to convert agricultural land into forest and receive compensation to offset the foregone annual

Table 2 Dynamics in emissions due to land-use change and forestry.[57]

$Mt\,C\,year^{-1}$	1990	1995	2000	2005	2010	2015	2020
Forest sink[a]	2.6	2.8	2.9	3.2–3.3	3.1–3.4	2.7–3.0	2.4–2.8
Planting since 1990[b]	0	0.2	0.3	0.5–0.6	0.6–0.8	0.9–1.2	1.2–1.6
Emission from LUC[c]	8.7	7.3	6.5–8	4.9–7.8	4.1–8.2	2.8–8.4	1.4–8.3

[a]carbon accumulation in biomass, soil, litter and in wood products.
[b]entries from woodlands planted since 1990, excluding in timber products.
[c]net emissions caused by LUCA. (The trends do not consider possible effect of climate change on forest productivity).

income.[37] In Scotland, as part of the Rural Development Programme (RDP), new grants have been introduced which aim to deliver targeted environmental, social and economic benefits from forests. The RDP brings together a range of formerly separate support schemes, including those covering farming, forestry and primary processing sectors, rural enterprise and business development, diversification and rural tourism. Grant support will now be delivered through a number of forestry-specific (*e.g.* SRTP of willow or poplar) and non-specific (*e.g.* support for renewable energy – forestry) options, including those of CCS.[37] The Bioenergy Infrastructure Scheme funded by DEFRA was set up to provide grants to farmers, foresters and businesses to help in developing the supply chain required to harvest, store, process and supply energy crops and wood fuel to end-users.[52]

The maximum rate at which the woodlands expanded during the 20th century was about 40 kha yr^{-1} in the early 1970s, with this taking place primarily in the uplands. However, the establishment of 1.3 Mha of mainly monoculture conifer plantations met with considerable objections, due to their perceived impact on social, ecological and environmental components of sensitive rural landscapes.[18] Partly as a result of this, and also because of low forest profitability and uncertainty over CAP reform, the average rate of forest expansion went down to 10 kha per year in the last decade. Nowadays, forests in the UK sequester nearly 3 Mt C yr^{-1}, with 0.5 Mt C by trees planted since 1990 (ref. 37). The largest carbon pools in the UK are in soils and litter.[58,59] However, carbon stock in vegetation is also high, with total above ground in woodland of around 120 Mt C, and total including roots close to 150 Mt C (ref. 37). In the future, forests will probably expand on average at *ca.* 8 kha yr^{-1} rate. Given these assumptions, the forest sink is expected to rise to about 3.1 Mt C yr^{-1} in 2020, storing by then an extra 50 Mt C (ref. 33).

For the UK to be carbon neutral using only afforestation as the mitigation measure, each year it would need about the same amount of woodland expansion as the current area.[37] However, this is a very static assertion, considering also that the age structure of forests, declining availability of land suitable for afforestation and public preferences for multifunctional LUC suggest that in recent times the net rate of carbon sequestration in forests peaked around 2005. The projections[60] show that potential carbon savings from forestry, timber production and bio-energy could enable avoidance of *ca.*

Table 3   Carbon sequestration potential of afforestation in the UK, kt
C per yr.[61]

| Year | Carbon sequestration from forest land | | Additional carbon sequestration potential |
	Baseline	Abatement	
2007	3909	3867	−42
2008	3761	3711	−50
2009	3528	3495	−32
2010	2939	2949	−10
2011	2921	2991	70
2012	2715	2853	137
2013	2444	2648	204
2014	2331	2598	268
2015	2137	2464	327
2016	2107	2489	382
2017	2113	2548	435
2018	2114	2599	486
2019	1851	2386	535
2020	1376	1960	584

$8 \, \text{Mt C yr}^{-1}$ over the next fifty years. Woodland expansion alone (6.2 Mha, 50% broadleaved and 50% conifer) could save $4 \, \text{Mt C yr}^{-1}$ for the second half of this century.[42] Some authors[18] provide even higher estimates of the potential carbon sequestration by forests in the UK. These projections are largely based on assumptions that concern the expansion of forest cover. Although they show the role of forests in climate change mitigation, they do not account for a broad range of uncertainties associated with forest CCS and soil carbon estimates.

Carbon uptake in trees is temporary overall. Nevertheless, our research supports the suggestion that forestry development is a relatively effective and low-cost option for the UK to mitigate climate change, especially if fast-growing species are planted on marginal land. Following DEFRA[33] we considered a planting rate of $30 \, \text{kha yr}^{-1}$ for conifers under the abatement scenario. The baseline scenario extrapolated the 2005 planting rate of $8 \, \text{kha yr}^{-1}$ until 2020. The carbon sequestration potential of afforestation in the UK is shown in Table 3 (ref. 52). The results imply that under the abatement scenario, on average in the UK, new forest plantations could annually offset about $4.9 \, \text{Mt CO}_2$.

The results suggest that the woodlands expansion policy measure will likely be competitive with other means of removing carbon from the atmosphere, and that choosing appropriate species and management regimes is important for saving economic costs.

3.2   A Focus on Transitional Countries of Ukraine and Slovakia

In transition countries, primarily as the result of economic recession in the early 1990s, CO_2 emissions have fallen much below these countries' KP commitments (the fall of 30.7% on average was observed in 2000; ref. 62). These countries, therefore, have reached their KP targets and stand to profit from the sale of "hot air". Although "hot air" trading may appear advantageous to these particular countries, they will not necessarily be able to rely on it, because "hot air" is a hot topic pertaining to the environmental effectiveness and economic efficiency of implementation of the KP.[31] The countries in transition to a market economy are therefore wide open to a range of opportunities for cleaner industrial and energy production. Due to substantial carbon emissions per unit of GDP, these countries have a high potential for cheap Joint Implementation.[63] However, they have an insufficient institutional capacity[28] for foreign investors to enter their business environment effectively. Hence, over and above the emissions reduction, an enhancement of CCS in forests is important.

Ukraine is among the largest by area of all the countries in Europe (58 Mha), but it is sparsely forested (16.5% of its land).[56] Its forests have mostly been turned to agricultural land.[64] Nevertheless, because of the vast territory, Ukraine's forests possess a total biomass of about 1.7 Gt, containing 600 Mt C only above ground,[56] with average actual net carbon uptake of 17 Mt C per annum. Approximately 2.29 Mha of land is suitable for tree-planting, and afforestation would allow a 23% increase in Ukraine's forest cover.[46] Thus through afforestation in Ukraine, total cumulative carbon sequestration in forests can be increased by 180 Mt (0% discount rate for carbon savings) over a 40-year time horizon. The maximum additional amount of carbon sequestered annually could reach 4.6 Mt (C savings not discounted), or about 1.0 Mt if C savings were discounted at 4%. This roughly corresponds to 4.6% of the Ukraine's annual CO_2 emissions. Primarily due to the diversity of conditions, the potential benefits vary considerably through the territory of the country (see Table 4; ref. 28,46).

The estimates differ substantially also across the territory in Slovakia. Over 1.9 Mha is classified as forest land in this country, with the wooded cover being 40.1% (ref. 56). Forests in Slovakia possess a total above-ground biomass of 334 Mt, containing 167 Mt C (ref. 56), with a forest sink of 1.9–4.7 Mt C per year.[65] More than 50% of forests are owned privately in Slovakia but are managed by state-owned companies, leasing forests from their owners.[66]

Table 4 Carbon sequestration for the above-ground biomass and its value per ha in Euros, at 4% discount.[28]

	Slovakia			Ukraine				
	Western	Central	Eastern	Polissja	Wooded Steppe	Steppe	Carpa-thians	Crimea
Total tonnes per year of C	432.2	492.6	315.1	1846.2	2372.5	143.0	676.7	584.3
Permanent tonnes	8.4	9.6	6.2	36.9	47.5	2.9	13.5	11.7
Value of C uptake	124.7	149.0	90.2	553.9	711.8	40.1	203.0	175.3

Approximately 0.41 Mha of largely marginal and abandoned land withdrawn from agricultural production in the last decade is suitable for tree-planting. Its afforestation would result in 0.9 Mt C sequestered annually (4% discount rate), and this counts for *ca.* 1.4% of Slovakia's CO_2 emissions.[28]

In Ukraine, where nearly 66% of the forest land (7.1 Mha) is publicly owned, tree-planting activities are under the execution and control of the State Committee of Forests. The weaknesses of a "command-and-control" mechanism of afforestation include the lack of flexibility and economic incentives for encouraging tree-planting. Control mechanisms, however, could be justified on efficiency grounds, if the savings in transaction costs exceeded the gains from using other co-ordination mechanisms.

Afforestation enlarges social benefits, primarily to agriculture, because of soil protection and improved hydrological forest functions, and to society in terms of climate change mitigation. Due to market failures, however, the social gains from afforestation (external benefits) could hardly be achieved, and welfare maximisation conditions could hardly be met without government regulation. The main reason is the discrepancy in the distribution of benefits and costs from forestry development. The establishment of forest plantations, including those for CCS, is executed in the forest sector, while climate-change mitigation benefits accrue to society. The problem "who pays and who receives the benefits" cannot be solved through the market. Hence, despite afforestation costs being relatively low in the transition countries, a large-scale tree-planting for CCS will not take place, without government subsidies or foreign investment.

This argument has been proven in Slovakia, where with the cancellation of the state law and state forestry framework, according to which by the year 2000 the area of afforested land would have been 50 kha, the area actually afforested appeared to be just 877 ha. During the transition process, when 42% of previously state-owned forests were given back to their former owners and more than 90% of claims were processed, the incentives in support of afforestation were ineffective. The afforestation process was negatively influenced by uncertain land ownership and by problems with the allocation of subsidies to land owners. An average size of a private plot of land in Slovakia is 0.45 ha, and such fragmentation of the land also hampers afforestation.[28] As long as forest land remains fragmented into small ownership parcels and there is a lack of long-term investment and appropriate incentives for tree-planting, the land owners are unlikely to undertake afforestation activities.

A very important task pertaining to reconciliation of sustainable forest management and climate change mitigation is to settle upon a proper structure of land (forest) ownership. The countries' legal documents have to redefine (where necessary) and enforce property rights on forest resources and wooded land, for instance through the introduction of community-managed forests and de-fragmentation of private plots of land. There is also a need for simplified forest management guidelines for various owners within their land. Privatisation of forests is not the only solution for transition economies to enhance sustainable forest management.[67] The forest inventories may comprise public (state) forests; forests of municipalities, farmers, enterprises, organisations and institutions; as well as privately owned wooded land and land that undergoes afforestation. An afforestation fund may include state, communal and private land to be sold or leased for forest development.[28]

To date, climate change mitigation opportunities that involve forestry are not viewed as priorities for the national climate policies, forestry policies and rural development strategies in transition countries.[66] However, as the above analysis demonstrates, carbon sequestration through afforestation represents an opportunity, given the decline in agricultural production and increase in abandoned land. Therefore, afforestation of non–forested areas, increasing the level and efficiency of wood utilisation, using biomass as a substitute for fossil fuels and the protection of existing carbon storage, could provide relevant policy measures in transition countries trying to cope with the changing climate.

3.3 A Focus on The Netherlands

The Netherlands has a more energy-intensive economy than the EU average. Therefore, after the burden-sharing process within the EU, its emission reduction commitment was fixed at 6%. This country considers GHG emissions reduction (see Table 5) as the priority measure of its climate policy, and it does much to achieve its emission-reduction targets. Concerning the role of forestry, a payment discount has been introduced into regulatory energy tax to accelerate afforestation. The objective of Dutch forestry is to expand wooded area by 75 kha until the year of 2020. Over 63.5 kha is to be planted with trees through governmental agencies, mainly in rural areas (54 kha, of which 30 kha are on agricultural and 15 kha on natural land), but also in urban areas.

The present rate of afforestation, however, is lagging behind (around 70%) the planned expansion rate due to an increased demand for space and, consequently, due to high land prices. Recently, therefore, the Netherlands has revised its forest policy into more general objectives of an expansion of its wooded area.[69] Being the country which is the smallest by area (3.4 Mha) among the countries analysed, and the most homogeneously and densely populated, the Netherlands has little room for coping with changing climate through forestry. Although a 20.5% increase in wooded cover is projected (see

Table 5 Projections for Netherlands' sector emission reduction.[68]

Sector	Emission in 2010 in Mt CO_2 eq.	% Reduction in 2010
Industry (incl. refineries)	89	11.2
Energy/Waste companies	61	13.1
Agriculture	28	7.0
Forestry	0	0
Traffic	40	7.4
Households	23	10.0
Trade, service, government	12	8.3
Other	6	–

Table 6   Carbon storage in forests across several countries in Europe and their potential for afforestation.[56]

Country	Wooded cover, % area	Forest area, Mha	Stock of forest above-ground biomass, Mt	C in forest, above-ground, Mt	C in forest soil, Mt	Projected increase of wooded cover, %
Netherlands	10.8	0.365	43	21	40	20.5
Slovakia	40.1	1.929	334	167	270	21.5
UK	11.8	2.845	190	95	719	25.0
Ukraine	16.5	9.575	1199	600	n.a.	23.9

Table 6), this country is to rely on markets as a governance structure and on common values and norms, rather than on domestic CCS forestry initiatives.[31]

The Dutch implementation plan calls for 50% of the KP commitments to be met internally. Thus, 50% of the required emissions reduction is to be achieved externally, using the flexibility mechanisms of Joint Implementation and Clean Development. Promising CCS initiatives are already in place. They concern voluntary carbon markets, including those available in Europe, such as pilot afforestation projects for a total of 5 kha in Ukraine where a project was designed to regenerate forests on the land affected by radioactive contamination after the Chernobyl nuclear accident.

However, woodlands development for multiple purposes, where an increase in amenity and other landscape values comes along with CCS, is viable for the Netherlands not only beyond its boundaries but also internally. Various agreements with stakeholders to maintain multi-functional rural-land use in a sustainable manner and plant new forests, including for the purpose of climate-change mitigation, have been made. The National Green Fund, for instance, issues certificates for the number of hectares for which CO_2 sequestration rights have been acquired. Any company or organisation wanting to acquire the right for CCS can deduct €4.5k from its energy tax bill, providing this amount to the Fund.[68]

To complete the analysis of CCS across several countries in Europe we present in Figure 4 carbon sequestration rates per hectare of the selected forests in the UK, the Netherlands, Ukraine and Slovakia,[28] showing that these rates are comparable. The results of the analysis[28,50,52] across the selected countries in Europe allow us to argue that marginal lands would be available for tree-planting until 2020 and that, over and above other climate policy measures, an enhancement of CCS in forestry, including by use of bio-energy, is likely to be applicable in all these countries except the Netherlands.

The most optimistic projections are for Ukraine.[46,50] They suggest that with afforestation for CCS in this country, 80 t C per ha on average can be sequestered in the subsequent 40 years. This might be important for Ukraine and other Annex B countries (the 39 emissions-capped industrialised countries, including countries in transition listed in Annex B of the KP)[41] and especially for those where GHG emissions reductions are costly. In view of the KP commitments (and prospective international carbon-trade agreements), the

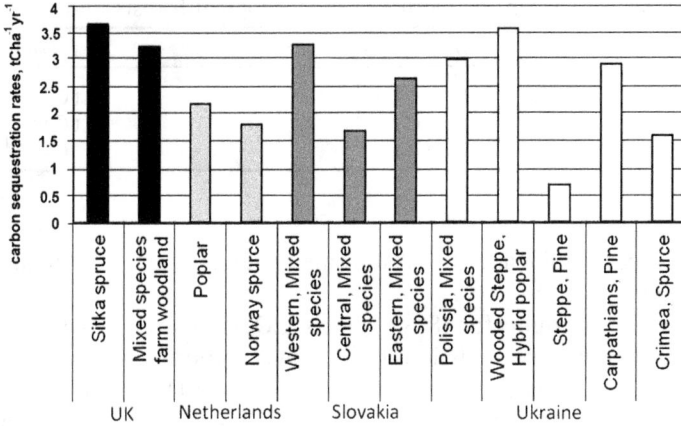

Figure 4 Comparison of averaged carbon sequestration rates across several European countries, $t\,C\,ha^{-1}\,yr^{-1}$ (ref. 11,33,38).

prospects of selling carbon-offset credits by Ukraine to these other countries is an issue that merits attention.

4 A Focus on Tropical Forests

In most major industrial wood-producing countries, such as Finland, Sweden and Canada, forest legislation stipulates reforestation of sites after forest harvesting. For these and other reasons, forests in temperate and boreal areas have expanded. During the last five years, the majority of countries in the Caribbean, North America, Oceania, and Western and Central Asia had no significant changes in forested area. In 2000–2005, there was a net gain in forest land in Asia (*ca.* 1 Mha per year), primarily as a result of large-scale (4.1 Mha per year) reforestation in China.[37] Policy and environmental drivers have resulted in a continuous growth of forest cover in Europe. Between 1990 and 2005, it increased by 13 Mha. However, approximately the same area of forestland got lost in tropics.[37]

In 2000–2005 the highest annual losses of *ca.* 4.3 Mha were observed in South America, followed by Africa which was losing about 4 Mha of forest land each year. Indonesia was losing *ca.* 1.9 Mha and Brazil over 3 Mha of forest each year.[56] Although the rates have decreased more recently, deforestation and forest fires are now responsible for nearly 75% of Brazil's GHG emissions. Land has been cleared for cattle and soya bean production, which is increasing, and due to illegal logging that continues.[70] Globally, therefore, the forest area is decreasing. To compare with forests in other regions, tropical forests have substantial carbon content of biomass (see Table 7). Therefore, when tropical forests are burned a huge amount of CO_2 is instantly released to the atmosphere. Given the rates of deforestation and the losses it brings, it is imperative to alleviate its occurrence and to save existing forests first of all, allowing for

Table 7   Carbon content of biomass across tropical forests and regions.[56]

Region	Wet Tropical Forest	Dry Tropical Forest
Africa	$187\,t\,C\,ha^{-1}$	$63\,t\,C\,ha^{-1}$
Asia	$160\,t\,C\,ha^{-1}$	$27\,t\,C\,ha^{-1}$
Latin America	$155\,t\,C\,ha^{-1}$	$27\,t\,C\,ha^{-1}$

their enhanced resilience to disturbances and their continual natural regeneration.[71]

Deforestation accounts for 18% of global CO_2 emissions, which is the second largest contributor to net emissions (after power stations, 24%) and a larger net emitter of CO_2 than agriculture (14%) and transport (14%). By slowing down deforestation and by increasing the rate of afforestation, the potential for increasing carbon storage in forests may reach 60–90 Gt C over 50 years, compared with current emissions from fossil fuels of 6.5 Gt C per year.[42] Reducing deforestation can be achieved by conserving and managing existing forests, *e.g.* by implementation of long rotations for carbon sequestration, which is cheaper than afforestation. However, whereas afforestation is likely to generate benefits from the double perspective of CCS in the forest and substitution benefits, the implementation of long rotations needs further investigation.[52] Apparently, this strategy is more pertinent for natural forests in the tropics or for primeval beech stands in the Carpathian Mountains, *i.e.* those which are valuable by their biodiversity and other ecosystem functions, including carbon storage. Lengthening rotations (and so slowing down the harvest rate) elsewhere may miss important opportunities of using wood for wood products and as a substitute for fossil fuels. Thus this issue needs to be analysed in detail and in connection with study areas under observation. Reducing deforestation can also be achieved through tackling its causes, reducing rates and decreasing the amounts of losses it brings, for example, by means of protection and enhancement of forest ecosystems services, including those of CCS, or by means of natural forest regeneration.

Forests in the tropics provide a range of ecosystem services, including biodiversity conservation, watershed protection and soil conservation, protection and enhancement of rural livelihoods, *etc.* The ecosystem services that forests in developing countries provide already necessitate their sustainable use and conservation. The countries, therefore, should have policies, capabilities and flexible mechanisms in place to combat deforestation. However, there is often a lack of resources, policies and institutions to pursue forest sustainability in developing countries. Tropical deforestation is not included in the Clean Development Mechanism (CDM) and there is now serious consideration that it should be included. Emissions reduction from deforestation is to be linked to efforts to tackle other types of emissions and the markets are to be created and developed to enhance efficiency in reaching climate policy targets. The problems to be solved include: (i) how developing countries might receive tradable carbon credits at the national level, and (ii) in which ways the benefits arising from meeting agreed national targets, should be shared by people so as to result in real changes in managing tropical forests at a local level.

Carbon markets are already starting to provide finance to support low-carbon development, including through the CDM. Although deforestation (*i.e.* forest conservation projects) has not been included in the scheme, the CDM allows eligible activities, which primarily comprise: tree-planting projects to enhance carbon sinks (generate certified emission reductions, CERs) and promote sustainable development in a developing country. However, under the regulatory scheme of the CDM, the percentage share of forestry projects in total expected CERs until 2012 comprises less than 0.5%. Only one project (in China) was registered, with seven projects under validation, 30 with detailed information and others under various levels of development.[72]

Implementation of regulatory trading schemes of the KP requires: accurate measurement of CCS and of the costs; reliable monitoring of the carbon sinks; addressing the problems of durability (permanence) of forestry projects; and alleviation of possible carbon leakages. The issues that make forestry-project implementation difficult under the CDM, and make them both time- and resource-consuming also include: assurance of project eligibility; demonstration that activities will contribute to sustainable development in a host country; arranging finance, property rights and legal matters; proper monitoring, verification and reporting; valuation of the baseline emissions; CER establishment and certification; registry, creation and acceptability of carbon trading; administration and adaptation assistance; establishment of an executive board; distribution and enforcement of responsibilities and designation of operational entities; provision of various kinds of support; and the introduction and imposition of penalties for non-compliance, *etc.*

Thus, the key obstacles for forestry projects to develop successfully under the CDM concern asymmetric information and high transaction costs. Consequently, to compare with the total of *ca.* 3000 projects eligible under the KP scheme, forestry activities are lagging behind. At the same time, voluntary carbon offsetting in forestry is performing successfully. It is popular in both developing and industrialised countries and comprises 37% of total voluntary transactions by sector.[73]

5 Economic Considerations of Carbon Sink and Storage in Forests

Afforestation is often considered to be among the low-cost methods for controlling the increase of CO_2 in the atmosphere.[74] However, a meta-analysis of various studies[75] estimated carbon-sequestration costs and identified their huge variability across countries worldwide. Under baseline conditions of forest conservation for CSS, the costs were in the range of €35–€199 per tonne of carbon and, when opportunity costs are taken into account, the cost range increases to €89–€1069 per tonne C. In the Netherlands, the costs of €199–€286 per tonne C are the highest of all the countries examined. The costs in the

Netherlands will not fall below €65–€202 per tonne C (ref. 11,48) even when the benefits are not discounted. These results can be explained by the high opportunity costs of land. Again, among the options for the Netherlands to contribute to joint efforts of coping with the changing climate is to establish forests outside its territory, *i.e.* in countries where land values are lower than in the Netherlands but the rates of carbon uptake may be even higher, for instance in Ukraine, as shown previously. Despite high cost-estimates of carbon uptake in some regions,[76] there is enough evidence to argue that, particularly in developing and transition countries with good tree-growing conditions, large amounts of carbon may be sequestered by forestry at costs of $30 per tonne C and lower.[28,42,50]

Several methodologies are in use[77] for estimating the cost of carbon sequestration in forests. They include econometric studies, optimisation models, and a bottom-up approach which is probably the most straightforward way to estimate costs. The estimates of marginal costs of carbon uptake provide benchmarks for cross-comparison of different measures and scenarios. To assess whether forestry offers an economic opportunity for CCS and in relation to which types of forests (their species composition, age structure and management), the marginal costs per tonne of sequestered carbon and the present value (PV) of costs per tonne of carbon have been computed for the UK, Ukraine and Slovakia.[28,50,52,78]

Since different methodologies and assumptions were applied in these countries, the cross-country comparison of the estimates cannot be precise. The stock-change approach (which consists of summing up carbon stocks over the length of the rotation) was used to estimate CCS in the UK. Under the assumptions and specific requirements of DEFRA,[78] the carbon sequestration costs range from £30.5 per tonne of carbon (afforestation of sheep grazing areas) to £174.9 per tonne of carbon (afforestation of wheat fields) for a discount rate of 3.5%. The estimates provide empirical evidence in support of prospective afforestation in the UK of some marginal land currently used for sheep grazing. These findings[52] are comparable with average costs of carbon sequestration of €72 to €116 per tonne C estimated by other authors.[80]

The costs that were taken into account in Slovakia and Ukraine included tree-planting costs (including soil preparation), care and protection costs, opportunity costs of land, replanting costs and the costs of timber harvesting. In Slovakia, the costs of carbon uptake (discounted at 4%) were estimated to be in the range of €8.5 to €14.2 per tonne (if carbon uptake benefits were not discounted). It is noticeable that the costs of CCS in forests are lower in Slovakia than in the UK (average country estimates). They are the highest in the Western region of Slovakia because of its more fertile soil. In the central region, the low value of the carbon sequestered results in a negative net PV of carbon uptake for the CCS policy scenario (strategy of carbon fixation).

In Ukraine, as shown in Figure 5, carbon uptake costs are €4.6–€78.5 per tonne (when carbon savings are not discounted). The estimates vary across the territory and depend on tree species, tree-growing conditions and forest-management practices. When the benefits of carbon uptake were discounted at 4%,

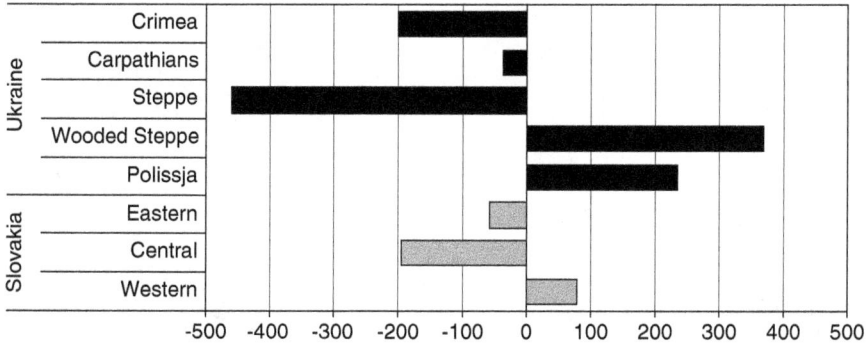

Figure 5 NPV of benefits of carbon storage through afforestation in Slovakia and Ukraine (€ per ha, 4% discount rate, carbon in permanent tonnes).[28]

at the same discount rate as for the costs, the PV of carbon uptake costs appeared to be €7.2–€173.3 per tonne.

For transition countries, the economics of wood products and renewable-energy scenarios were also considered and various discount rates were used to examine their influence on the results. The pilot analysis of renewable energy scenarios has shown that in Ukraine, the costs per tonne of carbon sequestered are more than €36.4 in the Polissja (wooded) region, €32.2 in the Wooded Steppe zone and €124.6 in the Steppe (at 4% discount rate). In Slovakia, the costs are €37–€48 per tonne, but become much higher when the costs for energy production from the planted trees are included.[28,50]

Commonly, in industrialised countries, even when all of the CCS is taken into account (but product sinks are not accounted under the KP), it is unlikely that 'additional' forest management will be a cost-effective and competitive means for sequestering carbon.[79] This is because on less-productive land, which is usually considered for afforestation, the growth rates of trees are usually low. At the same time, in the EU countries and, for example, in the USA and Canada, the land is anyway quite expensive. Afforestation, therefore, does not pay, especially when multiple benefits are not taken into account. Tree planting is often costly in these countries and returns that accrue to forestry in the distant future make the investment unprofitable. However, if SRTP are established, carbon-offset credits competitive with emission reductions might be created. In Canada, for example, hybrid poplar plantations on marginal land will likely be a cost-effective and competitive option.[31]

The costs of carbon sequestration in forests in transitional countries are commonly lower than elsewhere in Europe. For instance, in Ukraine and Slovakia they are lower than in the UK,[52] and even more so than in the Netherlands.[11] However, in some regions of Ukraine and Slovakia the relatively low PV costs of CCS are nevertheless higher than the value of the land. Furthermore, in areas that are strongly affected by the decline in agricultural production and land abandonment, the market prices of land can be significantly lower than the prices set by the government (which are based on the

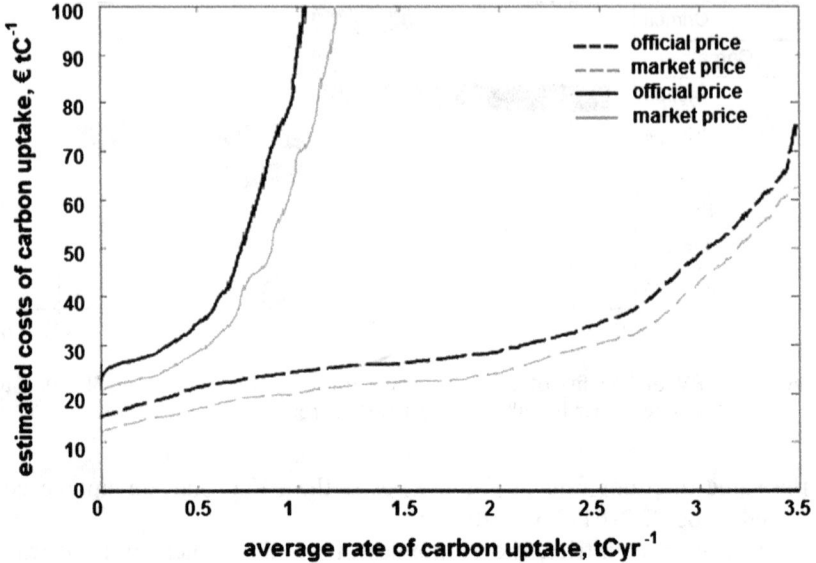

Figure 6   Estimated costs per tonne of carbon uptake for afforestation (dashed line) and short rotation bio-energy plantations (costs of energy production are included, continuous line) in Slovakia, by using land prices set by government (official price) and market price (€ per tonne, 4% discount).[28]

physical characteristics of the land). This phenomenon is reflected in the estimates of the costs presented in Figure 6 (ref. 81).

The flexible mechanisms of the KP were designed to help Annex B countries to meet their emissions reduction targets at least cost.[1] It provides opportunities for (non-EU) transition countries to sell carbon offsets to industrialised countries. The opportunities arise not only due to the decline of CO_2 emissions during the economic transition compared to their levels in 1990, *i.e.* "hot air", but also due to selling carbon offsets from newly established woodlands. Therefore, carbon sequestration in Ukraine's forests might be beneficial for Annex B countries in view of stabilising their collective emissions in the cheapest possible way *via* the trading of carbon-offset services.

The KP provides opportunities for countries to cope with the changing climate from an economic perspective. The Joint Implementation mechanism presumes attaining GHG emission reductions in another Annex B country, including by CCS (ERU: emission reduction units). The Clean Development Mechanism presumes adding to emission reductions in developing countries, including through tree-planting (the use is limited each year to 1% of 1990 Annex B country's emissions).[41] However, the analysis indicates that under the CDM and JI flexible mechanisms it appears unlikely that credit and permit (allowance) trading, and particularly regulatory carbon offset trading, will occur on a large scale internationally or even nationally. The trading schemes fail not because of a lack of interest from the involved parties, but primarily from a breakdown in the necessary economic and market conditions, such as imperfect information and high transaction costs.

Among the reasons for difficulties (market and governance) of the countries to meet their KP targets are their proclivity to rely primarily on administrative measures and voluntary actions. Further, some countries have low capacity of social capital and inadequate institutions to develop regulatory markets for CO_2 trading. Consequently, the costs of appear to be higher than they need to be, and these high costs lower the efficiency of climate policy implementation. Moreover, the KP addresses only a small proportion of potential global emissions, and it has no effective penalty for non-compliance.

In order to use terrestrial carbon sinks as a flexible mechanism for addressing the KP (and future international agreements), it is important to measure carbon uptake and release, as well as to develop economic and market conditions for creating and trading terrestrial carbon credits. Proper carbon measurement and accounting, reliable monitoring and economic incentives are also required for making bio-energy and wood-products policy scenarios feasible. Various CCS projects that adapted voluntary carbon offsetting schemes have been developed successfully in many countries. The funders are governments and NGOs, businesses and individuals. The projects include reforestation of land and conservation of forests and, in the majority of cases, they offer cheap carbon savings. For expanding internationally and beyond voluntary forest-based carbon markets, it is important to examine the economic efficiency of CCS, determine which scenarios are economically sound and suggest which of them should be implemented, where and how.

The KP "cap-and-trade" system that includes carbon offsets from forestry faces serious challenges in the creation of carbon markets and acceptability of carbon-trading exchanges. The costs per tonne of carbon locked-up and removed by terrestrial ecosystems must be compared with the costs of decreasing the stocks of carbon in ways other than through forestry. When only CO_2 emissions are considered, the emissions cap is set equal to the KP target. When, in addition, carbon offsets are the matter of concern, a cap is required also on permissible offsets.[31] In the light of carbon-trade negotiations, a conversion factor or exchange rate needs to be set. The fact that the sequestered carbon remains in the forest for many years into the future is to be ensured. Evaluation and inclusion of carbon offset credits in a trading system is difficult because of the high transaction costs associated with assessing and monitoring of terrestrial CCS, and due to its temporary and ephemeral nature.[31]

Problems also arise due to the so-called carbon "leakages" which happen when the CO_2 emissions which a project is meant to sequester are displaced beyond its boundaries, so that the full benefit of the project is, in broader terms, reduced. There are ways to cope with "leakages", for example by expanding the scope of the system to "internalise" the "leakages" or to design the project so as to be "leakages" neutralising. Overall, major concern about the cost-efficiency of CSS in forests and numerous problems with the inclusion of carbon offsets from forestry in regulatory emission trading schemes are caused by the following challenges:[31,52]

- Setting up the level of baseline emissions.
- Assurance of "additionality" of the projects.
- Assurance of durability (permanence) of the projects.
- Accuracy in measurement of carbon sequestration and of costs.
- Reliable monitoring of carbon sinks.
- Avoidance of "leakages" and of double counting (which means that no credits ought to be sold twice to a final customer).[82]
- Creation and acceptability of carbon trading.
- Establishment of proper carbon-offsets certification and of their "conversion" into emission permits.
- Assurance that actual carbon sequestration has taken place.
- Development of property rights and institutions for exchanging carbon offsets.
- Various legal aspects.
- Verification of sustainable development requirements, *etc.*

Therefore, more attention is to be paid to analysing the role of carbon-uptake credits from forestry in climate-change negotiations, and how to bring terrestrial carbon offsets into regulatory emission-trading schemes. It is important to identify carbon-sequestration projects which will be coherent, effective, cost-efficient, widely acceptable by the public, and consistent with other aspects of sustainable development.

6   Uncertainties Pertaining to Carbon Sink and Storage in Forests

The role for forestry to mitigate climate change comes with a great deal of uncertainty. Firstly, marginal damages from atmospheric carbon over time are uncertain. There is also uncertainty when trying to correctly ascertain the benefits to future generations of carbon-control strategies. Even if the vote of up-coming generations over climate-change mitigation strategies can be taken into account, future preferences are unknown.[83] Difficulties in estimating the future benefits of CCS will also arise due to uncertainties about the dynamics of carbon in forests. It is necessary not only to determine how much carbon is actually sequestered and stored and for how long, but how much carbon will be sequestered in the future, in conditions of changing climate.

   The uncertainties also concern causes, magnitudes and permanence of forest carbon. Scientific evidence suggests that the terrestrial carbon sink is increasing. However, complex relationships between climate and the terrestrial carbon cycle result in uncertainties about CCS future projections, in conditions of increasing temperatures and their effects on photosynthesis. Forest ecosystems are vulnerable to the effects of changing climate. For example, there are predictions that large areas of tropical forest in the Amazon may die back from 2040, releasing carbon to the atmosphere[42] and multiplying the effect of global warming.

Although it is now possible to quantify the amount of carbon being sequestered and stored in forests, measuring carbon uptake remains a difficult task, especially if the carbon sink is short-lived.[76] It is now possible to monitor carbon sequestration and to project possible CCS scenarios. However, the results are case-specific and there are considerable variations between sites and forest types.[42] Furthermore, existing methods are labour- and time-intensive and, as they are usually based on measurements in sample plots, the scaling-up of results often leads to significant errors. Recent advances in remote-sensing techniques look promising.

The rates of possible carbon uptake, or loss, depend upon forest management and land-use changes (*e.g.* forest clearance, species substitution for peat bog, heathland, ancient-woodland restoration, increased uptake of silvicultural systems) and on climate change, itself a causal influence on carbon sequestration, due to changes in precipitation, temperature or the frequency of extreme events. The concern over non-permanence is tangible, as removed carbon could be released before the accounting period ends. The losses that result in non-permanence of CSS are associated with: increasing forest fires and diseases due to climate change; the losses of dissolved organic carbon in rainwater drainage and runoff caused by climate change and anthropogenic pressures; and reduced soil inputs through stump harvesting, plus increased losses involved in de-stumping soil disturbance.[52]

Discussions of the uncertainties pertaining to terrestrial carbon sink and storage, and of the mechanisms for assuring that the associated emissions reductions in forestry are long-lived and are not double-counted, are available in the literature.[31,82,84] The temporary nature of terrestrial CCS can be addressed through: partial credits which account for the perceived risk of carbon release; by insurance coverage against the destruction or degradation of forest sinks; by assurance that the temporary activity will be followed by one that results in a permanent emissions reduction; and by using a conversion factor to translate years of temporary carbon storage in a forest into a permanent equivalent, *etc.* Although CCS in trees is carbon neutral in the long-run (at 0% discount rate for carbon-uptake benefits), because the sequestered carbon is eventually released back to the atmosphere through wood decay, some temporary forest carbon may become permanent.[85] The prevailing vision is that CCS in forests is important as it may be a relatively low-cost option and it postpones and reduces climate change, allowing time for adaptation, learning and technological innovation.[86]

However, in addition to the already discussed "pros and cons" of climate-change mitigation through forestry projects, there are various scientific, technological and socio-economic uncertainties pertaining to terrestrial CCS. For example, the assumption that forestry-based carbon sequestration may be considered as a universal remedy discourages our efforts to address emissions reduction. The CCS policy measure does not always complement economic growth. Large-scale afforestation, particularly the establishment of vast areas of fast-growing monoculture plantations, may result in negative environmental and social consequences. These challenges are often multiplied by a great deal of institutional uncertainty associated, for instance, with land (forest) tenure in some developing and transitional countries, and with property rights on carbon offsets, as well as with some managerial aspects in forestry, particularly with those that concern afforestation. Changes in government policies, markets fluctuations, and changes in social norms and behavioural patterns contribute to uncertainties. Analysis of planting trees for CCS, therefore, runs into uncertainties, and the extent to which the strategies can be justified on efficiency grounds also depends on the rate of discounting employed in the analysis.

The discounting of carbon uptake benefits at 0% suggests that the value of marginal carbon damages in the future will increase at the real rate of discount. This implies that all of the carbon sequestered is valued equally, no matter when it is captured.[31] Given this assumption, consider fast-growing poplar stands with initially high rates of carbon uptake, but which shortly decelerate. Consider also spruce stands that grow much more slowly but for longer, and in 100 years from now they can accumulate up to 300 t C per ha. If the costs of carbon sequestration (largely afforestation costs) are equal for these two types of forest, at a 0% discount rate for carbon savings spruce stands are to be chosen for planting. However, this might not be the solution when terrestrial carbon sequestration is considered. Therefore, approaches to CCS project-evaluation are case-specific. The economic way of reasoning, for instance, suggests that in long-term projections in forestry, the setting of a 0% discount rate for carbon savings is a very specific assumption. Cumulative carbon uptake and other benefits in forests would be available in the future. However, in economics, it is important to compare costs and benefits according to their present value. Therefore, to justify the cost-efficiency of carbon-sequestration forestry projects, positive discount settings for carbon uptake benefits could also be advocated.[50]

Among the challenges of CCS in forests is the meeting of the "additionality" condition.[82] In principle, credit should be given only for carbon sequestration above and beyond what occurs in the absence of CCS incentives. For example, if it can be demonstrated that the forest would not otherwise has been established (*e.g.* to provide higher returns to forest owner), the "additionality" condition is met. Similarly, afforestation projects are additional if they provide ecosystem services not captured by the landowner, and would not be undertaken in the absence of economic incentives, such as subsidy payments or an ability to sell carbon credits.[82] Moreover, carbon sink in forests has to take into account the carbon debit from LUC and timber harvesting; carbon stored in wood product sinks (currently, not considered under the KP); various carbon "leakages"; and additional carbon sequestered as a result of forest management activities, including fire control.[31]

To conclude, our major concerns with relation to using terrestrial CCS as a climate policy measure are, as follows:

- Wider use and promotion of CCS in forests, under the KP and upcoming international agreements, may distract attention of policy makers and practitioners from emissions reduction and from novel means of climate change alleviation.
- CCS in forestry tends to be ephemeral and thus not equivalent to emissions reduction. Often, terrestrial carbon sinks are short-lived, and this makes it particularly difficult to compare them with emissions reduction (but the techniques exist).[31]
- The 'value' of sinks to a country is tied to the land use existing in 1990 as the base year. Identification of both, a baseline scenario and additional CCS activities, is usually difficult.
- Measurement of carbon flux, its monitoring, and enforcement of durability of projects are costly in the case of carbon uptake in terrestrial ecosystems.
- Accounting for CCS in wood products needs to be resolved.
- Incentives and mechanisms to combat deforestation in some regions of the world, particularly in the tropics, are to be addressed.

7 Social Considerations of Carbon Sink and Storage in Forests

Forestry has been viewed as the basis for timber production, outdoor recreation and support of wildlife habitat, a means of watershed protection, a sink for pollution and an opportunity for carbon sequestration. However, forestry practices that sequester the most carbon may compromise projects that reduce net emissions of CO_2 to the atmosphere. Carbon sequestration activities could lead to changes in fossil-fuel use and could cause LUC that further impact the atmospheric CO_2 pool. Climate policy measures need, therefore, to be effective and feasible, and well embedded in land use and environmental policy strategies. If this is achieved, and economic instruments and flexible policy mechanisms are implemented, there is a scope for multi-functional countryside where CCS, production of sustainable energy, sustainable forestry and agriculture are combined, and where climate policies are connected to the strategies that promote integrated sustainable land use.

Integration of various sectors of the economy should be accompanied by reconciliation of carbon sequestration policies with the strategies dealing with remote rural areas affected by regional disparities. Support for afforestation on marginal land is important, and more attention is being given to agricultural and environmental linkages to climate-change related measures, and to the inclusion of carbon-sequestration forestry activities into rural and regional development schemes. It is anticipated that CCS forestry projects will be more successful if they are consistent with wider programmes of sustainable rural development, and if they focus on multiple (*i.e.* social, economic and environmental) components of sustainability, including on climate-change dimensions and various aspects of land-use policy. This particularly concerns remote rural areas in Europe, where CCS through forestry could target "win-win" situations by bringing together sustainable development and climate-change mitigation, and by combining the socio-economic objectives, *e.g.* those of increasing welfare of communities with the enhancement of nature and rural landscape.

In such a land-rich country as Canada, for instance, which now focuses on afforestation of agricultural land to meet a significant component of its KP commitment, the problem of woodland development is related to the will-ingness of landowners to create carbon credits rather than to the biophysical possibilities for carbon uptake.[31] This necessitates the development of oppor-tunities and incentives by which the capabilities of landowners to create carbon credits and to market carbon offsets would be enhanced. Landowner pre-ferences for carbon sequestration methods are imperative with this respect, and they probably are influenced by the available information, institutions and uncertainties concerning landowners' potential profits and their eligibility for subsidies.

In England, for example, tree-planting for multiple purposes rather than solely for CCS commonly enlarges social benefits and helps to prevent potential conflicts relating to trade-offs, *e.g.* between biodiversity and carbon seques-tration, or between landscape amenity values and those of climate-change mitigation. The Regional Development Agencies are taking a lead in inter-sectoral integration.[87] Although multi-functional forestry may result in lower rates of carbon sequestration, it is expected to be more attractive to people, because in the majority of cases, it will provide additional benefits and con-tribute to sustainable development.[28] Existing incentives in forestry need to be scanned to assess their influence on both carbon fluxes and climate-change adaptation, and measures enhancing forest sinks need to be based on principles of sustainable forest management and recognition of the multi-functional role of forests.[5] Afforestation for multiple purposes[37] can be seen as a sustainable way of restoring the productivity of abandoned land, whilst utilisation of woody biomass for energy could create new options for land development.

Carbon sequestration strategy[6] in complement with a replacement of fossil fuels by bio-energy is becoming a priority for forestry when coping with climate change. The rising importance of renewable energy could be explained by the fact that the avoidance of carbon release through the replacement of wood for fossil fuels is repeatable, as long as a continual process of tree growing, har-vesting and regeneration, with the use of harvested woody biomass for energy, is maintained. Therefore, the social benefits of renewable energy projects and those arising from the substitution of energy intensive materials with wood are likely to be higher, in the long run, than the benefits of the strategy of carbon fixation alone. Carbon sinks in wood products is an issue that merits further attention. This policy option provides multiple benefits by enlarging the supply of wood and adding to the total carbon sink. Wood substitutes for various raw materials implicated in GHG releases and can be used in construction, engineering and production of household goods. In all these cases, wood acts as a sink of carbon (beyond its storage in trees) and with the duration of the sink equivalent to the life of the goods.

A systematic promotion of wood products and renewable energy strategies offers opportunities for innovation and development of energy markets with locally and regionally oriented value chains, and thereby provides new employment opportunities and enhances rural development.[7] Policies and measures for CCS should, therefore, be integrated not only within spatial planning, and agricultural, forest and rural policies, but also with the policies and measures for sustainable energy systems.[51] This will enhance energy savings and will assist in coping with local environmental problems associated, for example, with health risks pertaining to changing climate.

Enhancement of carbon sequestration on marginal land, especially in combination with an increased use of bio-energy, will more likely represent a sound opportunity for rural areas and communities when there is synergy between different, and sometimes contrasting, policy areas, and when attention is given to the provision of long-term initiatives and infrastructure (*e.g.* markets) in support of forest-sector-based CCS. Integrated land-use systems, however, are often problematic because they require fiscal and other incentives outside forestry; they have high transaction costs and tend to deny self-interested landowners the right to determine their preferences which are often not economically optimal (*i.e.* amenity-related landowning).[52] So the institution of private ownership and the extent of amenity land purchase must be understood too. The optimum carbon-offset projects will likely be those which link long-term CCS with long-term substitution opportunities,[18] and which successfully develop capabilities to bridge gaps between rural policy priorities, and those of climate change and other issues of sustainable development.[34]

The bridging of gaps between climate policy and practice, and between various sectors of the economy will improve governance capacity, whether it is based on markets (as in the UK) or on the authority of government (as in Ukraine). Moreover, social and economic policies pertaining to CCS are to be worked out and implemented in collaboration between stakeholders. Substantial efforts are to be made to develop environmental awareness of farmers, forest managers and decision-makers concerning various aspects of climate change. This work is ongoing in the analysed European countries, including transition economies. For example, the Centre of Climate Change Initiatives in Ukraine[90] focuses on institutional strengthening, on increased involvement of NGOs and practitioners in climate-policy activities, and on knowledge transfer. However, there is a need for more information campaigns, training facilities and schemes to demonstrate CCS forest possibilities and make them attractive for end-users and, most of all, for forest and land owners, and managers.

In the UK, a high level of participatory democracy is already manifested in the development and implementing of CCS land use and LUC initiatives (*e.g.* through planting trees); in the involvement of the public in environmental and forestry decision-making and policy implementation (*e.g.* through consultations on Forestry Strategies across Britain, and on Forestry Commission

Climate Change Action Plan); and in the extension of information and education pertaining to climate change.[37] It is crucial to involve key forestry stakeholders and local communities in CCS policy-making, decision-taking and policy implementation. It is important to consult people on what climate policy alternatives are desirable for them and to get to know why it is so, as well as to develop our understanding of public perspectives on the role and place of forestry in the mitigation of climate change.

8 Conclusions

The activities enhancing carbon sequestration and sink in forestry overall contribute modestly to the removal of CO_2 from the atmosphere. The carbon sink of forests that could be accounted under the KP is relatively low. For instance, in 2000, carbon uptake by forests in the UK reached almost 3 Mt, whereas it was only 0.3 Mt from planting since 1990, which is eligible under the KP.[57] There are many uncertainties pertaining to terrestrial CCS as a climate-policy measure. Nevertheless, forestry projects have considerable relevance for national carbon budgeting in individual countries where wooded cover has a potential to grow. They are also important in view of reduction of collective carbon emissions at least cost, by trading of carbon offsets across countries. Important carbon sequestration activities involving forestry include afforestation, an increase of area of SRTP and forest regeneration. Even more pertinent are these activities when combined with substitution of wood for fossil fuels and non-wood materials (especially for energy-intensive materials in construction), and these policy options are to be considered further. Tropical deforestation requires in-depth consideration, particularly the development and implementation of economic incentives to cope with this problem more effectively.

In terms of climate-change mitigation priorities for forestry, the approach that includes the principles of sustainable forest management is advocated.[37] According to this approach, carbon sequestration forest policy measures must be socially and environmentally acceptable and based on sustainable-development principles.[6] Enhancement of carbon sequestration on marginal land in combination with increasing use of bio-energy represents a promising opportunity. It is important, however, to develop and provide knowledge of how to translate sustainability requirements for forestry and biomass production into policy guidelines; how to overcome market limitations and institutional obstacles for terrestrial CCS; how to implement flexibility mechanisms for more effective and cost-efficient use of forestry opportunities to mitigate climate change; and how to develop incentive mechanisms and good governance to implement CCS projects on the ground.

In transition countries, large-scale agriculture under the socialist regime supported the conversion of forest or grassland to agricultural land. Currently, the decreasing agriculture[56] will likely cause an increase of abandoned land. In the EU, policy changes under CAP reform might promote forestry development. Further expansion of woodlands in Europe and a rising role of forestry in mitigating climate change may be predicted. The analysis suggests that over and above other climate-policy measures, the enhancement of carbon 'sinks' and 'reservoirs' by forests is meaningful, and their inclusion in climate-policy activities is logical and viable. It is important, however, to develop our understanding of policy options which are most acceptable by people, and to identify places where CCS projects, including those for biomass production, will function most effectively and could be most appropriately integrated into the general context of land use, where multi-functional forestry and contemporary rural change are currently being observed.

In Europe, the regulations and national programmes supporting the conversion of agricultural land back to forest focus largely on remote areas. Hence, CCS land-use and forest-policy measures should be incorporated in regional schemes of integrated sustainable rural development, where socio-economic, environmental and climate-change-related components of LUC are to be considered jointly (where possible). The effective measures should aim at "win-win" situations,[4] which would benefit the development, environment, people and the economy all together, both at national levels and internationally.

However, carbon uptake in trees is temporary, and this needs to be addressed. Moreover, at some point in time there will be no more land available for tree-planting, and carbon sequestration through afforestation will no longer be applicable. Therefore, CCS in forestry alone (without a consideration of further use of wood after timber harvesting) can hardly be considered as a primary solution. Emissions reduction is seen as the priority for climate policy. Nevertheless, climate-change mitigation through forestry is deemed to be amongst the effective and relatively low-cost (complementary) climate-policy options for many countries, especially if fast-growing species are planted on marginal land, or when forestry projects deliver multiple benefits.

Carbon-sequestration rates per hectare of forest across the several European countries observed are relatively high, and marginal lands are available for tree-planting at least until 2020. The carbon sequestration potential of afforestation is deemed to be substantial in some of the countries. The results of the economic assessment of opportunities for CCS through forestry suggest that this policy measure is likely to be competitive with other means of removing carbon from the atmosphere, and that choosing the right locations for forestry development, and the most appropriate tree species and management regimes to be applied, are important factors for saving economic costs.

Although the potential to sequester carbon through afforestation is high in some countries, the potential for carbon-offsets selling is smaller and is limited to carbon balances resulting from the eligible climate-change-mitigation forestry projects which are subject to the cap set on emissions, as well as by the costs of GHG inventory preparation.[1] Moreover, a rising number of carbon offsets available to buy will likely lead to problems pertaining to the environmental effectiveness and economic efficiency of implementation of international agreements, such as the KP, and to contradictions associated with possible attitudes of policy makers to not restrict the emission targets for a polluting industry.[31] Nevertheless, the development of forestry capabilities for CCS, along with many other climate-policy activities, will contribute to the offsetting of CO_2 emissions and could allow European countries to improve their climate-policy performance.

There is a difference, however, between benefits provided by forest development and the benefits that accrue to a forest developer. Establishment of forest plantations to offset CO_2 emissions requires an appropriate institutional setting, incentives and sources of investment. In the EU, intra-European credits from the activities enhancing carbon sequestration will not be included in the trading schemes.[89] Sound incentives for afforestation are therefore required in individual Member States, with proper subsidies to be given to landowners for planting trees. In the transition and developing countries, for instance, the question of whether tree-planting for carbon-uptake should be on the national or EU-based agenda remains unresolved. European investors are clearly showing the interest to invest in JI and CDM projects.[88] The potential gains from international projects are not seen, however, as priorities for land use and climate policies in host countries.[63] Therefore, unless the necessary institutional infrastructure is developed and the barriers for investment are identified and addressed, the Annex B countries cannot expect to benefit widely from crediting JI and CDM systems.

In order to utilise most effectively and efficiently the potential of forests to contribute to the mitigation of climate change, it is imperative to clarify international agreements and rules on forest CCS accounting; to increase its technical effectiveness and accuracy; and to develop further CCS policies, tenure rights (*e.g.* forest carbon ownership), incentives and carbon markets. Among motivating research topics for socio-economists to consider are: who is responsible for carbon sinks after the KP commitment period to 2012; what is the value of (temporary) terrestrial carbon; and how will the value of terrestrial carbon change as markets develop and institutions evolve to handle the numerous uncertainty aspects affecting both its capture and storage.

References

1. UNFCCC, *The Kyoto Protocol to the Convention on Climate Change*, UNEP/IUC, Bonn, 1998.
2. Corporate Europe Observer, *COP-7: Widening the Loopholes*, 10, 2001.
3. European Commission, *Reform of the Common Agricultural Policy*, 1992.
4. European Commission, *Environment 2010: Our Future, Our Choice, The Sixth Environment Action Programme of the European Community*, 2002.
5. ECCP, Second ECCP Progress Report-*Can We Meet our Kyoto Targets? Conclusions and Recommendations Regarding Forests Related Sinks and Climate Change Mitigation*. www.europe.eu.int/comm/environment/climate/forest relatedsinks.htm2002, accessed 14/12/2005.

6. European Commission, *Communication from the Commission to the Council and European Parliament on a Forestry Strategy for European Union*, Brussels, 1998.
7. European Commission, *Energy for the Future: Renewable Sources of Energy, White Paper for Community Strategy and Action Plan*, Brussels, 1997.
8. IBN-DLO, *Resolving Issues on Terrestrial Biospheric Sinks in the Kyoto Protocol, Dutch National Research Programme on Global Air Pollution and Climate Change*, Wageningen, 1999.
9. European Commission, *Mid-Term Review of Common Agricultural Policy*, Brussels, 2002.
10. G. C. van Kooten, W. Thompson and I. Vertinsky, in *Forestry and the Environment: Economic Perspectives*, ed. W. Adamowicz, W. White and W. Phillips, Wallingford, 1993, 227–247.
11. L. Slangen, G. C. van Kooten and P. F. van Rie, *Economics of Timber Plantations on CO_2 Emissions in the Netherlands,* Tijdschrift van de Landbou, 1997, **12**(4), 318–33.
12. CA. Peng and M. Apps, *Global Biogeochem. Cycles*, 1998, **12**, 393–402.
13. D. Adams, R. Alig, B. McCarl, J. Callaway and S. Winnett, *Land Economics*, 1999, **75**(3), 360–374.
14. D. Pettenella, *On the Role of State Forest Institutions in the Sustainable Development of the Sector*, Lviv. http://www.tesaf.unipd.it/people/pettenella/index.html 2000, accessed 5/07/2002.
15. I. Bateman and A. Lovett, *Estimated Value of Carbon Sequestered in Softwood and Hardwood Trees, Timber Products and Forest Soils in Wales.* http://www.uea.aca.uk/∼e089/carbfinl.pdf 2001, accessed 24/03/2009.
16. M. Broadmeadow and R. Matthews, *Forests, Carbon and Climate Change: The UK Contribution,* Forestry Commission, Edinburgh, 2003, 48.
17. R. Matthews and K. Robertson, *Carbon Sequestration in the Global Forest Sector*, ed. T. Karjalainen and M. J. Apps, IUFRO Task Forces on Environmental Change State of the Art Report, 2003.
18. M. G. R. Cannell, *Biomass Bioenergy*, 2003, **24**, 97–116.
19. Royal Society, *Economic Instruments for the Reduction of Carbon Dioxide Emissions.* http://royalsociety.org/displaypagedoca.asp?id = 122002, accessed 4/07/2006.
20. The Stern Review, *The Economics of Climate Change.* http://www.hm-treasury.gov.uk/stern_ review_report.htm 2006, accessed 7/12/2006.
21. The Carbon Trust, *Biomass Sector Review.* www.thecarbontrust.co.uk/Carbontrust/about/publication 2005, accessed 23/02/2007.
22. AEA Technology, *Renewable Heat and Heat from Combined Heat and Power Plants – Study Analysis*, A Report to DTI and DEFRA, 2005.
23. E. Booth, J. Booth, P. Cook, B. Ferguson and K. Walker, *Economic Evaluation of Biodiesel Production from Oilseed Rape Grown in North and East Scotland,* Scottish Agricultural College, Consultancy Division, 2005.

24. W. Towers, R. Birnie, E. Booth, K. Walker and P. Howes, *Energy from Crops, Timber and Agriculture Residue*, SEERAD Report, 2004.
25. ENDS, *Climate Review 2006/07*, Haymarket, London, 2007.
26. D. Galbraith, P. Smith, N. Mortimer, R. Stewart, M. Hobson, G. McPherson, R. Matthews, P. Mitchell, M. Nijnik, J. Norris, U. Skiba, J. Smith and W. Towers, *Review of Greenhouse Gas Life Cycle Emissions, Air Pollution Impacts and Economics of Biomass Production and Consumption in Scotland*, SEERAD, 2006, 265.
27. E. Krcmar, I. Vertinsky and G. C. van Kooten, *Int. Trans. Operat. Res.*, 2003, **10**(5), 483–498.
28. M. Nijnik and L. Bizikova, *Forest Policy Economics*, 2008, **10**, 257–269.
29. Sustainable Development Commission for Scotland, *Wood Fuel for Warmth*. http://www.sd-commission.org.uk/publications/downloads/050626-SDC-Wood-Fuel-for-Warmth.pdf 2005, accessed 19/03/2006.
30. Fraser of Allander Institute, *The Economic Impact of Wood Heat in Scotland*, A Report to Scottish Enterprise Forest Industries Cluster. www.forestryscotland.com/download.asp?file = Fraser%20of%20Allander%20Woodheat 2006, accessed 29/07/2008.
31. G. C. van Kooten, *Climate Change Economics*, Edward Elgar, Cheltenham, 2004.
32. IPCC, *Climate Change 2007 Synthesis Report, An Assessment of the Intergovernmental Panel on Climate Change*, presented at IPCC Plenary XXVII, Valencia, Spain,12–17 November 2007. http://www.ipcca.ch/pdf/assessment-report/ar4/syr/ar4_syr.pdf 2007, accessed 22/03/2009.
33. DEFRA, *The UK's 3rd National Communication under the UN Framework on Climate Change*, London, 2001.
34. DEFRA, *Climate Change Programme*, London. http://www.defra.gov.uk/environment/climatechange/uk/progress/index.htm 2006, accessed 4/03/2009.
35. DEFRA, *Progress Towards National and International Targets, Statistical Release: UK Climate Change SD Indicators and GHG Emissions Final Figures*, DEFRA, London. http://www.defra.gov.uk/environment/climatechange/uk/progress/index.htm 2007a, accessed 20/03/2009.
36. Forestry Commission, *Forestry Statistic*, Forestry Commission, Edinburgh. www.forestry.gov.uk/sfs 2001, accessed 12/03/2009.
37. Forestry Commission, *Climate Change Action Plan 2009–2011*, Forestry Commission, Edinburgh. www.forestry.gov.uk/sfs 2008, accessed 9/03/2009.
38. R. Crabtree, in *Climate Change Mitigation and European Land-Use Policies*, ed. W. N. Adger, D. Pettenella and M. Whitby, CAB International, Wallingford, 1997.
39. R. Milne, *Carbon Sequestration in Vegetation and Soils*, DETR Contract, April 2000.
40. P. H. Freer-Smith, M. S. J. Broadmeadow and J. M. Lynch, *Forestry and Climate Change*, CABI, Wallingford, 2007, 253.
41. IPCC, *Good Practice Guidance for Land Use, Land-Use Change and Forestry*, Special Report of the Intergovernmental Panel on Climate Change, Cambridge University Press, UK, 2000, 599.

42. ECCM, *Impacts of Climate Change on Scotland, Adaptation to Climate Change, Mitigation of Scotland's Greenhouse Gas Emissions*, Submission ECCM to Environment and Rural Development Committee, 2004.
43. P. Lakida, S. Nilsson and A. Shvidenko, *Estimation of Forest Phytomass for Selected Countries of the Former European USSR*, IIASA, Austria, 1995.
44. Royal Society, *The Role of Land Carbon Sink in Mitigating Climate Change*, Policy Document 10/01, July 2001. http:/www.royalsoca.aca.uk 2001, accessed 24/03/2009.
45. J. Randerson, *Tree Planting Warning over Global Warming*, New Scientist. http://www.newscientist.com/channel/earth/mg18725116.2001, accessed 18/01/2004.
46. M. Nijnik, *To Sustainability in Forestry*, Wageningen University, The Netherlands, 2002, 156.
47. G. Nabuurs, A. Dolman and E. Verkaik, *Resolving Issues on Terrestrial Biospheric Sinks in the Kyoto Protocol*, Dutch National Research Programme on Global Air Pollution and Climate Change, 1999.
48. G. C. van Kooten and E. Bulte, *The Economics of Nature: Managing Biological Assets*, Blackwell, Oxford UK, 2000.
49. A. Pussinen, T. Karjalajnen, S. Kellomaki and R. Makipaa, *Biomass Bioenergy*, 1997, **13**(6), 377–387.
50. M. Nijnik, *Climate Policy*, 2005, **4**, 319–336.
51. M. Nijnik and L. Bizikova, in *Sustainable Forestry: From Monitoring and Modelling to Knowledge Management and Policy Science*, ed. K. M. Reynolds, CAB International, 2006, pp. 56–66.
52. M. Nijnik, G. Pajot, A. Moffat and B. Slee, *Exploring Opportunities of British Forests to Mitigate Climate Change*, European Association of Environmental and Resource Economics Proceedings, Amsterdam, 2009, 1–19.
53. DEFRA, *Demonstrable Progress. The United Kingdom Report on Demonstrable Progress under the Kyoto Protocol. Tomorrow's Climate Today's Challenge*. http.www.gov.uk/environment 2007b, accessed 19/03/2009.
54. DEFRA, *Climate Change Act*. http://www.defra.gov.uk/environment/ climatechangeuk/legislation/provisions.htm 2008, accessed 23/03/2009.
55. DEFRA, *New Bill and Strategy lay Foundations for Tackling Climate Change*, News release. http://www.defra.gov.uk/news/2007/070313a.htm 2007c, accessed 14/01/2009.
56. FAO, *FORSTAT database*. www.fao.org/forestry 2006, accessed 3/03/ 2009.
57. DEFRA, *Climate Change: Draft UK Programme*, Sections 1–4, Annex G: Carbon Sequestration, London, 2002.
58. S. Chapman, W. Towers, B. Williams, M. Coull and E. Paterson, *Review of the Contribution to Climate Change of Organic Soils under Different Land Uses*, Scottish Executive Central Research Unit, 2001, 62.
59. P. Smith, J. U. Smith, M. Wattenbach, J. Meyer, M. Lindner, S. Zaehle, R. Hiederer, R. Jones, L. Montanarella, M. Rounsevell, I. Reginster and S. Kankaanpää, *Can. J. Soil Sci.*, 2006, **86**, 159–169.

60. R. Tipper, R. Carr, A. Rhodes, A. Inkinen, S. Mee and G. Davis, *The UK's Forest: A Neglected Resource for the Low Carbon Economy*, Scottish Forestry, 2004, 58.

61. A. M. Thomson and M. van Oijen, *Inventory and Projections of UK Emissions by Sources and Removals by Sinks due to Land Use, Land Use Change and Forestry*, DEFRA Annual Report, 2007.

62. E. Petkova and G. Faraday, *Good Practices in Policies and Measures for Climate Change Mitigation. A Central and Eastern European Perspective*, Regional Environmental Center for CEE and World Resource Institute, Szentendre, 2001.

63. S. Fankhauser and L. Lavric, *Climate Policy*, 2003, **3**, 417–434.

64. M. Nijnik and C. A. van Kooten, *Forestry Policy Economics*, 2000, **1**(2), 139–153.

65. MoE, *Third Report on Climate Change*, Ministry of Environment, Bratislava, 2001.

66. GRA, *Green Report*, Agriculture, Ministry of Agriculture, Bratislava, 2004.

67. L. Carlsson, *Towards a Sustainable Russian Forest Sector*, FR, IIASA, Laxenburg, Austria, Panel 3-3, June 11, 1999.

68. Ministry of Housing, *Spatial Planning and the Environment*, 3rd Netherland's National Communication On Climate Change Policy, Directorate Climate Change and Industry, Den Haag, 2001.

69. LNV, *Nature for People, People for Nature; Memorandum Nature. Forests and Landscape in the 21st Century*, Ministry of Agriculture, Nature and Fisheries of the Netherlands, The Hague, 2000.

70. G. Daffy, *Pressure Build on Amazon Jungle*, BBC News, World News, 14/01/2008.

71. C. S. Papadopol, *Climate Change Mitigation. Are there any Forestry Options?* http://www.eco-web.com/editorial/05934-03.html 2000, accessed 12/09/2001.

72. UNEP, *Tropical Forest Update: Climate Change in Tropical Forest, Climate and Atmosphere*, report UNEP. www.iisd.ca/publications_resources/climate_atm.htm 2008, accessed 20/03/2009.

73. K. Hamilton, R. Bayon, G. Turner and D. Higgins, *State of the Voluntary Carbon Market 2007*, Washington D. C. and London. http://ecosystemmarketplace.com/documents/acrobat/2007, accessed 3/04/2008.

74. D. Binkley, M. G. Ryan, J. L. Stape, H. Barnard and J. Fownes, *Ecosystem*, 2002, **5**, 58–67.

75. G. C. van Kooten, A. J. Eagle, J. Manley and T. Smolak, *Environ. Sci. Policy*, 2004, **7**, 239–251.

76. G. C. van Kooten and A. J. Eagle, in *Sustainability, Institutions and Natural Resources: Institutions for Sustainable Forest Management*, ed. S. Kant and R.A. Berry, Springer, 2005, 233–255.

77. R. Savins and K. Richard, *The Cost of U.S. Forest-based Carbon Sequestration*, Arlington, VA, Pew Center on Global Climate Change, 2005.

78. D. Moran, M. MacLeod, E. Wall, V. Eory, G. Pajot, R. Matthews, A. McVittie, A. Barnes, B. Rees, A. Moxey and A. Williams, *2008 Final Report to the Committee on Climate Change*, 2008, **152**.

79. J. Caspersen, S. W. Pacala, J. CA. Jenkins, G. CA. Hurtt, P. R. Moorcroft and R. A. Birdsey, *Science*, 2000, **290**(5494), 1148–1151.

80. Global Atmosphere Division, *Paper to the British Government Panel on Sustainable Development Sequestration of Carbon Dioxide*, Sequestration of Carbon Dioxide, London. http://www.sd-commission.org.uk/panel-sd/position/co2/index.htm 1999, accessed 20/02/2009.

81. L. Bizikova, *Carbon Sequestration Potential on Agricultural Land-Use Management in Slovakia,* Institute for Forecasting, Slovak Academy of Sciences, Bratislava, 2004.

82. K. M. Chomitz, *Evaluating Carbon Offsets from Forestry and Energy Projects: How do they Compare?* Development Research Group, The World Bank, 2357. www.worldbank.org/research 2000, accessed 14/03/2009.

83. N. Hanley and C. A. Spash, *Cost-Benefit Analysis and the Environment,* Edward Elgar Publishing, UK–USA, 1998.

84. S. Subak, *Climate Policy*, 2003, **3**, 107–122.

85. G. Marland, K. Fruit and R. Sedjo, *Environ. Sci. Policy*, 2001, **4**(6), 259–268.

86. UNFCCC, *Japan Marks the Protocol's Entry into Force on 16 February 2004*, Bonn. http://unfccca.int/2860.php, 2006, accessed 1/02/2007.

87. Regional Development Agencies, *Tackling Climate Change in the Regions.* http://www.englandsrdas.com/files/2008/2/28/policy_rda_climate_change.pdf UTH 2007, accessed 16/12/2007.

88. R. Schwarze, *Ecolog. Economics*, 2000, **32**, 255–267.

89. European Commission, *Proposal for a Directive of the European Parliament amending the Directive Establishing a Scheme for GHG Emission Allowance Trading within the Community, in the respect of the Kyoto Protocol's Project Mechanisms*, Brussels, 2003.

90. Ukraine's Centre of Climate Change Initiatives, Kiev. http://www.climate.org.ua/ 2007, accessed 1/02/2008.

Carbon Uptake, Transport and Storage by Oceans and the Consequences of Change

C. TURLEY,* J. BLACKFORD, N. HARDMAN-MOUNTFORD, E. LITT, C. LLEWELLYN, D. LOWE, P. MILLER, P. NIGHTINGALE, A. REES, T. SMYTH, G. TILSTONE AND S. WIDDICOMBE.

1 Summary

This chapter explores the greatest biospheric reservoir of carbon on planet Earth – the oceans. When in balance, there is a large flux of CO_2 between the oceans and the atmosphere of almost $90 \, Gt \, C \, yr^{-1}$ due to a combination of primary production and particle sinking (the biological pump) and ocean circulation and mixing (the solubility pump). Climate change will tend to suppress ocean-carbon uptake through reductions in CO_2 solubility, suppression of vertical mixing by thermal stratification and decreases in surface salinity. It is envisaged that climate-driven changes in any of these physical mechanisms will have a subsequent impact on the phytoplankton and their ability to draw carbon from the atmosphere and into the ocean. This will increase the fraction of anthropogenic CO_2 emissions that remain in the atmosphere this century and produce a positive feedback to climate change.

Increased burning of fossil fuel, cement manufacturing and land-use change since the industrial revolution has increased atmospheric CO_2 and caused an imbalance in the exchange of CO_2 between atmosphere and ocean, resulting in more ocean uptake. Oceans have taken up around 25% of the anthropogenic CO_2 produced in the last 200 years and through this have buffered climate change. However, this has already lead to a profound change in ocean carbonate chemistry (a 30% increase in hydrogen ions), coined "ocean acidification", and this change will increase in magnitude in the future as anthropogenic

---

* Corresponding author

Issues in Environmental Science and Technology, 29
Carbon Capture: Sequestration and Storage
Edited by R.E. Hester and R.M. Harrison
© Royal Society of Chemistry 2010
Published by the Royal Society of Chemistry, www.rsc.org

CO_2 emissions increase and more CO_2 dissolves in the surface of oceans. The atmospheric CO_2 increase alone will lead to continued uptake by the ocean, although the efficiency of this uptake will decrease as the carbonate buffering mechanism in the ocean weakens. Research so far indicates that these changes to ocean pH, and bicarbonate and carbonate ion saturation, will have a profound impact on ocean biology, both in pelagic (free-floating) and benthic (sea-floor) realms.

Ocean productivity is far from uniform and may cause impacts when vast numbers of phytoplankton cells are concentrated in high-biomass, sometimes harmful or toxic, algal blooms. The most significant harm caused by high-biomass blooms is oxygen depletion, usually caused when dead phytoplankton cells sink down the water column and are decomposed by bacteria, using oxygen to do so. The degree of depletion is determined by the quantity of organic matter accumulated, the stability of the water column, and the bathymetry (depth of the water column), the first two being sensitive to climate change.

2 Carbon Uptake by Oceans

2.1 Air–Sea Exchange of Carbon Dioxide and the Chemistry of Carbon in Seawater

2.1.1 Global Air–Sea Fluxes of Carbon Dioxide

The large natural annual flux of CO_2 between the ocean and the atmosphere of almost $90 \, Gt \, C \, yr^{-1}$ was, pre-1800, believed to be almost in balance.[1] The ocean is responsible for 58% of global carbon fixation, and land 42%.[2] This huge influx and efflux is due to a combination of marine productivity and particle sinking (the biological pump) and ocean circulation and mixing (the solubility pump) (see Section 3). Phytoplankton growth consumes dissolved inorganic carbon (DIC) in the surface seawater, causing an under-saturation of dissolved CO_2 and uptake from the atmosphere. The re-equilibration rate for CO_2 is slow (typically taking several months) due to the dissociation of CO_2 in seawater. Ocean circulation also results in air–sea exchange of CO_2 as the solubility of CO_2 is temperature-dependent. Warming decreases the solubility of CO_2 and promotes a net transfer of CO_2 to the atmosphere, whereas cooling results in a flux from the atmosphere to the ocean. *Anthropogenic* CO_2 modifies the fluxes due to the solubility pump as CO_2 availability does not normally limit biological productivity in the world's oceans. Hence there is little potential for increased biological production sequestering anthropogenic CO_2. However, the observation that the net oceanic uptake of *anthropogenic* CO_2 is only about 2% of the total CO_2 cycled annually across the air–sea interface ought to be of major concern. The significant perturbations arising from this small change in flux imply that the system is extremely sensitive. Any resulting changes in the biogeochemistry of the mixed layer could have a major impact on the magnitude (or even sign) of the total CO_2 flux and hence on the Earth's climate.[3]

There has been an increase in atmospheric carbon dioxide from 280 ppm in AD1800 to 380 ppm at the present day. This increase is due to a supply of anthropogenic CO_2 to the atmosphere which is currently estimated at 7 Gt C yr^{-1} (ref. 4). The observed annual increase in atmospheric CO_2 represents 3.2 Gt C yr^{-1}, the balance being removed from the atmosphere and taken up by the oceans and land. There is now generally good agreement that the ocean absorbs 1.7 ± 0.5 Gt C yr^{-1} (ref. 4). The rate-limiting step in the long-term oceanic uptake of anthropogenic CO_2 is not air–sea gas exchange, but the mixing of the surface waters with the deep ocean.[1] The ocean can theoretically absorb 70–80% of the projected production of anthropogenic CO_2. So, whilst there is in principle sufficient capacity in the oceans to uptake 70–80% of anthropogenic CO_2, achieving this would take hundreds of years longer than the production of anthropogenic CO_2; it would take many centuries to do so.[5]

2.1.2 Seawater Carbon Chemistry

The chemistry of carbon dioxide in seawater has been the subject of considerable research and has been summarised by Zeebe and Wolf-Gladrow.[6] Dissolved inorganic carbon can be present in any of four forms: dissolved carbon dioxide (CO_2), carbonic acid (H_2CO_3), bicarbonate ions (HCO_3^-) and carbonate ions (CO_3^{2-}). Addition of CO_2 to seawater, by air–sea gas-exchange due to increasing CO_2 in the atmosphere, leads initially to an increase in dissolved CO_2, see Equation (1). This dissolved carbon dioxide reacts with seawater to form carbonic acid, see Equation (2). Carbonic acid is not particularly stable in seawater and rapidly dissociates to form bicarbonate ions, see Equation (3), which can themselves further dissociate to form carbonate ions, see Equation (4). At a typical seawater pH of 8.1 and salinity of 35 the dominant DIC species is HCO_3^- with only 1% in the form of dissolved CO_2. It is the relative proportions of the DIC species that control the pH (that is, the H^+ ion concentration) of seawater on short-to-medium timescales.

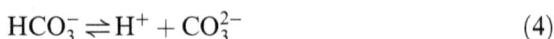

$$CO_{2(atmos)} \rightleftharpoons CO_{2(aq)} \tag{1}$$

$$CO_2 + H_2O \rightleftharpoons H_2CO_3 \tag{2}$$

$$H_2CO_3 \rightleftharpoons H^+ + HCO_3^- \tag{3}$$

$$HCO_3^- \rightleftharpoons H^+ + CO_3^{2-} \tag{4}$$

It is also important to consider the interaction of calcium carbonate with the inorganic carbon system. Calcium carbonate ($CaCO_3$) is usually found in the environment either as calcite or, less commonly, aragonite. Calcium carbonate dissolves in seawater forming carbonate ions (CO_3^{2-}) which react with carbon dioxide as follows:

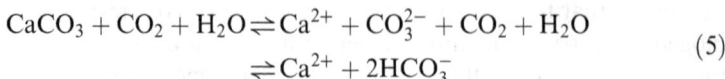

$$CaCO_3 + CO_2 + H_2O \rightleftharpoons Ca^{2+} + CO_3^{2-} + CO_2 + H_2O$$
$$\rightleftharpoons Ca^{2+} + 2HCO_3^- \tag{5}$$

This reaction represents a useful summary of what happens when *anthropogenic* carbon dioxide dissolves in seawater. The net effect is removal of carbonate ions and production of bicarbonate ions and hence a lowering in pH due to the reaction shown in Equation (4). This in turn will encourage the dissolution of more calcium carbonate. Indeed, the long-term sink for anthropogenic CO_2 is dilution in the oceans and reaction with carbonate sediments. As can clearly be seen above, formation of calcite [the reverse of Equation (5)] actually produces CO_2. Seawater at current pH levels is highly buffered with respect to carbon dioxide and has a great capacity to absorb carbon dioxide, as most of the CO_2 added will rapidly be converted to bicarbonate ions. It can be shown that if the atmospheric CO_2 levels doubled, dissolved CO_2 would only rise by 10%, with most of the remaining 90% being converted to bicarbonate ions. However, if bicarbonate ions increase, then the equilibrium of Equation (4) will be forced forwards and hence the pH of the seawater will be reduced. This is of great importance both for seawater chemistry and for the buffering capacity of seawater as it reduces the ability of seawater to buffer further CO_2 increases,[6] *i.e.* as the partial pressure of carbon dioxide increases, the buffering capacity of seawater decreases.

2.2 Carbon Fixation and Controlling Factors

2.2.1 Light and Photosynthesis

Photosynthesis is the light-driven biochemical transformation of water and CO_2 into O_2 and carbohydrates, the cellular building blocks. In the marine environment, photosynthesis is a complex system controlled by a number of environmental factors, including underwater irradiance, nutrient content, temperature, phytoplankton type and ecosystem trophic status.[7,8] The interplay between these factors, and the air–sea flux of CO_2, regulates the capacity of the oceans to take up carbon. Chlorophyll-a (Chla) and irradiance are key factors in marine primary production, but their effectiveness can be subtly altered by the physiological state of phytoplankton which, in turn, is dependent on temperature, nutrients and the composition of the phytoplankton community present.

Light in the marine environment is determined by two factors: first, by the incident light flux at the ocean surface (apparent optical properties) and second, by the optical properties (known as inherent optical properties) of dissolved and suspended (particulate) substances, of the seawater itself and air bubbles.[9] Incident light at the sea surface is affected by latitude, sky conditions, sea state, time of year and day. Inherent optical properties of oceanic waters are generally dominated by phytoplankton light absorption[10] which can be modified by variations in cell size, species type and pigment type and concentration within the cell.[11] In oceanic waters there is an almost constant background of light-absorbing (*i.e.* coloured) dissolved organic material (CDOM), which is coupled with variations in phytoplankton biomass and is modified by microbial and photochemical degradation.[12,13] In coastal regions CDOM and suspended material, originating from riverine run-off and re-suspension of bottom sediment, are highly variable. They often have a larger impact on light penetration in the water column than phytoplankton[14] and may therefore limit phytoplankton photosynthesis and primary production.[15]

Phytoplankton have complex light-harvesting systems, with pigment arrays arranged in asymmetrical membranes that are orientated to maximise light capture by the cell.[16] Each light-absorbing pigment captures light over a specific wavelength range and channels it through the photosystems. The principal absorbing phytoplankton pigment, which captures light and channels it through the photosystems, is Chla. It has absorption peaks in the blue (440 nm) and red (675 nm) regions of the spectrum. Other key marker pigments and accessory pigments, such as the carotenoids which are coupled with the Chla unit, also absorb blue, green and yellow light and channel it through the photosystems. Little work has been conducted on the energy efficiency conversion of sunlight into carbon by different phytoplankton pigments (see Section 2.2.2). Some of the carotenoids have a photo-protective role;[17] contrasting results on the effects of non-photosynthetic carotenoids on primary production and the capacity of the oceans to absorb CO_2 have been reported.[18,19]

Radiochemically-labelled carbon in the form of ^{14}C sodium bicarbonate ($NaH^{14}CO_3$) has traditionally been used as a tracer for carbon uptake by phytoplankton. Steeman Nielsen[20] first used the technique to measure carbon fixation in the oceans and it has become the universal benchmark method of estimating water-column integrated primary production (PP). The technique was refined to measure carbon uptake as a functional response to light,[21] to indicate the photosynthetic response of phytoplankton. PP was firstly calculated using broad-band photosynthetic parameters and light.[22] These models assume that all light wavelengths in the photo-synthetically active radiation (PAR) part of the spectrum (400 to 700 nm) are absorbed by phytoplankton with the same efficiency. Absorption of light by phytoplankton, however, is strongly wavelength-dependent.[23,24] The use of spectral models greatly improved the calculation of primary production.[25,26] The extent to which temporal variations in hydrography affect broad-band and spectral models of PP, particularly in relation to light limitation, is poorly understood. Spectrally-dependent phytoplankton light-absorption models estimate the maximum quantum yield of photosynthesis from photosynthetic useable radiation to be between 25 to 50% higher than from models based on PAR.[25,27–29]

Ship-borne ^{14}C uptake measurements are limited, both spatially and temporally, and therefore provide only a snap-shot of the total ecosystem carbon uptake. Modelling of carbon fixation in specific biogeochemical provinces and improving estimates of PP from remotely-sensed data requires an understanding of the variations in photophysiology in relation to changes in environmental conditions, and an assessment of the variable(s) that influence PP.[31,32] Using satellite ocean-colour chlorophyll-a and radiance data, it is now possible to construct accurate maps of PP and provide spatially and temporally resolved global and regional estimates of PP.[33–36] For example, recent regional trends in PP in the Atlantic, derived from satellite data from 1998 to 2005, show

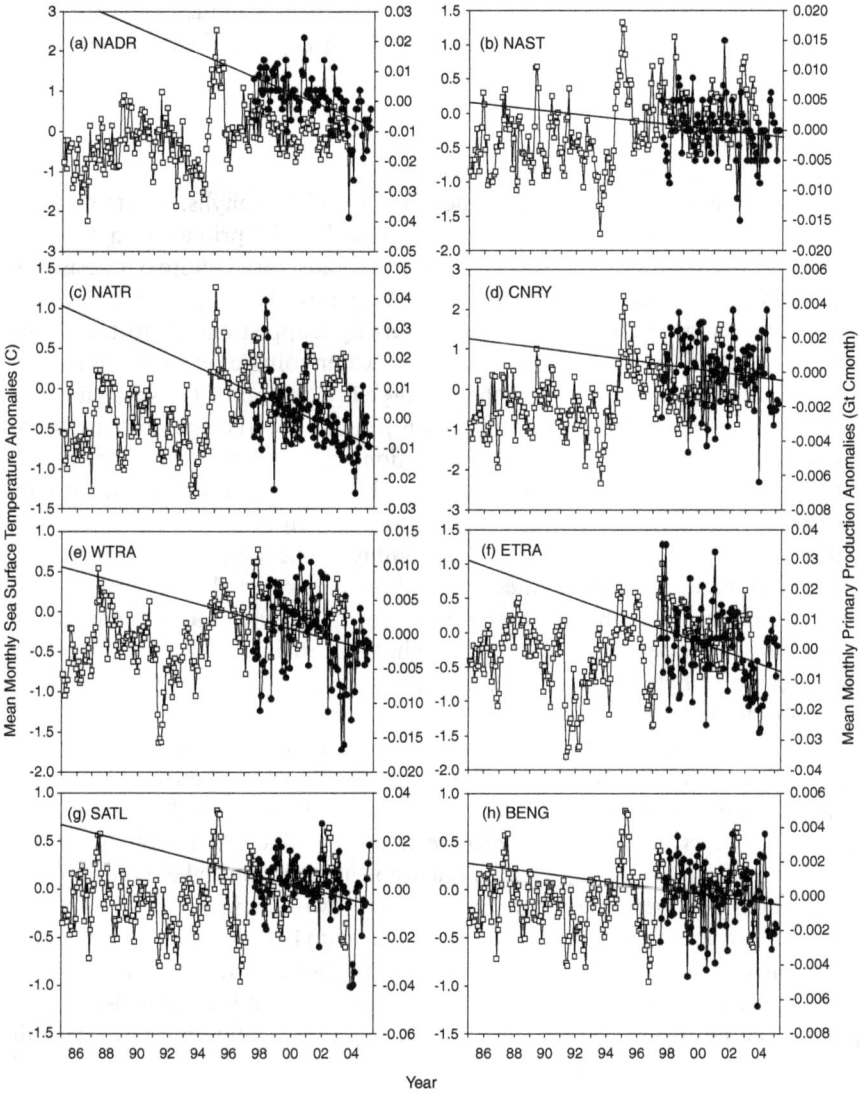

Figure 1 Mean monthly anomaly in primary production using SeaWiFS data from 1998 to 2005, and in AVHRR sea-surface temperature from 1985 to 2005 in nine Atlantic provinces. PP anomaly (closed circles) and linear regression (solid line), SST anomaly (open squares) and linear regression (dashed line). (a) North Atlantic Drift: NADR, (b) North Atlantic Subtropical Gyre Province: NAST, (c) North Atlantic Tropical Gyre: NATR, (d) Canary Current Coastal upwelling: CNRY, (e) Western Tropical Atlantic: WTRA, (f) Eastern Tropical Atlantic: ETRA, (g) South Atlantic Subtropical Gyre: SATL, and (h) Benguela Current Coastal: BENG. (Source: Tilstone *et al.*)[30]

that with increasing sea-surface temperature there is a reduction in PP in the northern hemisphere (see Figure 1), suggesting a negative feedback on carbon capture in relation to global warming.[30]

2.2.2 Adaptation using Different Pigments

Central to the process of photosynthesis are the chlorophylls, carotenoids and phycobiliprotein pigments. Chlorophylls absorb light primarily in the blue-violet and red regions of the spectrum. Carotenoids make photosynthesis more efficient over a broader range of wavelengths by absorbing light primarily in the blue-green region of the spectrum, transferring trapped energy to the chlorophylls. In addition, cyanobacteria contain water-soluble accessory phycobiliprotein pigments that absorb in the green-yellow region of the spectrum.

The ability of photosynthetic plankton to adapt to the changing light conditions (intensity and chromatic adaptation) through the water column by altering pigment type and composition, enables maximum use of the available light energy. There is building evidence of the importance of pigments and photosynthetic characteristics in determining niche opportunity, promoting biodiversity and ultimately affecting carbon uptake in the oceans. The following paragraphs highlight the many recent discoveries about pigments and the adaptive consequences in a range of photoautotrophic organisms, including the prochlorophytes, cyanobacteria, anoxygenic phototrophic bacteria and proteobacteria.

Prochlorococcus, which dominates phytoplankton in tropical and sub-tropical oceans, is responsible for a significant fraction of global photosynthesis and is unique in containing divinyl (rather than monovinyl) chlorophyll-a and -b. These unique pigments enable *Prochlorococcus* to absorb blue-green light efficiently at low-light intensies characteristic of the deep euphotic zone where they are most abundant.[37] Furthermore, there are at least two distinct ecotypes of *Prochlorococcus*: low-light adapted, with a high divinyl chl-b to divinyl-chl-a ratio predominating in the deeper portion of the euphotic zone where nutrients are abundant, and high-light adapted, with a low divinyl chl-b to divinyl-chl-a ratio predominating in the surface where nutrients are typically limiting.[38] This distribution of multiple *Prochlorococcus* ecotypes in the same water column results in a greater integrated production than could be achieved by a single ecotype and permits survival of the population as a whole over a broader range of environmental conditions than would be possible for a homogenous population.[39]

Chromatic adaptation has been observed in two closely related picocyanobacteria strains of *Synechococcus* where striking differences in pigment composition was found in competition experiments, with one strain a blue-green colour due to phycocyanin which absorbs photons in the red-orange part of the spectrum (620–630 nm), and the other strain red due to the pigment phycoerythrin which absorbs photons in the green-yellow part of the spectrum (560–570 nm).[40] This chromatic adaptation of pigment divergence allows for efficient utilisation of light energy, with partitioning of the light spectrum favouring coexistence of different strains of the same species.[40]

Recent evidence indicates that chlorophyll-d (chl-d) is globally distributed and that its contribution to photosynthesis should be properly evaluated in estimating global primary production.[41] Chlorophyll-d absorbs at longer wavelengths (up to 30 nm red-shifted) than chl-a and is primarily known to be associated with the cyanobacterium *Acaryochloris marina* which thrives in environments with low visible-light intensity but high near-infrared intensity, where no other photosynthetic organisms absorb strongly.[42] Recent evidence indicates that chl-d has a widespread distribution in oceanic environments covering a wide range of temperatures and salinities that receive near-infrared light, thus providing new insight into utilisation of the near-red infrared light energy.[41]

Aerobic anoxygenic phototrophic bacteria (AAPs), traditionally associated with being heterotrophic, are also capable of photosynthetic CO_2 fixation under DOM deficient conditions. AAPs have been found widely with uniform distribution in the euphotic zone in the oligotrophic ocean[43] where they can comprise at least 11% of the total microbial community,[44] with vertical distribution closely correlating with chl-a. AAPs contain high levels of carotenoids relative to bacteriochlorophyll-a (ratio 8 : 1) and are able to utilise carotenoids as an efficient auxiliary pigment.[45]

The diversity of pigments associated with photosynthesis in the marine environment has been widened with the recent discovery of rhodopsin (absorbance λ_{max} 520 nm) in bacteria of the phyla Proteobacteria and Bacteroidetes.[46,47] These bacteria are widely distributed in oceanic surface waters and add to the growing recognition of the diverse range of organisms and pigments involved in photoautotrophy.

The ability of photosynthetic plankton to maximise use of incident energy according to intensity and wavelength distribution is directly related to the uptake of carbon and global primary productivity. Recent discoveries about photosynthesis and pigments and on the contribution that new types of photoautotrophy have on carbon uptake in the oceans have yet to be evaluated. However, despite this, the oceans are estimated to contribute to over half of global primary productivity,[2,48] with the vast majority carried out by this diverse range of microscopic algae and bacteria. Unlike on land, these phototrophs have a turnover of hours to days.

2.2.3 Nutrient Availability

In order for phytoplankton to be able to fix CO_2 into organic material during photosynthesis, plants have an absolute requirement for other nutrient elements in order to support the energetic and physiological processes which control the growth and activity of the organism. In the marine realm, the rate and mechanism of the supply of nutrients is inherently variable, largely as a result of geographic features, and it is this supply which is often the factor which controls primary productivity. There are several autochthonous mechanisms by which nutrients can be supplied which are cyclical in nature and occur over different scales of time (hours to millenia) and space (μm to km).

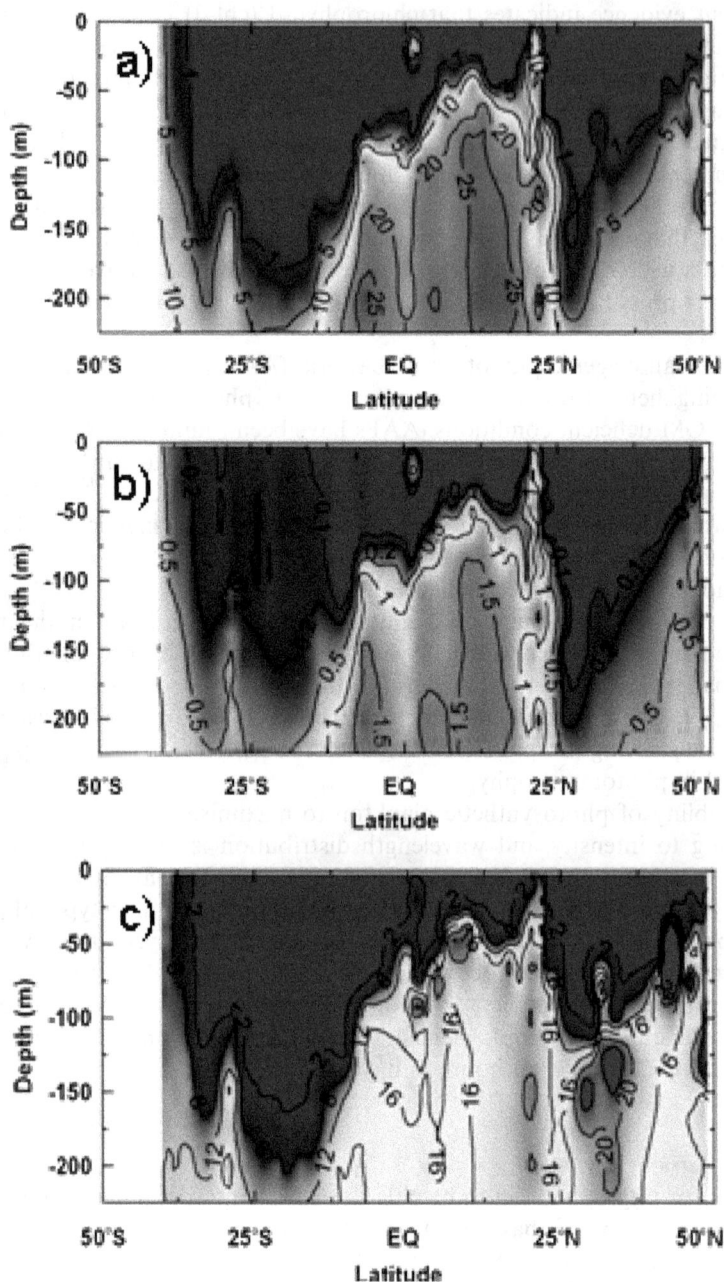

Figure 2   (a) Nitrate, (b) phosphate and (c) their N : P ratio contours along a transect through the Atlantic Ocean. Data were collected during Atlantic Meridional Transect cruise: AMT15 (ref. 51) between the UK and South Africa in October 2004, and highlight the relative differences in nutrient conditions between the North and South Atlantic gyres.[52]

Within the surface layers of the oceans – the euphotic zone – the rapid uptake and regeneration of nutrients contributes to the maintenance of primary productivity. Meanwhile the organic material which sinks out of these waters may undergo re-mineralisation within the deep ocean or benthic sediments, prior to being re-introduced as inorganic nutrients to the euphotic zone following vertical mixing or diffusion across the thermocline. The rate at which these processes occur is largely under the control of physical conditions which vary from global phenomena, such as the ocean conveyor belt,[49] which drives the ocean circulation, to more local conditions which may be influenced by topographic changes (*e.g.* islands) or wind-induced upwelling (*e.g.* eastern boundary currents), both of which are characterised by high nutrient conditions and enhanced productivity. In contrast, the conditions found in the mid-oceanic gyres are characterised by a physically stratified water column with low exchange of material, depleted nutrient concentrations and low primary production. In coastal regions and on the continental shelves, riverine inputs and atmospheric deposition play major roles in the supply of nutrients and these inputs have been increased by human activities.[50]

Nutrients are often described as being either macro- or micro-nutrients and are defined according to the concentrations at which they occur and are utilised. Nitrogen, phosphorus (see Figure 2) and silicate are generally considered to be the major macro-nutrients, while micro-nutrients are trace elements which include iron, zinc and cobalt. Oceanic phytoplankton which live in the euphotic zone have been found to contain nutrient concentrations at largely the same $C : N : P$ ratio as found in the deep-oceans, *i.e.* $106 : 16 : 1$ (ref. 53), and it is considered that the supply of N and P at these ratios is often the controlling or limiting factor for phytoplankton carbon fixation. Silicate is a key nutrient for the important phytoplankton group, the diatoms. Without silicate, other non-silicate requiring phytoplankton will grow if there is sufficient N and P. In the early 1990s, John Martin proposed his hypothesis that in vast tracts of the ocean it was actually the supply of iron which is the limiting nutrient.[54] These regions are commonly referred to as HNLC (High Nitrogen Low Chlorophyll) regions, describing their characteristic mis-match of replete macro-nutrient concentrations and low phytoplankton biomass. Since that time there has been significant research activity to assess which, if any, of the nutrient species provide the ultimate control over primary productivity. This is far from being a simple exercise; while we know that phytoplankton require nitrogen for a host of structural, genetic and metabolic requirements, phosphorus is needed for nucleic acids and for the intracellular transfer of energy and the production of phospholipid membranes, and iron, which is essential to several enzyme structures which are fundamental to phytoplankton activity. Each nutrient is required to varying degrees by different algal groups, and is supplied by different routes and in different forms (*e.g.* nitrogen may be utilised as NO_3^-, NH_4^+, urea, N_2, *etc.*).

The supply of nutrients due to anthropogenic loading is becoming increasingly more important and predictions of change to the oceanic nutrient regime are wide-ranging, but in all cases are likely to result in alterations to nutrient loading and stoichiometry. Changes in nutrient supply will ultimately result in changes to phytoplankton communities and carbon fixation.

3 Carbon Transport and Storage by Oceans

The oceans play a crucial role in controlling atmospheric carbon dioxide through a variety of physical, chemical, and biological processes (see Figure 3).

3.1 The Solubility Pump

The physical or solubility pump is the term generally given to the thermo-dynamic exchange of CO_2 between the surface ocean and the atmosphere. CO_2

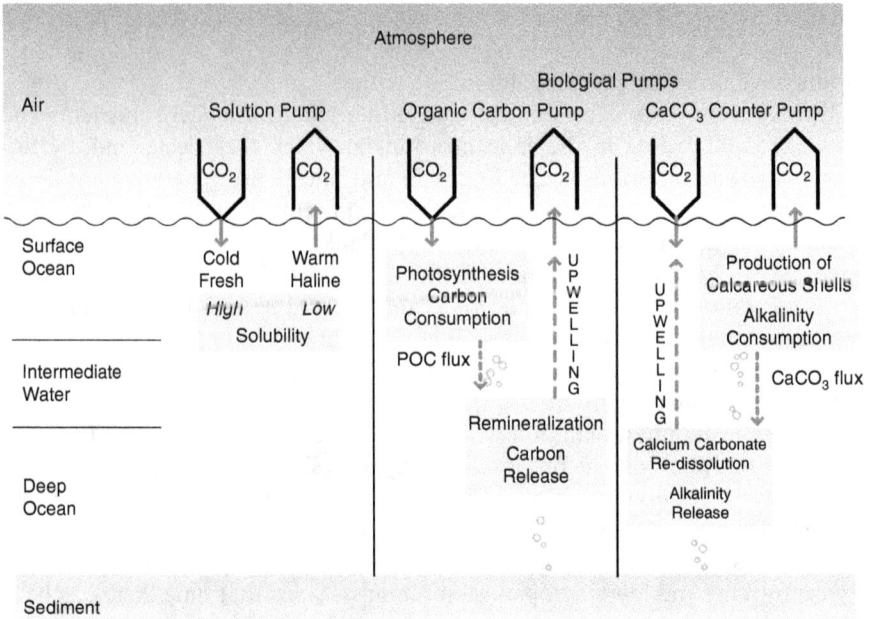

Figure 3 Three main ocean carbon pumps govern the regulation of natural atmospheric CO_2 changes by the ocean: the solubility pump, the organic carbon pump and the $CaCO_3$ 'counter pump'. The oceanic uptake of anthropogenic CO_2 is dominated by inorganic carbon uptake at the ocean surface and physical transport of anthropogenic carbon from the surface to deeper layers. For a constant ocean circulation, to first order, the biological carbon pumps remain unaffected because nutrient cycling does not change. If the ocean circulation slows down, anthropogenic carbon uptake is dominated by inorganic buffering and physical transport, as before, but the marine particle flux can reach greater depths if its sinking speed does not change, leading to a biologically-induced negative feedback that is expected to be smaller than the positive feedback associated with a slower physical downward mixing of anthropogenic carbon.[57]

is a weakly acidic gas and is highly soluble in seawater, itself being slightly alkaline due to its strong mineral content.[55–57] This solubility is highly tem-perature-dependant, and this temperature dependence is accurately known: a 1 K increase in seawater temperature will decrease its solubility to CO_2 by *ca.* 4%; likewise a 1 K decrease in temperature will increase solubility by the same amount.[58,59]

The chemical buffering of CO_2 in seawater is, quantitatively, the most important process contributing to the oceanic uptake of anthropogenic CO_2. Carbon dioxide entering the ocean is buffered due to scavenging by CO_3^{2-} ions and conversion to HCO_3^-, that is, the resulting increase in gaseous CO_2 con-centration in seawater is smaller than the amount of CO_2 added per unit of seawater volume.[57,60] This buffering effect is quantified by the Revelle factor, relating the fractional change in the partial pressure of CO_2 in seawater (pCO_2) to the fractional change in total DIC after re-equilibration:[6,61]

$$\text{Revelle factor (or buffer factor)} = (\Delta p[CO_2]/p[CO_2])/(\Delta[DIC]/[DIC]) \quad (6)$$

The lower the Revelle factor, the larger the buffer capacity of seawater. Variability of the Revelle factor in the ocean depends mainly on changes in pCO_2 and the ratio of DIC to total alkalinity. Uptake of CO_2 perturbs the pH and equilibrium concentrations of HCO_3^- and CO_3^{2-}. In the present-day ocean, the buffer factor varies between 8 and 13 (ref. 57,62).

3.1.1 Role of Ocean Circulation, including Upwelling and Deep Water Formation

The key mechanism by which the solubility pump operates is ocean circulation, especially the thermohaline circulation and latitudinal and seasonal changes in ocean ventilation.[63] Indeed, when integrated over the whole ocean and periods of a decade, the ocean sink is controlled mainly by the rate of vertical mixing and overturning of the ocean, *i.e.* how quickly surface waters penetrate into the interior. However, on shorter timescales, it is the interaction of the solubility pump (and biological pump) with the ocean circulation that determines surface levels of CO_2, hence the air–sea exchange.[59]

Heat loss from the poleward transport of warm water in western boundary currents (*e.g.* Gulf Stream, Kuroshio Current) increases the solubility of surface waters, forming sinks for atmospheric CO_2. The strongest, most intense and most consistent natural sink is the North Atlantic Ocean (including the Nordic Seas), where an overall northward drift and cooling of water as a result of the Atlantic meridional overturning circulation contributes to a net under-saturation in CO_2 throughout the region north of the sub-tropical gyre.[59] This water then sinks at high latitudes, in regions of deep-water formation, is sequestered to depths greater than 500 m (ref. 63) and is then transported southwards in the North Atlantic Deep Water. This thermodynamic drawdown of CO_2 is seen also in the Southern Indian Ocean and the Brazil–Falklands current convergence region of the South Atlantic, both areas of Antarctic

Intermediate Water formation, which is the major source of thermocline waters for the world ocean.[59] Once CO_2 has been entrained in these deeper waters it is effectively prevented from re-equilibrating with the atmosphere until transported back to the surface decades-to-centuries later.[63]

The solubility pump also drives the release of CO_2 into the atmosphere from oceanic source regions. Sub-surface waters have relatively higher carbon content than the surface.[59] The strongest net source regions are tropical and sub-tropical upwelling zones, such as in the Equatorial Pacific, where DIC-rich sub-surface water is vigorously upwelled, rapidly warming as it rises to the surface, decreasing the solubility of CO_2 and thus releasing it to the atmosphere.[59,63] This out-gassing has also been described in the Arabian Sea during the south west monsoon[64] and along the coast of Peru and Chile.[65] In the Southern Ocean, south of the Polar Front, is a site of strong upwelling which might be expected to be a source zone, although less than in lower latitudes because the water does not warm as it rises; in fact, it probably cools in winter.[59] However, measurements suggest this region to be neutral with respect to the atmosphere.[59]

3.1.2 The Shelf Pump

Continental shelf seas, including coastal and marginal seas, are thought to play a key role in the global carbon cycle, linking the terrestrial, oceanic and atmospheric carbon pools.[66,67] The term "continental-shelf pump" was coined by Tsunogai *et al.*[68] to account for the net CO_2 uptake they observed along a transect in the East China Sea. It has since been used more widely, usually to describe the net uptake of atmospheric CO_2 through primary production in stratified shelf-sea regions,[69] where the CO_2 exported below the thermocline is then carried off the shelf to the open ocean at depth, contributing to the higher DIC concentrations in sub-surface waters (seasonally well-mixed shelf regions tend to act as a weak source of CO_2 to the atmosphere). Globally, shelf seas are thought to be net sinks for CO_2, with inner estuaries acting as net sources for CO_2 (ref. 70). Thomas *et al.*[69] calculated the North Sea to be a highly efficient continental shelf pump, exporting approximately 93% of atmospheric CO_2 taken-up in this region off the NW European Shelf into the deep waters of the North Atlantic.

3.1.3 Inter-Annual and Decadal Scale Variability

Inter-annual variability in ocean physical processes, particularly circulation and thermal variability, have been linked to fluctuations in regional-to-basin-scale ocean biogeochemistry and air–sea CO_2 fluxes.[59,71–79]

Ocean inter-annual variability is modulated primarily by the major atmosphere and atmosphere-ocean climate modes.[71,80] Globally the largest signals are seen in the tropical Pacific, associated with the El Nino–Southern Oscillation (ENSO). Regional changes in atmospheric convection and the trade winds over the Pacific influence the upwelling of sub-surface water containing high DIC and pCO_2, while remotely-forced Kelvin waves modulate the depth of the

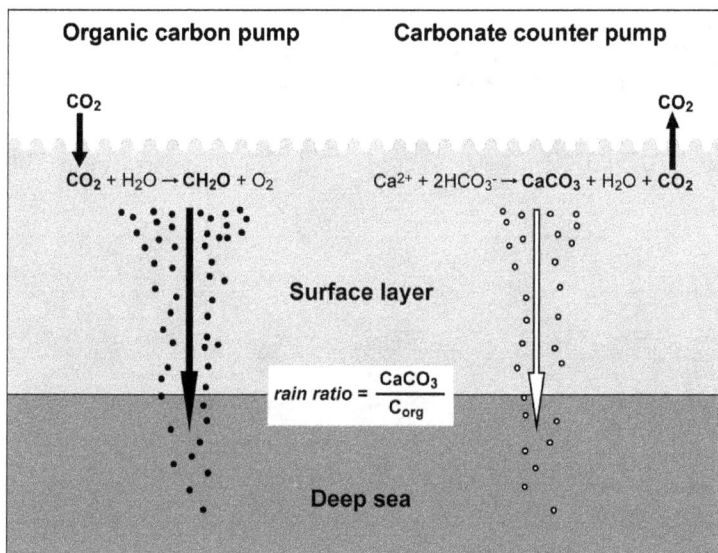

Figure 4   The biological carbon pumps.[91] (with kind permission of Springer Science + Business Media).

thermocline and the concomitant biogeochemical concentrations of source waters.[36,72,81] ENSO-related variability extends over much of the globe because of ocean-wave propagation from the tropical Pacific and atmospheric teleconnections.[80] There is a strong correlation between reduced CO_2 outgassing and the onset of El Nino events, in line with reduced upwelling and a deeper pycnocline. Conversely, La Nina events show a corresponding increase in CO_2 outgassing, in line with enhanced upwelling and a shallower pycnocline.[81,82]

Outside of the tropics, three major climate modes cause inter-annual variability: the North Atlantic Oscillation (NAO[75,77,83]) the Pacific Decadal Oscillation (PDO[84,85]) and the Southern Annular Mode (SAM) in the Southern Ocean.[86–89] All three modes involve atmospheric pressure oscillations that drive substantial changes in the strength and location of the surface winds, ocean upwelling, ocean convection patterns, sea-surface temperature (SST) and air–sea heat and freshwater fluxes on regional scales. The physical impact on air–sea CO_2 fluxes depends upon the interaction of several, often competing, climatic factors, such as thermal solubility, upwelling/mixing of nutrient- and DIC-rich waters, net surface-freshwater fluxes (through dilution of DIC and alkalinity) and wind speed. Gruber *et al.*[75] suggest a correlation of negative NAO index years with deeper mixed layers, lower SST, increased entrainment and biological production, and enhanced CO_2 uptake. A positive SAM phase is associated with increased surface wind stress, enhanced upwelling of carbon-rich, sub-surface circumpolar deep water, and outgassing of ocean CO_2 to the atmosphere.[86–88] Although models correlate variability in air–sea CO_2 fluxes

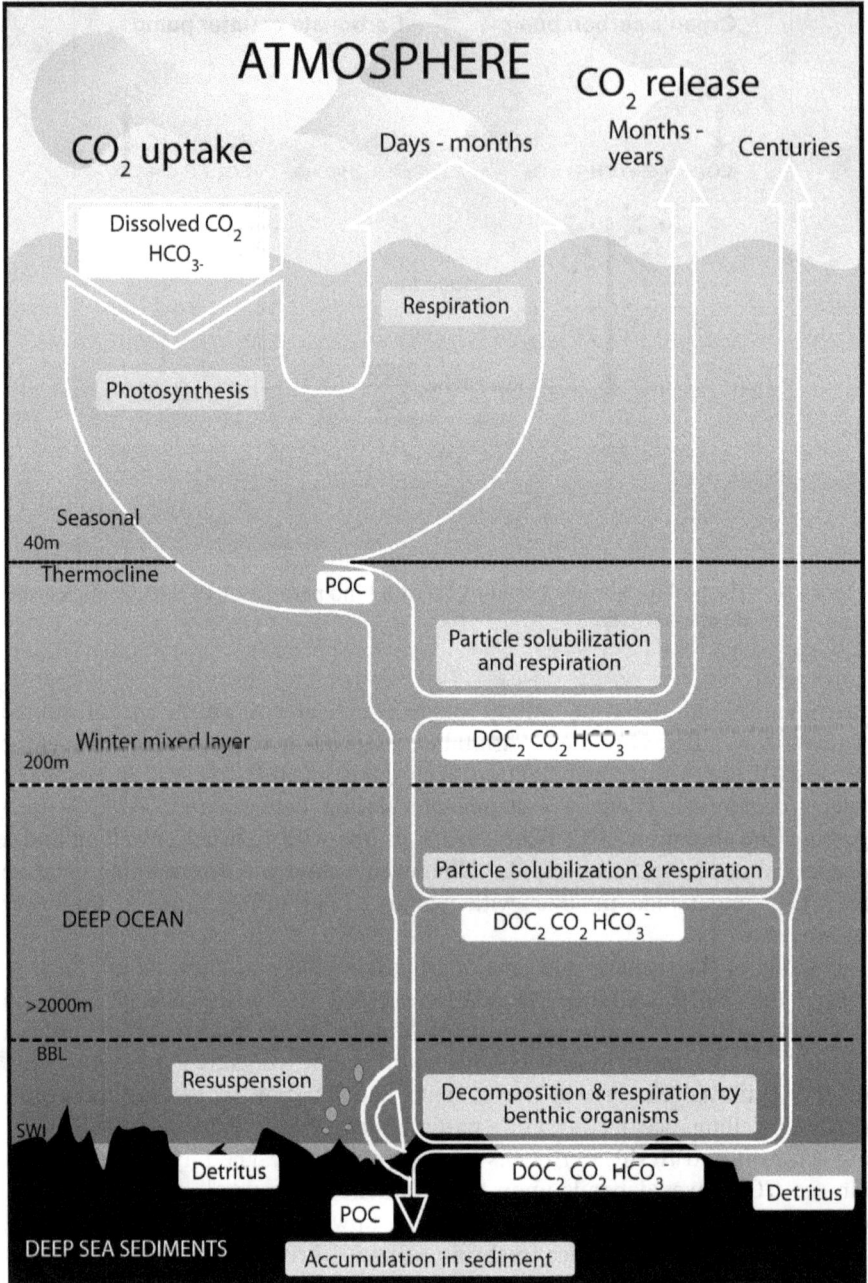

Figure 5 The organic carbon pump.[94]

with the PDO, the total projection is small because attendant changes in SST, DIC and alkalinity have opposing effects on surface CO_2 concentration.[71]

It should be noted that inter-annual variability in ocean uptake of CO_2 is also influenced by variability in biological uptake and in the gas-transfer velocity (*e.g.* through variability in storms and hurricane frequency).[90]

3.2 The Biological Pumps

The biological pumps consist of two counteracting pumps: the organic carbon pump, which depends on a combination of marine productivity and particle sinking, and the carbonate counter pump, which depends on biogenic calcification (see Figure 4).[91]

Integrated over the global ocean, the biologically-driven, surface-to-deep DIC pool amounts to *ca.* 2500 Pg C (ref. 92). This amounts to around 3.5 times the atmospheric carbon pool. Thus small changes in this pool, *e.g.* due to changes in the biological pump, would have a strong impact on atmospheric CO_2.

3.2.1 The Organic Carbon Pump

A key process responsible for about three quarters of the surface-to-deep-ocean gradient in DIC is the organic carbon pump.[93] This process transports carbon fixed through photosynthesis by microorganisms in the sunlit surface layer into the deep ocean and is a key component of the biological pump.

Around half of primary production on Earth is carried out in the oceans by photosynthetic planktonic micro-organisms[48] which require sunlight, nutrients (phosphate, nitrate, silicate and micronutrients such as iron; Section 2.2) supplied from deep waters and dissolved inorganic carbon (DIC). As these photosynthetic plankton consume DIC in the sunlit surface ocean, they cause an undersaturation of dissolved CO_2 and further CO_2 uptake from the atmosphere.

The majority of the carbon fixed through this process will be respired within the upper mixed layer through processing by the marine food web within days to months (see Figure 5). However, a small but significant proportion will fall from the surface waters into the ocean depths, removing organic carbon from the surface waters and hence from short-term exchange with the atmosphere. Repackaging of ingested photosynthetic cells into faecal pellets, discarded mucus feeding webs by zooplankton or aggregation of smaller cells into mucopolysaccharide macroaggregates called "marine snow", can result in even more rapid removal than single cells alone, due to increased sinking rates.[95–97] Particles weighted with additional ballast in the form of, for instance, $CaCO_3$ or clay, will sink faster and therefore enhance the organic carbon pump. Much of the organic carbon in the particles is remineralised (oxidised to DIC and other inorganic compounds through the action of bacteria attached to the aggregates), primarily in the upper 1000 m of the oceanic water column, and their carbon removed from reacting with the atmosphere for decades to centuries.[98] Particles that reach these deep ocean depths will essentially remove carbon, either in the form of organic or inorganic particles or DIC, from the atmosphere for centuries. Particles that arrive on the deep sea bed are subjected to an intense period of remineralisation by feeding deep-sea benthic fauna and are subject to dissolution within the benthic boundary layer.[94] "Marine snow" can result in a visible layer of detrital material or "fluff" on the deep-sea bed,[99] and can result in considerable biological activity until its ingestion and remineralisation.[100] It is only a very small fraction of the more recalcitrant material that accumulates within the sediments, that is essentially removed from the atmosphere for millions of years.

3.2.2 The Carbonate Counter Pump

Oceanic organisms can alter atmospheric CO_2 concentration in another way, through the process called calcification. Calcification involves the formation of biogenic calcium carbonate ($CaCO_3$) material, in the form of shells in the case of planktonic foraminifera, and pteropods or small platelets called liths in the case of coccolithophores. This is called the carbonate counter pump (see Figure 4) as, in contrast to the biological pump, calcification consumes total and carbonate alkalinity and releases CO_2 through the following reaction:

$$Ca^{2+} + 2HCO_3^- \rightarrow CaCO_3 + H_2O + CO_2 \qquad (7)$$

In coccolithophores, there is a tight and contrasting coupling of the biological and carbonate counter pumps. Through primary production in the surface layer of the ocean and removal of organic matter by sinking into deeper waters, they draw down CO_2 via the organic carbon pump. In contrast, through calcification during lith production they also produce CO_2. The relative strength of these two pumps (the ratio of particulate inorganic to organic carbon in exported biogenic matter, the so-called *rain ratio*) mainly determines the biogenic exchange of CO_2 between the surface ocean and the overlying atmosphere. The $CaCO_3$ may also act as ballast to the biological pump (see Section 3.2.1).

Coccolithophores are among the main carbonate producers in today's oceans and, together with other calcifying organisms (mainly planktonic foraminifera), they have been considered to be responsible for generating the ocean's vertical distribution of total alkalinity (TA) and for regulating the atmospheric pCO_2 since the Mesozoic era. Today, *Emiliania huxleyi,* one of the most abundant and widespread calcifying organisms on Earth, forms massive blooms, sometimes covering $100\,000\,km^2$ (ref. 101) in temperate and sub-polar continental margins and shelf seas.

4 Consequences of Too Little Uptake

4.1 Slow Down of the Physical Ocean Sink and Feedbacks to Climate

Observed changes in global climate are extensive: global average land and ocean surface temperatures have increased at a rate of about $0.2\,°C\,decade^{-1}$

over the last few decades;[102] sub-surface ocean heat content increasing over the upper 300 m since at least 1955;[103] temperatures rising in bottom waters of Antarctic origin;[104,105] higher precipitation rates observed at mid-high latitudes and lower rates in the tropics and sub-tropics, with corresponding changes measured in surface-water salinities; surface winds intensifying and moving poleward over the Southern Ocean, associated with a shift toward a more-positive state of the SAM;[106,107] and a decline in summer Arctic sea-ice extent, with September ice cover in 2007 and 2008 about 40% lower than pre-1980 conditions.[108] Many of these observed trends are projected to continue and even accelerate over the next several decades.[109]

In light of these changes, ocean CO_2 uptake can also be expected to change, with most mechanisms leading to a decline in the efficiency of the ocean carbon sink and a positive feedback on atmospheric CO_2 concentrations. Doney *et al.*[109] highlight the following mechanisms as being of particularly importance:

- Ocean warming will act to reduce ocean-carbon storage by reducing the seawater solubility of CO_2 (thermal and buffering effects), with a positive feedback on atmospheric CO_2 concentrations.
- Enhanced ocean stratification, due to upper-ocean warming globally and increased freshwater inflow in temperate and polar latitudes, will reduce vertical mixing, thereby slowing northern hemisphere intermediate and deep-water formation. As a result, the effectiveness of ocean uptake will decline.
- A strengthening and poleward contraction of westerly winds in the Southern Ocean is expected, which may increase vertical upwelling of old CO_2-rich circumpolar deep water. Enhanced upwelling would act to increase both the release to the atmosphere of metabolic CO_2 and the uptake of anthropogenic CO_2. In the near term the net effect would be an overall decline in ocean CO_2 uptake, though the effect could reverse on a century timescale.[110]

A decline in the efficiency of the ocean carbon sink has already been reported. Atmospheric CO_2 levels suggest that the Southern Ocean CO_2 sink (south of 45°S) did not increase between 1981 and 2004, despite increasing atmospheric CO_2 levels.[88] Furthermore, the substantially higher rate of accumulation of atmospheric CO_2 in the high-latitude North Atlantic compared to atmospheric rates of increase,[111] suggest this intense sink may become increasingly saturated, leading to a further decline in efficiency. Thomas *et al.*,[112] however, caution from a modelling study that much of the variability can be explained by decadal scale oscillations linked to the North Atlantic Oscillation, overlying a secular trend. Unravelling these signals is a key challenge. Globally, no significant change has been detected in the fraction of anthropogenic CO_2 emissions taken up by the oceans.[113]

4.2   Changes in Net Primary Productivity

Marine net primary productivity (NPP) accounts for approximately half of the total global biosphere and, as such, represents an important link in the carbon

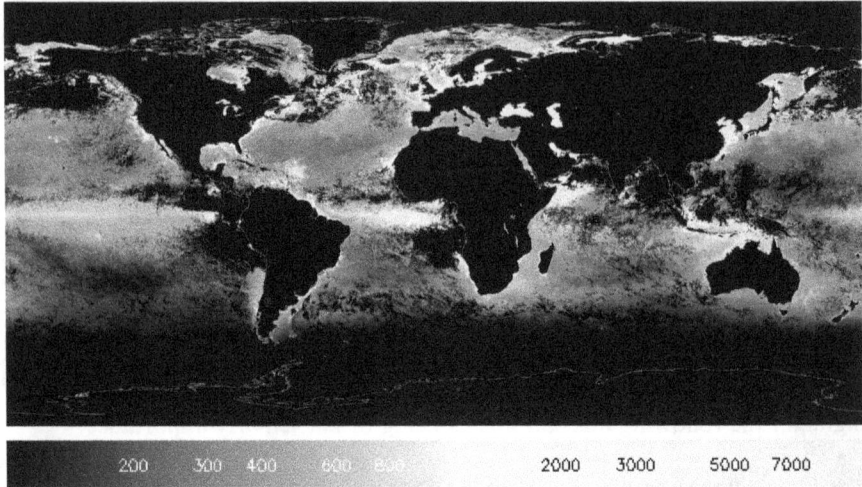

Figure 6 Global distribution of phytoplankton primary production from satellite Earth observation data. An example monthly composite map for August 2007, derived from NASA's Sea-viewing Wide Field-of-view Sensor (SeaWiFS) chlorophyll-a data together with photosynthetically available radiation and sea-surface temperature by coupling photosynthesis and radiative transfer models.[34] Black areas indicate persistent cloud cover and land; units in $mg\,C\,m^{-2}\,d^{-1}$. (Sea WiFS data courtesy of NASA Ocean Colour Web).

cycle. As on land, there are highly productive areas (such as upwelling areas along some coasts, the North Atlantic and some shelf seas) and areas of lower productivity, such as the central oceanic gyres (see Figure 6). The turnover of global phytoplankton biomass is also rapid, with the entire stock being consumed within two to six days. The main controls of marine primary production by phytoplankton are light and nutrients (Section 2.2). The supply of these nutrients to the surface-lit layers of the ocean is regulated by mixing processes such as wind speed, upwelling, ocean-circulation and convection. The atmosphere also has a role to play, with the deposition of aerosols containing nitrogen, phosphorous and iron. It is envisaged that climatic changes in any of these physical mechanisms will have a subsequent impact on the phytoplankton and their ability to draw carbon from the atmosphere and into the ocean. This can be either a positive or a negative feedback.

The dominant global mechanism for limiting NPP is thermal stratification; phytoplankton in the surface mixed layer rapidly deplete the supply of nutrients. Further productivity is hampered as the nutrient-rich waters below the thermocline become disconnected from the surface layer. This climate-plankton link is found primarily in the tropics and mid-latitudes, and it is envisaged that in a warmer world the degree and extent of thermal stratification will increase, thereby reducing NPP. In the high latitudes, however, the situation is more complicated. The retreat of sea-ice (this being well documented in the Arctic) obviously opens up more water for phytoplankton to inhabit, but the impact of

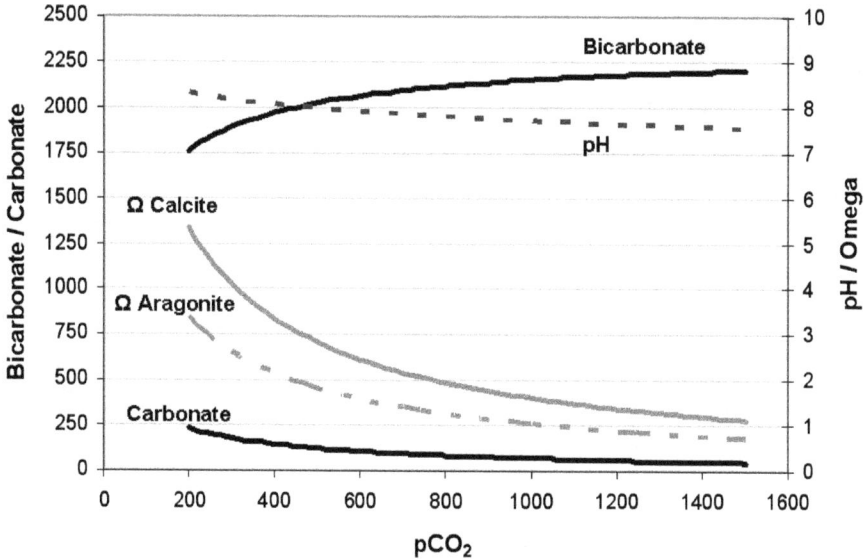

Figure 7 The variation in bicarbonate and carbonate ion concentration (left axis, both μmol kg^{-1}) and pH and the saturation states of calcite and aragonite (right axis) with changing pCO_2 (water) at 10 °C, 36 psu, with an alkalinity of 2324 μmol kg^{-1} for surface waters. The saturation states are the product of the concentrations of the reacting ions divided by the product of those ions at equilibrium, hence saturation states < 1 indicate that mineral carbonate will dissolve into the surrounding seawater.

fresh meltwater upon deep-water mixing and the increased absorption by phytoplankton are unclear.

Recent studies using satellite data to drive models of NPP (see Section 2.2.1, Figure 1) have shown that there are likely to be reductions in NPP in the tropics and mid-latitudes and increases at the poles. Behrenfeld *et al.*[36] found a strong correlation between a multi-variate El Nino/Southern Oscillation Index (MEI) and satellite anomalies of NPP. They showed that an increase in the MEI (towards warmer conditions) resulted in a decrease in NPP and *vice versa*. Tilstone *et al.*[30] showed decreases in satellite-derived NPP over the past decade in all provinces of the Atlantic Ocean sampled as part of the Atlantic Meridional Transect (AMT), this being linked to warming.

At the high latitudes, Smith and Comiso[114] found significant increases in annual productivity of the entire Southern Ocean since 1998 (again using satellite data), but the causes of the trend were unclear. Arrigo *et al.*[115] found a link between NPP and changes in sea-ice cover and postulated that the annual totals of NPP could increase as stronger winds increase nutrient upwelling. Pabi *et al.*[116] looked at trends in the Arctic over the past decade and found increases in productivity as the sea-ice has retreated.

It should be noted that all of the above studies have relied upon satellite estimates of NPP and that there is a large number of models in the literature

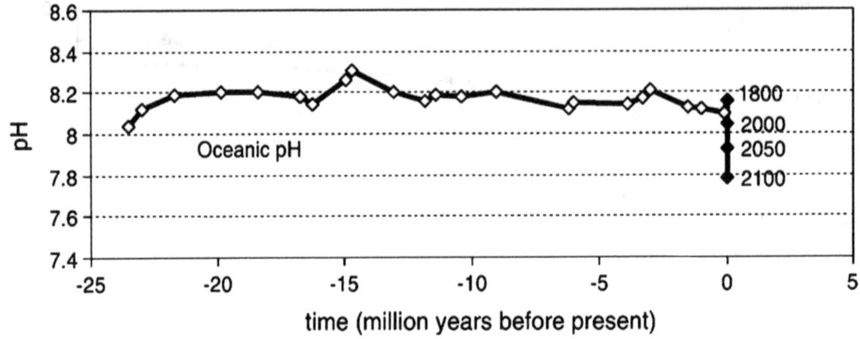

Figure 8   Past (white diamonds, data from Pearson and Palmer)[4] and contemporary
variability of marine pH (grey diamonds with dates). Future predictions are
model-derived values based on IPCC mean scenarios. (Reproduced courtesy
of Blackford & Gilbert).[121]

that derive NPP.[35] The trends represented in the studies are likely to be
reproduced in the majority of the models, however, the absolute magnitudes
will be different.

5   Consequences of Too Much Uptake

5.1   Ocean Acidification

5.1.1   Global and Regional Predictions of Omega and pH
The carbonate system is, in contrast, well constrained; equilibrium constants
are extensively published and, although there are variations in particular
constants emerging from different studies, a consistent, robust approach is
generally possible (see ref. 6, among others; Section 2.1.2). This enables predictive models of future carbonate chemistry to be developed with reasonable
certainty. Thus for a given set of environmental conditions (temperature,
pressure, salinity) the relationship between pCO_2 in water and pH, carbonate
ions and saturation state is straightforward to calculate (see Figure 7).

Uncertainties about rates of air–sea exchange of CO_2 are diminishing (see ref.
117 and refs therein) and sensitivity analysis of alternative parameterisations
has identified little impact on water carbonate chemistry if considered at broad
spatial and temporal scales.[118,119] Thus large-scale predictions of the oceanic
uptake of atmospheric CO_2 and the consequences for the marine carbonate
system are tractable and reasonably robust. It is predicted that global mean
ocean pH will fall by a maximum of 0.77 pH units to around pH 7.5 (ref. 120),
although clearly this is dependant on emission scenarios and mitigation. This
fall represents an extreme and rapid perturbation from estimated marine pH
values over at least 20 million years[4] (see Figure 8).

Many marine organisms depend on synthesising calcium carbonate structures and their ability to perform this synthesis depends at least partly on the

Table 1 Mineral composition of calcareous parts of key calcifying marine pelagic and benthic organisms Note that many benthic species have juvenile stages that spend time in the pelagic realm. (Amended from the Royal Society).[56]

Organisms	Form of calcium carbonate	Habitat
Foraminifera	Calcite	Benthic
	Calcite	Planktonic
Coccolithophores	Calcite	Planktonic
Macroalgae	Aragonite or calcite	Benthic
Coralline algae	High-magnesium calcite	Benthic
Corals: warm water cold water	Aragonite	Benthic
	Aragonite	Benthic
Pteropod molluscs	Aragonite	Planktonic
Non-pteropod molluscs	Aragonite + calcite	Benthic and Planktonic
Crustaceans	Calcite	Benthic and Planktonic
Echinoderms	Calcite	Benthic

carbonate saturation state, omega (Ω) (see below). Because of the dependence of saturation state on temperature and pressure, under-saturation is a property of deeper waters and the depth of the saturation horizon ($\Omega = 1.0$ at the saturation horizon which divides oversaturated waters above from under-saturated waters below) is an important diagnostic of the marine environment. The temperature-dependency creates a latitudinal variation in the saturation horizon depth, such that this depth is significantly shallower in polar waters. Increasing CO_2 has the effect of decreasing the carbonate ion concentration (see Figure 7), promoting dissolution and lowering the saturation state, thereby impeding the formation of carbonate minerals. Global predictions of saturation state project that polar surface waters will become under-saturated within decades.[122,123]

However, if one takes a regional or, especially, a shelf perspective, predicting the precise behaviour of the carbonate system is far more complex. Shelf systems are an important consideration, not only from their productive capacity and economic importance but also because they provide the interface between the terrestrial and ocean-carbon compartments. Typically models use total inorganic carbon and a parameterisation of total alkalinity to drive the carbonate equations and derive pH and pCO$_2$. For open-ocean situations, the derivation of alkalinity from modelled salinity is robust.[124,125] Unfortunately this relationship is, at best, highly approximate in shelf seas, and linear relationships with salinity break down.[112,126]

Partially, this is due to significant riverine input of dissolved inorganic carbon and alkalinity.[127] Furthermore, both of these may change over time; for example, changing rainfall and land-use patterns have increased the alkalinity of the Mississippi by $> 50\%$ over the last half century.[128] The close coupling of the benthic system and euphotic zone in shelf systems provides another uncertainty. Recently, Thomas *et al.*[112] have demonstrated the significance of anaerobically generated alkalinity in the North Sea and estimate that this process could be responsible for up to 60% of carbon uptake in marginal seas. Hence, if the aim is to accurately represent the carbonate system of shelf systems, it is important to consider a range of physical, boundary and biological processes.

5.1.2 Sensitivities of Pelagic Systems

The decrease in the saturation state of $CaCO_3$ in the ocean (Section 5.1) is likely to have a future impact on pelagic calcifying organisms, especially those in waters with lower than optimum saturation or under-saturation, such as the polar and sub-polar waters and areas with upwelling of waters high in carbon dioxide.[56,122,129,130] The major planktonic calcifiers are coccolithophores, foraminifera and pteropods (see Table 1). Additionally, many benthic organisms have pelagic juvenile stages which may be particularly sensitive to ocean acidification.[131] As aragonite is more soluble than calcite, those pelagic organisms with $CaCO_3$ structures made of aragonite, such as shelled pteropods, will be affected earlier than those such as coccolithophores and foraminifera which make their $CaCO_3$ structures with calcite. Indeed, when incubated at double the pre-industrial CO_2 concentration, the aragonite rods of the shells of pteropods from the Southern Ocean started dissolving.[122] According to a model experiment based on the IPCC Scenarios 1992a (IS92a) emission scenario, bio-calcification will be reduced by 2100, in particular within the Southern Ocean,[122] and by 2050 for aragonite-producing organisms It is expected that the Arctic Ocean, where pteropods can also be an important part of the food web, will be similarly sensitive to low pH and $CaCO_3$ under-saturation, but that this may occur earlier than projected for the Southern Ocean.[114,126,132]

A 4–14% decrease in shell mass of the planktonic foraminifera *Orbulina universa* and *Globigerinoides sacculifer* was observed in laboratory experiments with reduced CO_3^{2-} concentration and calcite saturation state.[133,134] Both these foraminifera have photosynthetic symbiotic microalgae which may partly reduce the effects of ocean acidification as a result of CO_2 removal during their photosynthesis, implying that foraminifera without symbionts may be more vulnerable.[135] These authors also suggest that the symbiont-free, post-zygote stage of foraminifera, during which calcification is weak or absent, may also be particularly vulnerable to ocean acidification. Both these and long-term changes in foraminiferal calcification rates need clarification by further research.

The greatest number of experiments has been carried out on the sensitivity of coccolithophores to ocean acidification, especially on the most important calcifier, *E. huxleyi*. The experiments have been mainly on laboratory cultures on single strains, but shipboard and mesocosm experiments have also been carried out. The great majority of experiments have shown a reduction in coccolithophore calcification, with one or two showing little change and one an increase in calcification with increase in carbon dioxide.[136–138] There is currently discussion about whether this variability may be due to different laboratory conditions, use of strains isolated decades ago, or the first indications that there could potentially be strains that may be able to adapt to high carbon dioxide.[139,140] Changes to oceanic calcification could have a profound impact on the organic carbon pump and the $CaCO_3$ counter pump (Section 3.2).

Coccolithophores are also a major producer of dimethyl sulfide (DMS)[141] which may have a role in climate regulation *via* the production of cloud-condensation nuclei.[142] A reduction in the occurrence of the coccolithophore blooms that occur in large areas of the global oceans could lead to a reduced flux of DMS from the oceans to the atmosphere and hence further increases in global temperatures *via* cloud changes.

The availability of marine nutrients is important in determining both the rate of primary production and the composition of the primary producers and, therefore, the structure of pelagic ecosystems (Section 2.2). The chemical forms, or speciation, of the key macro- nutrients (phosphorus, nitrogen and silicate) are theoretically very sensitive to changes in pH around 8.0 (ref. 6). For example, lower pH results in a reduction in the available form of phosphate (PO_4^{3-}) and a decrease in ammonia (NH_3) with respect to ammonium (NH_4^+). If such changes occurred in nature, as indicated by preliminary research,[143] this could alter the energetics of cellular acquisition for members of phytoplankton, archaea and bacteria and therefore the competition between them for these important growth substrates.

Other consequences of lower pH on the nitrogen cycle may be changes to the important metabolic processes of nitrification, denitrification and nitrogen-fixation. For example, Huesemann *et al.*[144] found rates of microbial nitrification (conversion of NH_3 and NH_4^+ to nitrate (NO_3^-)) were reduced by *ca.* 50% at pH 7, with inhibition at pH 6, both levels of pH far lower than is likely to occur as a result of ocean acidification. However, if nitrification is also sensitive at pH between 8.1 and 7.6 this may result in the future accumulation of ammonia instead of nitrate (that is, increasing the $NH_4 : NO_3$ ratio). A recent laboratory study has also shown that N-fixation rates increased significantly with elevated CO_2 in cultured *Trichodesmium*.[145]

Micro-nutrients are also important in the control of primary production and other important cellular processes involving metal-requiring enzymes.[56] For instance, phytoplankton growth rates are iron-limited in $>10\%$ of the ocean's surface[146] and one effect of reduced pH would theoretically be to increase the proportion of the soluble form Fe(II) to the insoluble form Fe(III).[147] Should this occur, and assuming no change in macro-nutrients, there could be enhanced productivity in these areas.

Unlike land plants, most marine phytoplankton are thought to have mechanisms to actively concentrate CO_2 so that changes in seawater CO_2 have little ($<10\%$) direct effect on their photosynthetic rate or their elemental composition.[148–151] However, whilst taxon-specific differences in CO_2 sensitivity have been observed in laboratory culture,[152] it is currently unknown whether a reduction of the advantage of possessing a CO_2-concentrating mechanism will impact phytoplankton species-diversity in the natural environment. This is a possibility and, should it occur, may impact the contribution

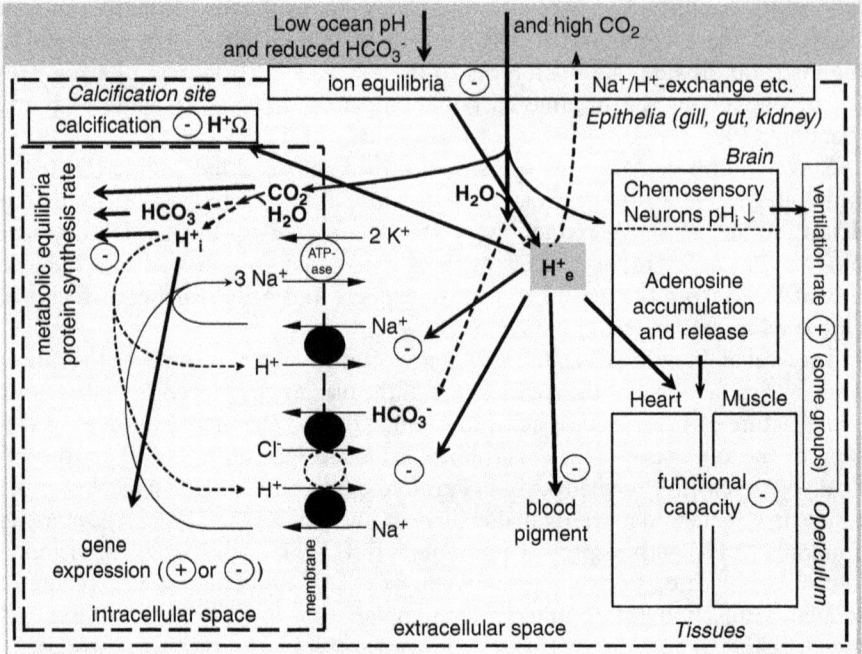

Figure 9   The impact of reduced pH and elevated CO_2 on physiological processes. (Source: Pörtner[155] and Pörtner et al.[158]).

of different functional groups, primary production, food-web structure and marine biogeochemical cycles. Exceptionally, the coccolithophore E. huxleyi increased its rate of photosynthesis in laboratory monocultures with elevated carbon dioxide.[91] However, this has not been found in mixed communities and, considering E. huxleyi has also been observed to decrease its calcification rate at higher carbon dioxide,[153] it is unsure whether this may offset the organisms which directly benefit from increasing CO_2. The Royal Society,[56] assessing this topic, reported that most of the experiments on marine phytoplankton have been short-term and did not provide sufficient time for any genetic modification that might enable them to adapt. Additionally, some were carried out by artificially altering pH and therefore do not mimic the situation in the real world, or were at pH levels unlikely to be seen in future scenarios.

   These pH-induced changes to carbonate, nutrient and trace-metal chemistry and metabolic processes could exert profound pressures on future ocean productivity and biogeochemistry and the structure of food webs and ecosystems. However, it must be stressed that the science of ocean acidification is young and our understanding of many of the possible consequences still limited to output from relatively limited research. A deeper understanding will be a great challenge to future marine scientists.

5.1.3 Sensitivities in Benthic Systems

5.1.3.1 Physiological Impacts of Hypercapnia. Recent efforts have been made to describe the theoretical impacts of elevated CO_2 levels on the physiology and function of benthic organisms and to review the corroborative evidence from experimental observations.[154–159] From these studies it is evident that ocean acidification (that is, elevated dissolved CO_2, H^+ and HCO_3^- concentrations, plus reduced CO_3^{2-} concentration, Figure 7) has the potential to disrupt a host of intracellular and extracellular physiological processes (see Figure 9 for a summary).

One of the primary effects of exposure to CO_2-"acidified" seawater is a decrease in the pH or "acidosis" of the body fluids. The physiological response of organisms to extracellular acidosis (blood or haemolymph pH falls) is broadly similar across a wide variety of marine animals where an increase in bicarbonate occurs, with near-full or partial pH compensation as a result of predominantly active ion-transport processes in the gills (see ref. 153 and refs therein). However, in some invertebrates, such as the mussel *Mytilus edulis,*[160] the crabs *Callinectes sapidus*[161] and *Chionoecetes tanneri*,[162] and the sea urchin *Psammechinus miliaris,*[163] studies have shown there is only partial, or no, compensation in hypercapnia-induced disturbance of extracellular acid–base balance. This lack of regulatory capacity is important because extracellular pH is usually regulated 0.5–0.8 pH units above intracellular pH[164] which, in turn, is important because the maintenance of intracellular pH is essential for countless cellular functions and regulations.[165] Maintenance of extracellular pH is not only linked to intracellular pH but is seen as important for the function of respiratory proteins. Both pCO_2 (*i.e.* a specific CO_2 effect)[166] and acidity can have pronounced effects on O_2 binding by respiratory pigments: hemoglobins,[167] hemocyanins,[168] but particularly the annelid pigments erythrocruorin and chlorocruorin.[167]

As organisms look to optimise the structural and kinetic coordination of molecular, cellular and systemic process, they are constrained to exist within a physiologically prescribed thermal window.[157] At the edges of these thermal windows an organism will experience decreased performance with respect to many functions: growth, reproduction, immune response. Environmental stressors, such as high CO_2, hypoxia and salinity change, can narrow an organism's thermal window and therefore increase its susceptibility to extreme temperature events. Some marine invertebrates inhabiting the highly variable inter-tidal environment are also known to produce an infusion of the neuro-modulator adenosine when exposed to elevated levels of carbon dioxide.[169] This adaptive strategy, known as metabolic depression, suppresses aerobic energy turnover rate, and whilst this may be beneficial in the short term, it may be detrimental to whole organism functions during long-term exposure.

5.1.3.2 Reproduction and Growth (including Impacts on Larval Development). The impact of a low-pH environment has been shown to reduce sperm motility in invertebrates that broadcast spawn, such as sea urchins,[170] resulting in reduced fertilisation success. Reduced egg-production has also been exhibited in copepods and decreased hatching success has been reported in gastropods.[56] Kurihara and Shirayama[171] reported severely impacted development in sea urchins at reduced pH and Kurihara and coworkers[172] also reported impairment of development in oysters. Reduced pH has been shown to lead to slower growth in corals, which could impact on reproductive success in those species where reproductive maturity is related to size and not age.[173,174] Kikkawa *et al.*[175] observed decreased reproductive success in bream; however, the pH values used for these studies were more related to a CCS-leakage scenario than sea-surface deposition. A low-pH environment has been shown to impact on the development of calcifying plankton, including the dissolution of the protective armour of coccolithophores,[176] forminiforera and pteropods. Additionally, impaired development and growth for a range of invertebrate species has also been demonstrated, including weakening of the shell structure,[177–179] and mortality of bivalve larvae is greatly increased during settling.[56,177] A general feature of many studies for a range of organisms is that the degree of the impact, in descending order, is greatest in eggs, followed by larvae and early life-stages, with adults being the least affected.[180] Part of the explanation for the sensitivity of eggs may be due to their untimely release into the environment, as laboratory studies for a range of benthic invertebrates have indicated that a reduction in pH to 7.8 or 7.6 can lead to accelerated gamete development and maturation.

5.1.3.3 Organism Health and Survival. Whilst many of the published studies to date on the impacts of reduced pH in the marine environment have concentrated on changes to calcification, growth, reproduction and survival, there is a paucity of evidence on impacts on general health status. Research by Beesley *et al.*[181] observed that within seven days of exposure to reduced levels of pH (7.7 and lower) in mussel, blood cells were exhibiting reduced membrane stability, which is indicative of reduced health status. Similarly, coelomocyte lysosomal membranes were observed to become fragile in urchins exposed to pH 7.2 and pH 6.8 during laboratory mesocosm studies. Studies by Bibby *et al.*[182] demonstrated a decline in immunocompetence in *Mytilus edulis* following exposure to reduced-pH environments (7.7 and lower) which would have consequences for disease susceptibility. Metzger *et al.*[183] report on a narrowing of thermal tolerance in the crab, *Cancer pagurus*, with increasing CO_2 concentrations, which has implications for longer-term climate-change consequences. *Amphiura filiformis*, a brittle star, exhibited muscle wastage following a period at pH 7.7 which was more severe at pH 7.3 and pH 6.8 and has health-status implications in that it reflects on reduced feeding potential.[184]

5.1.3.4 Impact on Communities, Biodiversity and Ecosystem Function. The current challenge facing scientists is to predict the long-term implications of ocean acidification for the diversity of marine communities and the

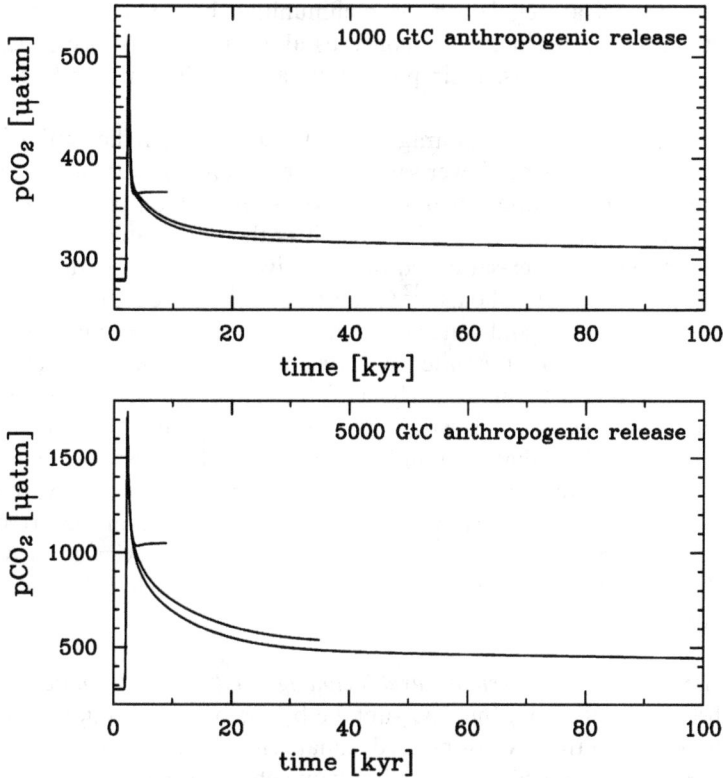

Figure 10 Model projections of the neutralisation of anthropogenic CO_2 for an ocean-only model, a model including dissolution of $CaCO_3$ sediment and a model including weathering of silicate rocks, (top) for a total of 1000 Gt C of anthropogenic CO_2 emissions and (bottom) for a total of 5000 Gt C of anthropogenic CO_2. Note that the y-axis is different for the two diagrams. Without $CaCO_3$ dissolution from the seafloor, the buffering of anthropogenic CO_2 is limited. Even after 100 kyr, the remaining pCO_2 is substantially higher than the pre-industrial value.[192] (Reproduced by permission of the American Geophysical Union).

ecosystem functions that this diversity sustains. This challenge is made more difficult as empirical data which directly quantify the impact of ocean acidification on marine biodiversity are currently lacking. Given the physiological impacts described above, it is implicit that ocean acidification has the potential to alter community structure and reduce biodiversity through the extinction of those species with a limited tolerance to high levels of CO_2. In addition, if sensitivity to acidification is a function of an organism's phylogeny (*e.g.* echinoderms being more vulnerable than polychaetes) or ecology (*e.g.* epifauna being more vulnerable than infauna), acidification may also reduce taxonomic and functional diversity, respectively. Finally, the loss of keystone or critical species, or a reduction in their activity (*e.g.* predation, grazing, bioturbation), could reduce habitat complexity. It could also affect the biological regulation of communities by reducing interspecific competition. There are increasing concerns about impacts of ocean acidification on marine organisms, their physiology and biodiversity and ecosystem function.[135,185–189]

Future projections of global aragonite-saturation state indicate that warmwater corals will experience lower saturation levels and are very likely to suffer from reduced calcification, such that bioerosion will outpace reef growth (reviewed in Kleypas et al.).[187] However, it is the cold-water corals that are likely to experience under-saturated or corrosive conditions as the aragonite-saturation horizon (ASH) shoals.[122] Guinotte et al.[190] have estimated that 70% of known scleractinian cold-water coral ecosystems will be in under-saturated water with respect to aragonite by 2100, although some will experience aragonite under-saturation as early as 2020. It would seem unlikely that scleractinian cold-water corals would be able to calcify under these conditions; it would be more likely that aragonitic structures would experience dissolution. Indeed, if cold-water corals respond in the same way as warm-water coral species, where a substantial decrease in calcification occurred with relatively small reductions in aragonite saturation state (reviewed in Kleypas et al.),[187] then their calcification rates may decrease well before aragonite under-saturation occurs.

5.1.4   Sensitivities on a System Level including Feedback to Climate

The substantial anthropogenic CO_2-uptake by the oceans (Section 5.1) is currently helping to mitigate CO_2-induced climate change. The major processes for neutralising anthropogenic CO_2 are inorganic chemical buffering and dissolution of marine $CaCO_3$ sediments, but this will take tens of thousands of years and will therefore not prevent a temporary build-up of high atmospheric CO_2 concentrations[191,192] (see Figure 10). Even after this period, atmospheric CO_2 will not return to pre-industrial levels.[192]

However, the ocean carbon cycle is also sensitive to climate. Climate change will tend to suppress ocean carbon uptake through reductions in CO_2 solubility, suppression of vertical mixing by thermal stratification and decreases in surface salinity (Section 4). This will increase the fraction of anthropogenic CO_2 emissions that remain in the atmosphere this century and produce a positive feedback to climate change. The atmospheric CO_2 increase alone will lead to continued uptake by the ocean, although the efficiency of this uptake will decrease as the carbonate-buffering mechanism in the ocean weakens. Future potential changes in large-scale circulation driven by climate change (e.g. a slowing down of the thermohaline circulation) could also affect the ocean carbon sink. All C4MIP models indicate a reduction in the ocean carbon sink by climate change of between -14 and $-60\,\mathrm{Gt\,C\,^{\circ}C^{-1}}$, implying a positive climate-CO_2 feedback.[57] An example of a negative feedback is that a reduction in sea-ice cover may increase the area for anthropogenic CO_2 uptake and act as a minor negative carbon feedback.[193]

Ocean ecosystem function depends strongly on climate, including the strength and timing of stratification of the upper layer of the ocean, ocean circulation and upwelling, temperature, salinity, wind strength and sea-ice cover. For instance, increased sea-surface temperatures could stimulate the physiology of photosynthetic organisms and lead to a temporary increase in fixation of CO_2. However, this could not be sustained, as associated reductions in vertical mixing and overturning circulation would decrease the return of required nutrients to the surface ocean and alter the vertical export of carbon to the deeper ocean *via* the biological pump (see Section 3.2).[57] Shifts in the structure of ocean ecosystems can influence the rate of CO_2 uptake by the ocean.[194]

A potentially complex feedback may involve iron limitation associated with large areas of the global ocean. Wind-borne dust deposited on oceans provides an important source of micronutrients, especially iron for the growth of phytoplankton,[195] and can enhance ocean-carbon uptake by providing ballast for the biological pump[196] (Section 3.2). Climate-induced changes to the wind strength and direction, and therefore mobilisation, transport and deposition of dust, could modify oceanic patterns of primary production and export of carbon-rich aggregates ballasted with clay. For example, a decrease in dust loads in a warmer climate[197] could result in a net positive feedback, further increasing atmospheric CO_2 through a weakening of marine production and export of aggregates due to clay ballast.[196] On the other hand, iron could become more bio-available in a future ocean with decreasing pH,[56] resulting in a strengthening of marine production and export and a negative feedback to climate.

The relationship between climate and the ocean carbon cycle is two-way, with ocean ecosystems affecting the composition of the atmosphere (*e.g.* CO_2, N_2O, O_2, dimethyl sulfide (DMS), sea salt and sulfate aerosol). Most of these components are expected to change with a changing climate and high atmospheric-CO_2 conditions.[57] DMS, for example, is produced by phytoplankton and linked to the formation of cloud droplets and albedo.[198] Coccolithophores, calcarious plankton, are thought to be important producers of DMS but are also sensitive to ocean acidification, implying a feedback between acidification and climate, although this may be small.[199] Microscopic plankton also influence the near-surface radiation budget through changes in marine albedo and absorption of solar radiation.[200,201]

These examples indicate that feedbacks between ocean ecosystems and climate change are highly complex and their magnitude and even their sign is still uncertain.

5.2 Oxygen Depletion and Harmful Algal Blooms (HABs)

Marine productivity is a primary mechanism for carbon uptake in the ocean (see Section 2.2). If this productivity were distributed evenly across the ocean there would be little negative effect on the ecosystem, even if the productivity and carbon uptake were to increase considerably above current levels. However, productivity is far from uniform (see Figure 6) and may cause detrimental

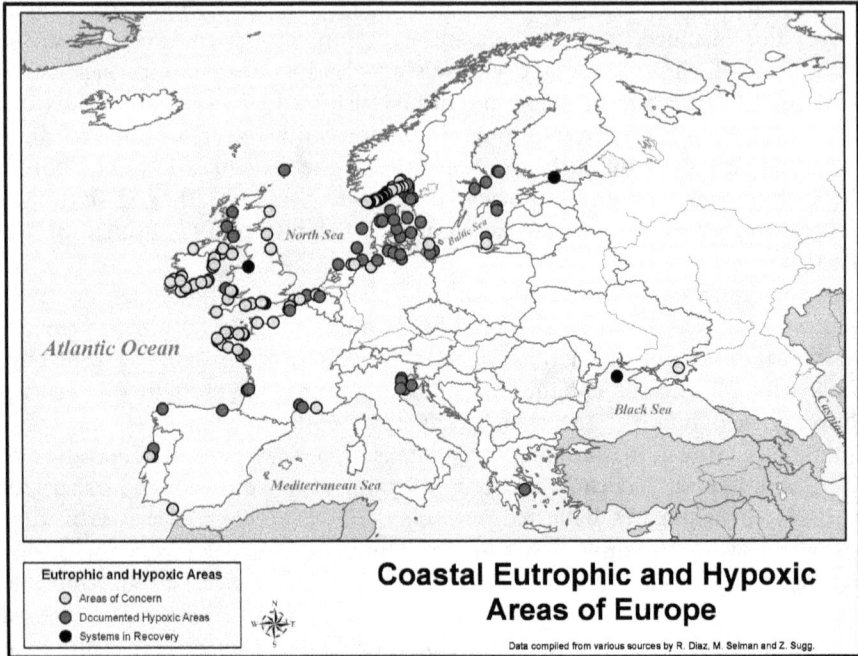

Figure 11 This map identifies 168 eutrophic and hypoxic coastal areas in Europe. Fifty-nine of these are documented hypoxic areas, 106 are areas of concern, and three are improved systems that are in recovery. This is based on multi-national surveys, such as those carried out by the Commission for the Protection of the Marine Environment of the North-East Atlantic.[208] (Reproduced from the Harmful Plankton Project website, University of Liverpool, courtesy of World Resources Institute).[207,215]

impacts when vast numbers of phytoplankton cells are concentrated in high-biomass algal blooms. Often only one or a few species dominate the bloom in concentrations of up to a billion cells per litre, covering a small region of an estuary or thousands of square kilometres of the shelf seas. The cause of such inhomogeneous phytoplankton growth is the horizontal distribution of nutri-ents that result from a complex relationship between the various inputs – deep ocean, riverine and atmospheric deposition – and mixing processes[202] (Section 2.2). Eutrophication of coastal zones due to increased riverine nutrient loads has often been blamed for increases in algal blooms and degradation of water quality,[203] though this a highly controversial topic, indicative of the difficulty in separating anthropological effects (*e.g.* agricultural run-off) from natural or climatic variations.[204]

Proliferation of phytoplankton can sometimes lead to harmful algal blooms (HABs). HABs can be broadly classified into two types: those that produce toxins that can contaminate seafood or kill fish even when their cell density is low, and those where high biomass causes impacts such as oxygen

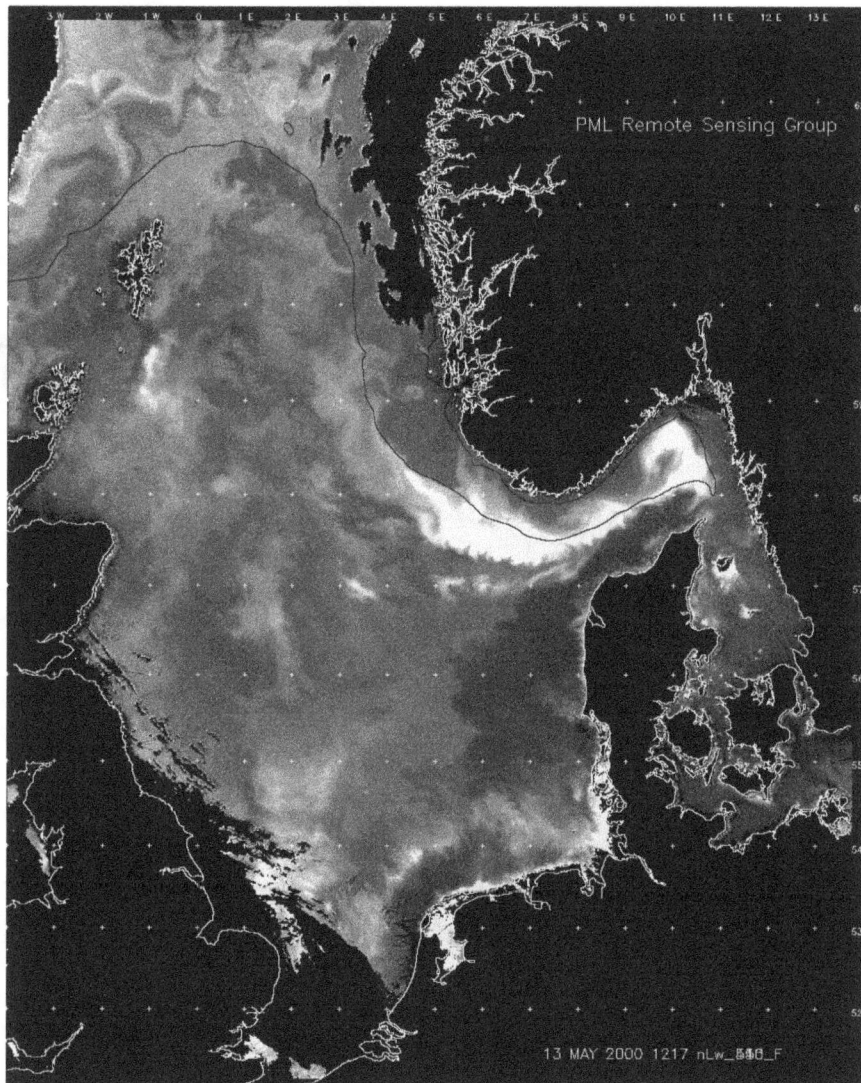

Figure 12 A harmful bloom of the flagellate *Chattonella* in the North Sea observed using SeaWiFS enhanced colour (using water-leaving radiance at 555, 510 and 443 nm) on 13 May 2000 1217 UTC (the bloom is the dark grey patch west off Denmark).[211] A similar bloom in 1998 killed 350 tonnes of farmed salmon in Norway.

depletion as the bloom decays, production of unpleasant scums, or other reduction of habitat for fish or shellfish.[205] Blooms are most likely to cause harm when they occur in restricted-exchange environments such as in estuaries or fjords.[206]

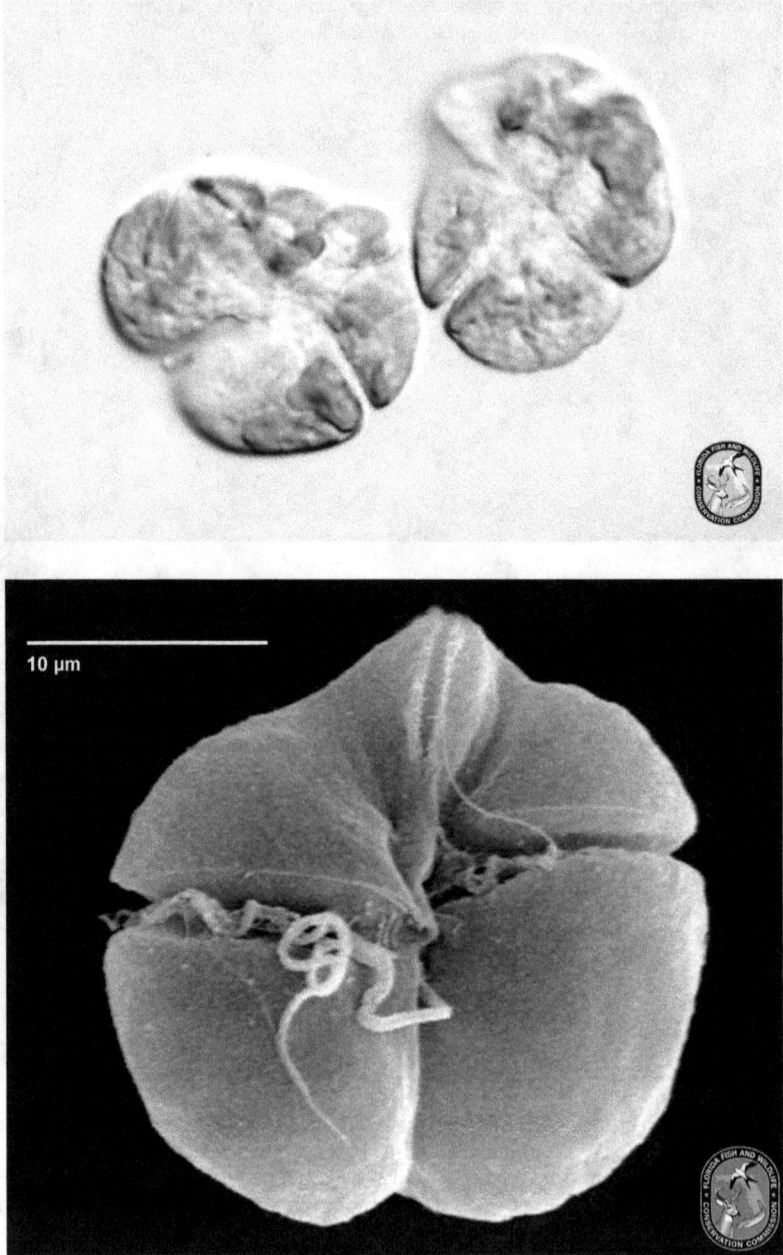

Figure 13   Harmful dinoflagellate algae *Karenia brevis* viewed under a microscope: Two cells under a light microscope (upper image) and one cell under a scanning electron microscope (lower). (Reproduced courtesy of the World Resources Institute).[207,215]

5.2.1 Oxygen Depletion

The most significant harm caused by high-biomass blooms is oxygen depletion, usually caused when dead phytoplankton cells sink down the water column and are decomposed by bacteria. The degree of depletion is determined by the quantity of organic matter accumulated, the stability of the water column, and the bathymetry.

Hypoxia is the state when the dissolved oxygen concentration is low enough to be detrimental to aquatic organisms (below 30% saturation), and anoxia when there is no oxygen. Hypoxia may have a drastic effect on the benthic environment and pelagic fish.[209,210] The World Resources Institute has identified 375 hypoxic coastal zones around the world, concentrated in coastal areas in Western Europe (see Figure 11), the Eastern and Southern coasts of the US, and East Asia.[207]

5.2.2 Toxic Algal Blooms

HABs that produce algal toxins are of great concern to public health and the fish-farming and aquaculture industries. The most toxic species are recorded among dinoflagellates; for example, several species of the genus *Alexandrium* which can have dramatic effects at barely detectable concentrations (10^2–10^3 cells l^{-1}); though a number of diatoms and cyanobacteria also produce neurotoxins that can endanger human health. The toxins may have direct impacts on wild and farmed fish and shellfish (*e.g.* Figure 12), though some non-toxic species may also kill fish by clogging their gills.

Human toxic syndromes caused by HABs include paralytic, diarrhetic, neurotoxic and amnesic shellfish poisoning;[205] these risks have instigated costly monitoring programmes and product-recalls to ensure the safety of seafood, for example, run by the Foods Standards Agency for the UK. There are many active areas of research relating to HABs, including attempts to understand environmental factors involved in their initiation, characterisation of toxins from *in situ* water samples,[212] and the early warning of potential HABs using satellite Earth-observation data.[211,213] Algal toxins such as those produced by *K. brevis* (see Figure 13) may even become airborne in aerosols under certain conditions of onshore winds and breaking surf, leading to respiratory and other health effects in humans and other mammals living near to the coast.[214]

5.2.3 Consequences of Change on Oxygen Depletion and HABs

Climate change may already be having an influence on HAB occurrence; for example, in the north-east Atlantic where Continuous Plankton Recorder surveys have reported increases in blooms of certain species since 1958, the increases are regional and are often associated with lower salinities or higher temperatures.[216] Higher temperatures and increase in rainfall and land run-off in the future may promote additional high-biomass blooms through better initial growth conditions and increased stability, the latter of which could also further increase the likelihood of hypoxia. Biological productivity is not normally limited by CO_2 availability (Section 5.2) and so an anthropogenic increase, or leakage from a carbon-storage system, would not directly affect HABs. As previously mentioned, the influence of eutrophication on algal bloom growth is a primary concern for the immediate future.

High-biomass blooms result in a significant export of carbon to the deep ocean *via* the biological pump (Section 3.2); for example, a peak uptake of *ca.* $2 \, \mathrm{g \, C \, m^{-2} \, d^{-1}}$ was measured during a diatom bloom in the north-east Atlantic.[217] As oxygen depletion is caused by a high carbon flux and subsequent remineralisation, those high-biomass blooms which achieve the greatest carbon uptake would also be most likely to cause detrimental impacts to the marine environment. Indeed, a recent global ocean-modelling study shows increased deoxygenation of intermediate depths in tropical waters with increased ocean acidification.[218] Future increases to marine primary production due to climate change, eutrophication or artificial fertilisation, while increasing carbon uptake would almost certain increase the severity of HABs and oxygen depletion.

References

1. J. L. Sarmiento and E. T. Sundquist, *Nature*, 1992, **356**, 589.
2. J. L. Sarmiento, T. M. C. Hughes, R. J. Stouffer and S. Manabe, *Nature*, 1998, **393**, 245.
3. P. D. Nightingale and P. S. Liss, in *The Oceans and Marine Goechemistry: A Treatise on Geochemistry*, ed. H. Elderfield, Elsevier, 2003, **vol. 6**.
4. P. N. Pearson and M. R. Palmer, *Nature*, 2000, **406**, 695.
5. E. Maier-Reimer and K. Hasselmann, *Climate Dyn.*, **2**, 63.
6. R. E. Zeebe and D. Wolf-Gladrow, *CO₂ in Seawater: Equilibrium, Kinetics, Isotopes,* Elsevier Oceanography Series 65, Elsevier, Amsterdam, 2001.
7. B. Wozniak, S. B. Wozniak, K. Tyszka, M. Ostrowska, D. Ficek, R. Majchrowski and J. Dera, *Oceanologia*, 2003, **45**, 171.
8. J. Aiken, N. J. Hardman-Mountford, R. Barlow, J. Fishwick, T. Hirata and T. Smyth, *J. Plankton Res.*, 2008, **30**, 165.
9. J. T. O. Kirk, *Light and Photosynthesis in Aquatic Ecosystems,* Cambridge University Press, Cambridge, 2000.
10. A. Morel and S. Maritorena, *J. Geophys. Res.*, 2001, **106**, 7163.
11. A. Morel and A. Bricaud, *Deep-Sea Res. II*, 1981, **28**, 1375.
12. R. Del Vecchio and N. V. Blough, *Marine Chem.*, 2002, **78**, 231.
13. J. I. Hedges, R. G. Keil and R. Benner, *Org. Geochem.*, 1997, **27**, 195.
14. A. Bricaud, A. Morel and L. Prieur, *Limnol. Oceanogr.*, 1981, **26**, 43.
15. G. H. Tilstone, T. J. Smyth, R. J. Gowen, V. Martinez-Vicente and S. B. Groom, *J. Plankton Res.*, 2005, **27**, 1127.
16. Z. Dubinsky, in *Primary Productivity in the Sea*, ed. P. G. Falkowski, Plenum Press, New York and London, 1980, 83.
17. A. J. Young, *Physiol. Plant.*, 1991, **83**, 702.
18. H. A. Bouman, T. Platt, G. W. Kraay, S. Sathyendranath and B. D. Irwin, *Mar. Ecol.: Prog. Ser.*, 2000, **200**, 3.

19. J. Marra, C. C. Trees, R. R. Bidigare and R. T. Barber, *Deep-Sea Res. II*, 2000, **47**, 1279.
20. E. Steeman Nielsen, *J. Conseil Int. Explorat. Mer*, 1952, **18**, 117.
21. T. Platt, C. L. Gallegos and W. G. Harrison, *J. Mar. Res.*, 1980, **38**, 687.
22. E. L. Smith, *Proc. Natl. Acad. Sci. U. S. A.*, 1936, **22**, 504.
23. D. A. Kiefer and J. B. Soohoo, *Limnol. Oceanogr.*, 1982, **27**, 492.
24. A. Bricaud and D. Stramski, *Limnol. Oceanogr.*, 1990, **35**, 562.
25. M. N. Kyewalyanga, T. Platt and S. Sathyendranath, *Mar. Ecol.: Prog. Ser.*, 1992, **85**, 171.
26. A. Morel, *Prog. Oceanogr.*, 1991, **26**, 263.
27. M. N. Kyewalyanga, T. Platt and S. Sathyendranath, *Mar. Ecol.: Prog. Ser.*, 1997, **146**, 207.
28. F. G. Figueiras, B. Arbones and M. Estrada, *Limnol. Oceanogr.*, 1999, **44**, 1599.
29. G. H. Tilstone, F. G. Figueiras, L. M. Lorenzo and B. Arbones, *Mar. Ecol.: Prog. Ser.*, 2003, **252**, 89.
30. G. H. Tilstone, T. J. Smyth, A. Poulton and R. Hutson, *Deep-Sea Res. II*, in press.
31. S. Sathyendranath, A. Longhurst, C. M. Caverhill and T. Platt, *Deep-Sea Res. I*, 1995, **42**, 1773.
32. M. J. Behrenfeld and P. G. Falkowski, *Limnol. Oceanogr.*, 1997, **42**, 1479.
33. T. Platt, C. Caverhill and S. Sathyendranath, *J. Geophys. Res.*, 1991, **96**(C8), 15147.
34. T. J. Smyth, G. H. Tilstone and S. B. Groom, *J. Geophys. Res.*, 2005, **110**(C10), C10014.
35. M.-E. Carr, M. A. M. Friedrichs, M. Schmeltz, M. N. Aita, D. Antoine, K. R. Arrigo, I. Asanuma, O. Aumont, R. Barber, M. Behrenfeld, R. Bidigare, E. Buitenhuis, J. Campbell, A. Ciotti, H. Dierssen, M. Dowell, J. Dunne, W. Esaias, B. Gentili, S. Groom, N. Hoepffner, J. Ishizaka, T. Kameda, C. LeQuere, S. Lohrenz, J. Marra, F. Melin, K. Moore, A. Morel, T. Reddy, J. Ryan, M. Scardi, T. Smyth, L. Turpie, G. Tilstone, K. Waters and Y. Yamanaka, *Deep Sea Res. II*, 2006, **53**, 741.
36. M. Behrenfeld, R. T. O'Malley, D. Siegel, C. McClain, J. L. Sarmiento, G. C. Feldman, A. J. Milligan, P. Falkowski, R. M. Letelier and E. Boss, *Nature*, 2006, **444**, 752.
37. A. Bricaud, K. Allali, R. Morel, D. Marie, M. J. W. Veldhuis, F. Partensky and D. Vaulot, *Mar. Ecol.: Prog. Ser.*, 1999, **188**, 21.
38. L. R. Moore, R. Goericke and S. W. Chisholm, *Mar. Ecol.: Prog. Ser.*, 1995, **116**, 259.
39. L. R. Moore, G. Rocap and S. W. Chisholm, *Nature*, 1998, **393**, 464.
40. M. Stomp, J. Huisman, F. de Jongh, A. J. Veraart, D. Gerla, M. Rijkeboer, B. W. Ibelings, U. I. A. Wollenzien and L. J. Stal, *Nature*, 2004, **432**, 104.
41. Y. Kashiyama, H. Miyashita, S. Ohkubo, N. O. Ogawa, Y. Chikaraishi, Y. Takano, H. Suga, T. Toyofuku, H. Nomaki, H. Kitazato, T. Nagata and N. Ohkouchi, *Science*, 2008, **321**, 658.

42. W. D. Swingley, M. F. Hohmann-Marriott, T. Le Olson and R. E. Blankenship, *Appl. Environ. Microbiol.*, 2005, **71**, 8606.
43. Z. S. Kolber, C. L. Van Dover, R. A. Niederman and P. G. Falkowski, *Nature*, 2000, **407**, 177.
44. Z. S. Kolber, F. G. Plumley, A. S. Lang, J. T. Beatty, R. E. Blankenship, C. L. Van Dover, C. Vetriani, M. Koblizek, C. Rathgeber and P. G. Falkowski, *Science*, 2001, **292**, 2492.
45. V. V. Yurkov and J. T. Beatty, *Microbiol. Mol. Biol. Rev.*, 1998, **62**, 695.
46. O. Beja, L. Aravind, E. V. Koonin, M. T. Suzuki, A. Hadd, L. P. Nguyen, S. Jovanovich, C. M. Gates, R. A. Feldman, J. L. Spudich, E. N. Spudich and E. F. DeLong, *Science*, 2000, **289**, 1902.
47. L. Gomez-Consarnau, J. M. Gonzalez, M. Coll-Llado, P. Gourdon, T. Pascher, R. Neutze, C. Pedros-Alio and J. Pinhassi, *Nature*, 2007, **445**, 210.
48. C. B. Field, M. J. Behrenfeld, J. T. Randerson and P. Falkowski, *Science*, 1998, **281**, 237.
49. W. S. Broecker, *Oceanography*, 1991, **4**, 79.
50. T. D. Jickells, *Science*, 1998, **281**, 217.
51. Atlantic Meridional Transect programme. http://www.amt-uk.org 2009, accessed 11/04/2009.
52. E. M. S. Woodward, Plymouth Marine Laboratory, personal communication.
53. A. C. Redfield, in *James Johnstone Memorial Volume*, ed. R. J. Daniel, University Press of Liverpool, 1934, 177.
54. J. H. Martin, K. H. Coale, K. S. Johnson, S. E. Fitzwater, R. M. Gordon, S. J. Tanner, C. N. Hunter, V. A. Elrod, J. L. Nowicki, T. L. Coley, R. T. Barber, S. Lindley, A. J. Watson, K. Vanscoy, C. S. Law, M. I. Liddicoat, R. Ling, T. Stanton, J. Stockel, C. Collins, A. Anderson, R. Bidigare, M. Ondrusek, M. Latasa, F. J. Millero, K. Lee, W. Yao, J. Z. Zhang, G. Friederich, C. Sakamoto, F. Chavez, K. Buck, Z. Kolber, R. Greene, P. Falkowski, S. W. Chisholm, F. Hoge, R. Swift, J. Yungel, S. Turner, P. Nightingale, A. Hatton, P. Liss and N. W. Tindale, *Nature*, 1994, **371**, 123.
55. E. T. Degens, S. Kempe and A. Spitzy, in *The Handbook of Environmental Chemistry*, ed. O. Hutzinger, Springer-Verlag, Berlin Heidelberg, 1984, 1(3), 127.
56. Royal Society, *Ocean Acidification due to Increasing Atmospheric Carbon Dioxide*, Policy document 12/05, June 2005, The Royal Society, London, 2005. http://www.royalsoc.ac.uk/document.asp?tip=0&id=3249.
57. K. L. Denman, G. Brasseur, A. Chidthaisong, P. Ciais, P. M. Cox, R. E. Dickinson, D. Hauglustaine, C. Heinze, E. Holland, D. Jacob, U. Lohman, S. Ramachandran, P. L. da Silva Dias, S. C. Wofsy and X. Zhang, in *Climate Change 2007: The Physical Science Basis. Contribution of Working Group I to the Fourth Assessment Report of the Intergovermental Panel on Climate Change*, ed. S. Solomon, D. Qin, M. Manning, Z. Chen, M. Marquis, K. B. Averyt, M. Tignor and H. L. Miller, Cambridge University Press, Cambridge, United Kingdom and New York, NY, USA, 2007.

58. T. Takahashi, J. Olafsson, J. G. Goddard, D. W. Chipman and S. C. Sutherland, *Global Biogeochem. Cycles*, 1993, **7**, 843.
59. A. J. Watson and J. C. Orr, in *Ocean Biogeochemistry: The Role of the Ocean Carbon Cycle in Global Change*, ed. M. J. R. Fasham, Springer-Verlag, Berlin Heidelberg, 2003, 123.
60. H. Elderfield, *Science*, 2002, **296**, 1618.
61. R. Revelle and H. E. Suess, *Tellus*, 1957, **9**, 18.
62. C. L. Sabine, R. A. Feely, N. Gruber, R. M. Key, K. Lee, J. L. Bullister, R. Wanninkhof, C. S. Wong, D. W. R. Wallace, B. Tilbrook, F. J. Millero, T. -H. Peng, A. Kozyr, T. Ono and A. F. Rios, *Science*, 2004, **305**, 367.
63. K. Richardson and R. B. Hanson, in *Global Change and the Earth System a Planet Under Pressure*, ed. W. Steffen, A. Sanderson, P. D. Tyson, J. Jäger, P. A. Matson, B. Moore III, F. Oldfield, K. Richardson, H. J. Schellnhuber, B. L. Turner II and R. J. Wasson, Springer-Verlag, Berlin, Heidelberg, 2004, 40.
64. C. L. Sabine, R. Wanninkhof, R. M. Key, C. Goyet and F. J. Millero, *Mar. Chem.*, 2000, **72**, 33.
65. R. Torres, D. R. Turner, N. Silva and J. Rutllant, *Deep-Sea Res. I*, 1999, **46**, 1161.
66. A. Yool and M. J. R. Fasham, *Global Biogeochem. Cycles*, 2001, **15**, 831.
67. M. A. Omar, T. Johannessen, A. Olsen, S. Kaltin and F. Rey, *Mar. Chem.*, 2007, **104**, 203.
68. S. Tsunogai, S. Watanabe and T. Sato, *Tellus*, 1999, **51B**, 701.
69. H. Thomas, Y. Bozec, K. Elkalay and H. J. W. De Baar, *Science*, 2004, **304**, 1005.
70. C. -T. A. Chen and A. V. Borges, *Deep-Sea Res. II*, in press.
71. S. C. Doney, I. Lima, R. A. Feely, D. M. Glover, K. Lindsay, N. Mahowald, J. K. Moore and R. Wanninkhof, *Deep-Sea Res. II*, in press.
72. F. P. Chavez, P. G. Strutton, G. E. Friederich, R. A. Feely, G. C. Feldman, D. G. Foley and M. J. McPhaden, *Science*, 1999, **286**, 2126.
73. C. Le Quéré, J. C. Orr, P. Monfray and O. Aumont, *Global Biogeochem. Cycles*, 2000, **14**, 1247.
74. C. Le Quéré, O. Aumont, L. Bopp, P. Bousquet, P. Ciais, R. Francy, M. Heimann, C. D. Keeling, H. Kheshgi, P. Peylin, S. C. Piper, I. C. Prentice and P. J. Rayner, *Tellus*, 2003, **55**, 649.
75. N. Gruber, C. D. Keeling and N. R. Bates, *Science*, 2003, **298**, 2374.
76. J. E. Dore, R. Lukas, D. W. Sadler and D. M. Karl, *Nature*, 2003, **424**, 754.
77. N. R. Bates, A. C. Pequignet, R. J. Johnson and N. Gruber, *Nature*, 2003, **420**, 489.
78. A. Corbière, N. Metzl, G. Reverdin, C. Brunet and T. Takahashi, *Tellus*, 2007, **59B**, 168.
79. U. Schuster and A. J. Watson, *J. Geophy. Res.*, 2007, **112**, C11006.
80. G. Wang and D. Schimel, *Annu. Rev. Environ. Resour.*, 2003, **28**, 1.

81. R. A. Feely, J. Boutin, C. E. Cosca, Y. Dandonneau, J. Etcheto, H. Y. Inoue, M. Ishii, C. Le Quéré, D. J. Mackey, M. McPhaden, N. Metzl, A. Poisson and R. Wanninkhof, *Deep-Sea Res. II*, 2002, **49**, 2443.

82. R. A. Feely, R. Wanninkhof, T. Takahashi and P. Tans, *Nature*, 1999, **398**, 597–601.

83. M. H. Visbeck, J. W. Hurrell, L. Polvani and H. M. Cullen, *Proc. Natl. Acad. Sci. U. S. A.*, 2001, **98**, 12876.

84. T. Takahashi, S. C. Sutherland, R. A. Feely and C. E. Cosca, *Science*, 2003, **32**, 852.

85. R. A. Feely, T. Takahashi, R. Wanninkhof, M. J. McPhaden, C. E. Cosca, S. C. Sutherland and M. E. Carr, *J. Geophys. Res.*, 2006, **111**, C07S03.

86. A. Lenton and R. Matear, *Global Biogeochem. Cycles*, 2007, **21**, GB2016.

87. N. S. Lovenduski, N. Gruber, S. C. Doney and I. D. Lima, *Global Biogeochem. Cycles*, 2007, **21**, GB2026.

88. C. Le Quéré, C. Rodenbeck, E. T. Buitenhuis, T. J. Conway, R. Lagenfelds, A. Gomez, C. Labuschagne, M. Ramonet, T. Nakazawa, N. Metzl, N. Gillett and M. Heimann, *Science*, 2007, **316**, 1735.

89. A. Verdy, S. Dutkiewicz, M. J. Follows, J. Marshall and A. Czaja, *Global Biogeochem. Cycles*, 2007, **21**, GB2020.

90. N. R. Bates, A. H. Knap and A. F. Michaels, *Nature*, 1998, **395**, 58.

91. B. Rost and U. Riebesell, in *Coccolithophores: From Molecular Processes to Global Impact*, ed. H. R. Thierstein and J. R. Young, Springer, 2004, 76.

92. N. Gruber and J. L. Sarmiento, in *The Sea: Biological-Physical Interactions in the Oceans*, ed. A. R. Robinson, J. J. McCarthy and B. J. Rothschild, John Wiley and Sons, 2002, **12**, 337.

93. T. Volk and M. I. Hoffert, in *The Carbon Cycle and Atmospheric CO₂: Natural Variations Archean to Present*, ed. E. T. Sunquist and W. S. Broecker, Geophys. Monogr. Ser. 32, Washington, DC, 1985, 99.

94. C. M. Turley, *FEMS Microbiol. Ecol.*, 2000, **1154**, 1.

95. A. L. Alldredge and E. O. Hartwig, *Aggregate Dynamics in the Sea*, Workshop Report, Asilomar Conference Center, Pacific Grove, California, September 22–24 1986, Office of Navel Research, American Institute of Biological Sciences, Washington DC, 1986.

96. C. M. Turley, *Arch. Hydrobiol. Beih. Ergebn. Limnol.*, 1992, **37**, 155.

97. A. Engel, S. Thoms, U. Riebesell, E. Rochelle-Newall and I. Zondervan, *Nature*, 2004, **428**, 929.

98. C. M. Turley, K. Lochte and R. S. Lampitt, *Phil. Trans. R. Soc. London, Ser. B*, 1995, **348**, 179.

99. R. S. Lampitt, *Deep-Sea Res.*, 1985, **32**, 885.

100. A. J. Gooday and C. M. Turley, *Phil. Trans. R. Soc. London, Ser. A*, 1990, **331**, 119.

101. P. M. Holligan, E. Fernandez, J. Aiken, W. M. Balch, P. Boyd, P. H. Burkill, M. Finch, S. B. Groom, G. Malin, K. Muller, D. A. Purdie, C. Robinson, C. C. Trees, S. M. Turner and P. van der Wal, *Global Biogeochem. Cycles*, 1993, **7**, 879.

102. J. Hansen, M. Sato, R. Ruedy, K. Lo, D. W. Lea and M. Medina-Elizade, *Proc. Natl. Acad. Sci. U. S. A.*, 2006, **103**, 14288.
103. S. Levitus, J. Antonov and T. Boyer, *Geophys. Res. Lett.*, 2005, **32**, L02604.
104. G. C. Johnson and S. C. Doney, *Geophys. Res. Lett.*, 2006, **33**, L14614.
105. G. C. Johnson, S. Mecking, B. M. Sloyan and S. E. Wijffels, *J. Climate*, 2007, **20**, 5365–5375.
106. D. W. J. Thompson and S. Solomon, *Science*, 2002, **296**, 895.
107. J. L. Russell, K. W. Dixon, A. Gnanadesikan, R. J. Stouffer and J. R. Toggweiler, *J. Climate*, 2006, **19**, 6382.
108. National Snow and Ice Data Centre. http://nsidc.org2009, accessed 10/04/2009.
109. S. C. Doney, B. Tilbrook, S. Roy, N. Metzl, C. Le Quéré, M. Hood, R. A. Feely and D. Bakker, *Deep-Sea Res. II*, in press.
110. K. Zickfield, J. C. Fyfe, M. Eby and A. J. Weaver, *Science*, 2008, **309**, 570.
111. U. Schuster, A. J. Watson, N. Bates, A. Corbière, M. Gonzalez-Davila, N. Metzl, D. Pierrot and M. Santana-Casiano, *Deep-Sea Res. I.*, accepted.
112. H. Thomas, A. E. F. Prowe, I. D. Lima, S. C. Doney, R. Wanninkhof, R. J. Greatbatch, U. Schuster and A. Corbière, *Global Biogeochem. Cycles*, 2008, **22**, GB4027. http://www.agu.org/journals/gb/gb0804/2007GB003167/.
113. J. G. Canadell, C. Quéré, M. R. Raupach, C. B. Field, E. T. Buitenhuis, P. Ciais, T. J. Conway, N. P. Gillett, R. A. Houghton and G. Marland, *Proc. Natl. Acad. Sci. U. S. A.*, 2007, **104**, 18886.
114. W. O. Smith and J. C. Comiso, *J. Geophys. Res.*, 2008, **113**, C05S93. http://www.agu.org/pubs/crossref/2008/2007JC004251.shtml.
115. K. Arrigo, G. L. van Dijken and S. Bushinsky, *J. Geophys. Res.*, 2008, **113**, C08004.
116. S. Pabi, G. L. van Dijken and K. Arrigo, *J. Geophys. Res.*, 2008, **113**, C08005.
117. A. V. Borges and R. Wanninkhof, *J. Mar. Syst.*, 2007, **66**, 1.
118. A. Merico, T. Tyrrell and T. Cokacar, *J. Mar. Syst*, 2006, **59**, 120.
119. J. C. Blackford, N. Jones, R. Proctor and J. Holt, *Mar. Pollut. Bull.*, 2008, **56**, 1461.
120. K. Caldeira and M. E. Wickett, *Nature*, 2003, **425**, 365.
121. J. C. Blackford and F. J. Gilbert, *J. Mar. Syst.*, 2007, **64**, 229.
122. J. C. Orr, V. J. Fabry, O. Aumont, L. Bopp, S. C. Doney, R. A. Feely, A. Gnanadesikan, N. Gruber, A. Ishida, F. Joos, R. M. Key, K. Lindsay, E. Maier-Reimer, R. Matear, P. Monfray, A. Mouchet, R. G. Najjar, G. -K. Plattner, K. B. Rodgers, C. L. Sabine, J. L. Sarmiento, R. Schlitzer, R. D. Slater, I. J. Totterdell, M. -F. Weirig, Y. Yamanaka and A. Yool, *Nature*, 2005, **437**, 681.
123. M. Steinacher, F. Joos, T. L. Frölicher, G. -K. Plattner and S. C. Doney, *Biogeosci. Discuss.*, 2008, **5**, 4353.
124. F. J. Millero, K. Lee and M. Roche, *Mar. Chem.*, 1998, **60**, 111.

125. K. Lee, L T. Tong, F. J. Millero, C. L. Sabine, A. G. Dickson, C. Goyet, G. H. Park, R. Wanninkof, R. A. Feely and R. M Key, *Geophys. Res. Lett.*, 2006, **33**, L19605.http://cdiac.ornl.gov/ftp/oceans/KitackLee_Alk_Climatology/GRL_Surface_alk.pdf.

126. R. G. J. Bellerby, A. Olsen, T. Furevik and L. G. Anderson, in *The Nordic Seas An Integrated Perspective*, ed. H. Drange, T. Dokken, T. Furevik, R. Gerdes and W. Berger, Geophys. Monograph Series 158, American Geophysical Union, 2005, 189.

127. J. Pätsch and H. J. Lenhart, *Ber. Zentr. Meeres- Klimaforsch. Reihe B: Ozeanogr.*, 2004, **48**, 159.

128. P. A. Raymond and J. J. Cole, *Science*, 2003, **301**, 88.

129. R. A. Feely, C. L. Sabine, K. Lee, W. Berelson, J. Kleypas, V. J. Fabry and F. J. Millero, *Science*, 2004, **305**, 362.

130. R. A. Feely, C. L. Sabine, J. M. Hernandez-Ayon, D. Ianson and B. Hales, *Science*, 2008, **320**, 1490.

131. S. Dupont, J. Havenhand, W. Thorndyke, L. Peck and M. Thorndyke, *Mar. Ecol.: Prog. Ser.*, 2008, **373**, 285.

132. J. C. Orr, S. Jutterström, L. Bopp, L. G. Anderson, V. J. Fabry, T. L. Frölicher, P. Jones, F. Joos, E. Maier-Reimer, J. Segschneider, M. Steinacher and D. Swingedouw, *Nature*, submitted.

133. J. Bijma, H. J. Spero and D. W. Lea, in *Use of Proxies in Paleoceanography: Examples from the South Atlantic*, ed. G. Fisher and G. Wefer, Springer-Verlag, New York, 1999, 489.

134. J. Bijma, B. Hönisch and R. E. Zeebe, *Geochem. Geophys. Geosyst.*, 2002, **3**, 1064.

135. V. J. Fabry, B. A. Seibel, R. A. Feely and J. C. Orr, *ICES J. Mar. Sci.*, 2008, **65**, 414.

136. U. Riebesell, I. Zondervan, B. Rost, P. D. Tortell, R. Zeebe and F. M. Morel, *Nature*, 2000, **407**, 364–367.

137. I. Zondervan, R. E. Zeebe, B. Rost and U. Riebesell, *Global Biogeochem. Cycles*, 2001, **15**, 507.

138. M. D. Iglesias-Rodriguez, P. R. Halloran, R. E. M. Rickaby, I. R. Hall, E. Colmenero-Hidalgo, J. R. Gittins, D. R. H. Green, T. Tyrrell, S. J. Gibbs, P. von Dassow, E. Rehm, E. V. Armbrust and K. P. Boessenkool, *Science*, 2008, **320**, 336.

139. A. Ridgwell, D. N. Schmidt, C. Turley, C. Brownlee, M. T. Maldonado, P. Tortell and J. R. Young, *Biogeosci. Discuss.*, 2009, **6**, 3455.

140. U. Riebesell, R. G. J. Bellerby, A. Engel, V. J. Fabry, D. A. Hutchins, T. B. H. Reusch, K. G. Schulz and F. M. M. Morel, *Science*, 2008, **322**, 1466.

141. G. S. Malin, P. Turner, P. Liss, P. Holligan and D. Harbour, *Deep-Sea Res. I*, 1993, **40**, 1487.

142. R. J. Charlson, J. E. Lovelock, M. O. Andreae and S. G. Warren, *Nature*, 1987, **326**, 655.

143. A. P. Rees, J. L. Dixon, S. Widdicombe and N. Wyatt, *CIESM Workshop Monograph*, 36, in press.

144. M. H. Huesemann, A. D. Skillman and E. A. Crecelius, *Mar. Pollut. Bull.*, 2002, **44**, 142.

145. D. A. Hutchins, F. X. Fu, M. E. Warner, Y. Fenh, K. Portune, P. W. Bernhardt and M. R. Mulholland, *Limnol. Oceanogr.*, 2007, **52**, 1293.

146. P. W. Boyd, A. J. Watson, C. S. Law, E. R Abraham, T. Trull, R. Murdoch, D. C. E. Bakker, A. R. Bowie, K. O. Buesseler, H. Chang, M. Charette, P. Croot, K. Downing, R. Frew, M. Gall, M. Hadfield, J. Hall, M. Harvey, G. Jameson, J. LaRoche, M. Liddicoat, R. Ling, M. T. Maldonado, R. M. McKay, S. Nodder, S. Pickmere, R. Pridmore, S. Rintoul, K. Safi, P. Sutton, R. Strzepek, K. Tanneberger, S. Turner, A. Waite and J. Zeldis, *Nature*, 2000, **407**, 695.

147. J. R. Andrews, P. Brimblecombe, T. D. Jickells, P. S. Liss and B. Reid, *An Introduction to Environmental Chemistry*, Blackwell Science, Oxford UK, 1996.

148. S. Burkhardt and U. Riebesell, *Mar. Ecol.: Prog. Ser.*, 1997, **155**, 67.

149. F. Gervais and U. Riebesell, *Limnol. Oceanogr.*, 2001, **46**, 497.

150. J. Beardall and J. A. Raven, *Phycologia*, 2004, **43**, 31.

151. M. Giordano, J. Beardall and J. A. Raven, *Ann. Rev. Plant. Biol.*, 2005, **56**, 99.

152. B. Rost, U. Riebesell, S. Burkhardt and D. Sültemeyer, *Limnol. Oceanogr.*, 2003, **48**, 55.

153. A. Engel, I. Zondervan, K. Aerts, L. Beaufort, A. Benthien, L. Chou, B. Delille, J. -P. Gattuso, J. Harly, C. Heemaan, L. Hoffmann, S. Jacquet, J. Nejstgaard, M. -D. Pizay, E. S. Rochelle-Newall, U. Schneider, A. Terbrueggen and U. Riebesell, *Limnol. Oceanogr.*, 2005, **50**, 493.

154. B. A. Seibel and P. J. Walsh, *Science*, 2001, **294**, 319.

155. H. O. Pörtner, *Mar. Ecol.: Prog. Ser.*, 2008, **373**, 203.

156. B. A. Seibel and P. J. Walsh, *J. Exp. Biol.*, 2003, **206**, 641.

157. H. O. Pörtner and A. P. Farrell, *Science*, 2008, **322**, 690.

158. H. O. Pörtner, M. Langenbuch and B. Michaelidis, *J. Geophys. Res.*, 2005, **110**, C09S10. http://www.agu.org/pubs/crossref/2005/2004JC002561.shtml.

159. S. Widdicombe and J. I. Spicer, *J. Exp. Mar. Biol. Ecol.*, 2008, **366**, 187.

160. B. Michaelidis, C. Ouzounis, A. Paleras and H. O. Pörtner, *Mar. Ecol.: Prog. Ser.*, 2005, **293**, 109.

161. C. M. Wood and J. N. Cameron, *J. Exp. Biol.*, 1985, **114**, 151.

162. E. F. Pane and J. P. Barry, *Mar. Ecol.: Prog. Ser.*, 2007, **334**, 1–9.

163. H. Miles, S. Widdicombe, J. I. Spicer and J. Hall-Spencer, *Mar. Pollut. Bull.*, 2007, **54**, 89.

164. H. O. Pörtner, M. Langenbuch and A. Reipschläger, *J. Oceanogr.*, 2004, **60**, 705.

165. R. W. Putnam and A. Roos, in *Handbook of Physiology, Cell Physiology*, ed. J. F. Hoffman and J. D. Jamieson, Oxford University Press, New York, 1997, 389.

166. C. P. Mangum and L. E. Burnett, *Biol. Bull.*, 1986, **171**, 248.

167. R. E. Weber, *Amer. Zool.*, 1980, **20**, 79.

168. C. P. Mangum, in *Handbook of Physiology, Section 13: Comparative Physiology II*, ed. W. H. Dantzler, Oxford University Press, Oxford, 1997.
169. A. Reipschläger, G. E. Nilsson and O. Pörtner, *Am. J. Physiol.*, 1997, **272**, 350.
170. J. N. Havenhand, F. R. Butler, M. C. Thorndyke and J. E. Williamson, *Curr. Biol.*, 2008, **18**, 651.
171. H. Kurihara and Y. Shirayama, *Mar. Ecol.: Prog. Ser.*, 2004, **274**, 161.
172. H. Kurihara, S. Kato and A. Ishimatu, *Aquat. Biol.*, 2007, **1**, 91.
173. K. Sakai, *Res. Popul. Ecol*, 1998, **40**, 287.
174. K. Sakai, *Biol. Bull.*, 1998, **195**, 319.
175. T. Kikkawa, J. Kita and A. Ishimatsu, *Mar. Pollut. Bull.*, 2004, **48**, 108.
176. U. Riebesell, I. Zondervan, B. Rost, P. D. Tortell, R. E. Zeebe and F. M. M. Morel, *Nature*, 2000, **407**, 364.
177. M. A. Green, M. E. Jones, C. L. Boudreau, R. L. Moore and B. A. Westman, *Limnol. Oceanogr.*, 2004, **49**, 727.
178. Y. Shirayama and H. Thornton, *J. Geophys. Res.*, 2004, **110**(C9), CO9S08.
179. P. M. Haugan, C. M. Turley and H. O. Pörtner, *Effects on the Marine Environment of Ocean Acidification resulting from Elevated Levels of CO_2 in the Atmosphere*, OSPAR Commission Report, 2006.
180. P. M. Haugan, *Possible Effects of Ocean Acidification*, Reports in Meteorology and Oceanography, University of Bergen Report Number, 2004, 2–224.
181. A. Beesley, D. M. Lowe, C. K. Pascoe and S. Widdicombe, *Clim. Change*, 2008, **37**, 215.
182. R. Bibby, P. Cleall-Harding, S. Rundle, S. Widdicombe and J. Spicer, *Biol. Lett.*, 2007, **3**, 699.
183. R. Metzger, F. J. Sartoris, M. Langenbuch and H. O. Pörtner, *J. Therm. Biol.*, 2007, **32**, 144.
184. H. L. Wood, J. I. Spicer and S. Widdicombe, *Proc. R. Soc. London, Ser. B.*, 2008, **275**, 1767.
185. J. Raven, K. Caldeira, H. Elderfield, O. Hough-Guldberg, P. Liss, U. Riebesell, J. Shepherd, C. M. Turley and A. Watson, *Ocean Acidification due to Increasing Atmospheric Carbon Dioxide*, Royal Society Policy document 2005.
186. C. Turley, J. Blackford, S. Widdicombe, D. Lowe, P. D. Nightingale and A. P. Rees, in *Avoiding Dangerous Climate Change*, ed. H. J. Schellnhuber, W. Cramer, N. Nakicenovic, T. Wigley and G. Yohe, Cambridge University Press, 2006, **8**, 65.
187. J. A. Kleypas, R. A. Feely, V. J. Fabry, C. Langdon, C. L. Sabine and L. L. Robbins, *Impacts of Ocean Acidification on Coral Reefs and Other Marine Calcifiers: A Guide for Future Research,* St Petersburg, Florida, 2006.
188. A. Fischlin, G. F. Midgley, J. T. Price, R. Leemans, B. Gopal, C. Turley, M. D. A. Rounsevell, O. P. Dube, J. Tarazona and A. A. Velichko, in *Impacts, Adaptation and Vulnerability. Contribution of Working Group II to the Fourth Assessment Report of the Intergovernmental Panel on Climate*

Change, ed. M. L. Parry, O. F. Canziani, J. P. Palutikof, P. J. van der Linden and C. E. Hanson, Cambridge University Press, Cambridge, 2007, 211.

189. S. Widdicombe and J. I. Spicer, *J. Exp. Mar. Biol. Ecol.*, 2008, **366**, 187.
190. J. M. Guinotte, J. Orr, S. Cairns, A. Freiwald, L. Morgan and R. George, *Frontiers Ecol. Environ.*, 2006, **4**, 141.
191. D. Archer, H. Kheshgi and E. Maier-Reimer, *Global Biogeochem. Cycles*, 1998, **12**, 259.
192. D. Archer, *J. Geophys. Res.*, 2005, **110**, C09S05. http://geosci.uchicago. edu/~archer/reprints/archer.2005.fate_co2.pdf.
193. ACIA, *Arctic Climate Impact Assessment*, Cambridge University Press, Cambridge, 1989.
194. L. Bopp, O. Boucher, O. Aumont, S. Belviso, J. L. Dufresne, P. Pham and P. Monfray, *Can. J. Fish. Aquat. Sci.*, 2004, **61**, 826.
195. P. W. Boyd, C. S. Law, C. S. Wong, Y. Nojiri, A. Tsuda, M. Levasseur, S. Takeda, R. Rivkin, P. J. Harrison, S. Strzepek, J. Gower, R. M McKay, E. Abraham, M. Arychuk, J. Barwell-Clarke, W. Crawford, D. Crawford, M. Hale, K. Harada, K. Johnson, H. Kiyosawa, I. Kudo, A. Marchetti, W. Miller, J. Needoba, J. Nishioka, H. Ogawa, J. Page, M. Robert, H. Saito, A. Sastri, N. Sherry, T. Soutar, N. Sutherland, Y. Taira, F. Whitney, S. K. E. Wong and T. Yoshimura, *Nature*, 2004, **428**, 549.
196. V. Ittekkot, *Global Planet. Change*, 1993, **8**, 17–25.
197. M. N. Mahowald and C. Luo, *Geophys. Res. Lett.*, 2003, **30**, 1903.
198. L. Bopp, O. Aumont, S. Belviso and P. Monfray, *Tellus Ser. B*, 2003, **55**, 11.
199. T. Tyrrell, P. M. Holligan and C. D. Mobley, *J. Geophys. Res.*, 1999, **104**(C2), 3223.
200. S. Sathyendranath, A. D. Gouveia, S. R. Shetye, P. Ravindran and T. Platt, *Nature*, 1991, **349**, 54.
201. P. Wetzel, E. Maier-Reimer, M. Botzet, J. Jungclaus, N. Keenlyside and M. Latif, *J. Climate*, 2006, **19**, 3973.
202. A. P. Martin, *Prog. Oceanogr.*, 2003, **57**, 125.
203. J. Heisler, P. M. Glibert, J. M. Burkholder, D. M. Anderson, W. Cochlan, W. C. Dennison, Q. Dortch, C. J. Gobler, C. A. Heil, E. Humphries, A. Lewitus, R. Magnien, H. G. Marshall, K. Sellner, D. A. Stockwell, D. K. Stoecker and M. Suddleson, *Harmful Algae*, 2008, **8**(1), 3.
204. P. Tett, R. Gowen, D. Mills, T. Fernandes, L. Gilpin, M. Huxham, K. Kennington, P. Read, M. Service, M. Wilkinson and S. Malcolm, *Mar. Pollut. Bull.*, 2007, **55**, 282.
205. P. Gilbert and G. Pitcher, *GEOHAB (Global Ecology and Oceanography of Harmful Algal Blooms) Science Plan, SCOR and IOC*, Baltimore and Paris, 2001.
206. B. G. Crespo, F. G. Figueiras, P. Porras and I. G. Teixeira, *Harmful Algae*, 2006, **5**, 770.
207. M. Selman, Z. Sugg, S. Greenhalgh and R. Diaz, *Eutrophication and Hypoxia in Coastal Areas: A Global Assessment of the State of Knowledge*, World Resources Institute, 2008.

208. OSPAR Commission, *OSPAR Integrated Report 2003 on the Eutrophication Status*, London, UK, OSPAR, 2003.
209. M. Plus, J. M. Deslous-Paoli and F. Dagault, *Aquat. Bot.*, 2003, **77**, 121.
210. M. Costantini, S. A. Ludsin, D. M. Mason, X. S. Zhang, W. C. Boicourt and S. B. Brandt, *Can. J. Fish. Aquat. Sci.*, 2008, **65**, 989.
211. P. I. Miller, J. D. Shutler, G. F. Moore and S. B. Groom, *Int. J. Remote Sensing*, 2006, **27**, 2287.
212. K. G. Sellner, G. J. Doucette and G. J. Kirkpatrick, *J. Ind. Microbiol. Biotechnol.*, 2003, **30**, 383.
213. C. R. McClain, *Ann. Rev. Mar. Sci.*, 2009, **1**, 19.
214. B. Kirkpatrick, L. E. Fleming, D. Squicciarini, L. C. Backer, R. Clark, W. Abraham, J. Benson, Y. S. Cheng, D. Johnson, R. Pierce, J. Zaias, G. D. Bossart and D. G. Baden, *Harmful Algae*, 2004, **3**(2), 99.
215. Harmful Plankton Project website, University of Liverpool, http://www.liv.ac.uk/hab/, accessed 11/04/2009.
216. R. Raine, M. Edwards, C. Reid, E. Bresnan and L. Fernand, *Scientific Review: Harmful Algal Blooms*, Annual Report Card 2007–2008, Marine Climate Change Impacts Partnership, 2008.
217. S. J. Bury, P. W. Boyd, T. Preston, G. Savidge and N. J. P. Owens, *Deep Sea Res. I*, 2001, **48**, 689.
218. M. Hofmann and H. J. Schellnhuber, *Proc. Natl. Acad. Sci. U. S. A*, 2009, **106**, 3017.

Methane Biogeochemistry and Carbon Stores in the Arctic Ocean: Hydrates and Permafrost

VASSILIS KITIDIS

1 Introduction

The Arctic marine environment holds large stores of methane (CH_4) in the form of methane hydrate and, potentially, permafrost. The dynamic interactions between these carbon (C) reservoirs and the Arctic CH_4 cycle are discussed in this chapter. A number of papers and books have recently reviewed diverse aspects of CH_4 biogeochemistry, methane hydrates and methane in permafrost, covering a regional to global range in scale and scope.[1-6] This chapter does not aim to replicate these efforts, but rather to provide an overview and links between the relevant aspects of methane biogeochemistry with particular reference to the Arctic marine environment. The Arctic marine environment is unique due to its seasonal or permanent sea-ice cover and extensive continental shelf areas, which receive substantial fluvial inputs from the surrounding land masses. In addition, large reservoirs of carbon (C), captured from the atmosphere over millennia, in the form of permafrost and methane hydrates are found on land and extend beneath the Arctic shelf. These characteristics give the Arctic Ocean a particular interconnection with the surrounding land masses. In reviewing the biogeochemistry of methane in the Arctic Ocean, it is therefore necessary to also consider the terrestrial Arctic environment, specifically in the light of increasing concern that C in permafrost and CH_4 hydrates may be mobilised rapidly by global warming or human exploitation.

Issues in Environmental Science and Technology, 29
Carbon Capture: Sequestration and Storage
Edited by R.E. Hester and R.M. Harrison
© Royal Society of Chemistry 2010
Published by the Royal Society of Chemistry, www.rsc.org

1.1 Methane: Marine Sources and Sinks

Methane is a long-lived greenhouse gas which contributes to hydroxyl radical (OH) and ozone (O_3) regulation in the atmosphere.[7] Per unit C, methane is much more effective as a greenhouse gas than carbon dioxide (CO_2), with a global warming potential of 25 over 100 years, compared to 1 for CO_2. The increase in the atmospheric mixing ratio of CH_4 from approximately 700 parts per billion by volume (ppbv) in the early 18th century to 1774 ppbv at present has contributed 30% of the net anthropogenic radiative forcing perturbation over the same period.[8] [For a compilation of past and present data on the atmospheric mixing ratio of CH_4 the reader is referred to http://cdiac.ornl.gov/products.html)].

Wetlands, agriculture and termite-mound emissions dominate the sources of atmospheric CH_4. The marine source is thought to contribute $4–19 \times 10^{12}$ g $CH_4 a^{-1}$, less than 4% of the total source.[9–11] Coastal waters account for the majority of marine CH_4 emissions to the atmosphere, due to the high concentrations of organic matter which may fuel methanogenesis and the low residence time for CH_4 in the water column due to the shallow nature of coastal seas.[9,12,13] Microbial CH_4 production by methanogenic archaea occurs in anoxic marine sediments and the water column in anoxic basins (*e.g.* Arabian Sea), as well as anaerobic micro-environments, such as zooplankton guts, faecal pellets and estuarine turbidity maxima.[13–19] Microbial CH_4 generation follows two distinct biochemical pathways: methylotrophic and autotrophic methanogenesis.[20] The latter reduces CO_2 with H_2 as the electron donor. It is conceivable that sub-sea floor carbon capture and storage (CCS) may provide additional substrate (CO_2) for this methanogenic pathway, though methanogenic archaea may be limited by the lack of H_2 in CCS reservoirs. In addition to microbial CH_4 production, cold seepage of thermogenic CH_4 and CH_4 hydrate dissolution may contribute to the marine source, but these are difficult to quantify due to their transient and episodic nature.[21–27] Terrestrial and marine thermogenic CH_4 sources to the atmosphere are thought to be in the order of $40–60 \times 10^{12}$ g $CH_4 a^{-1}$ and are currently not included in the Intergovernmental Panel on Climate Change (IPCC) global CH_4 budget.[10] Although microbial CH_4 is sometimes called '*biogenic*', in order to differentiate this from '*thermogenic*' CH_4, strictly this is incorrect as the latter is also derived from biologically sequestered carbon that was buried in sediments.

Microbial methane oxidation to CO_2, either aerobic or anaerobic coupled to sulfate reduction,[28,29] represents a major sink for marine methane, moderating the sediment–water and water–air fluxes in a range of aquatic and marine environments.[30–39] Oxidation consumes the majority of microbial and thermogenic CH_4 that is produced in marine waters before this is emitted to the atmosphere.

1.2 Arctic Ocean Methane and Global Warming

The Arctic Ocean is particularly sensitive to global warming through polar amplification where a disproportionate amount of warming is expected at high latitudes compared to the global average.[10,40] The dynamic links between CH_4 biogeochemistry and a changing climate are well documented, but it is worth mentioning a number of potential feedback mechanisms that are specific to the marine Arctic. Firstly, low seawater temperatures increase the solubility of CH_4 in seawater[41] and allow a shallowing of the CH_4 hydrate stability zone.[42,43] Therefore, a small increase in seawater temperature could potentially release CH_4 from the Arctic Ocean to the atmosphere and lead to hydrate dissolution. Secondly, the expected increase in seawater temperature for the Arctic Ocean is also likely to increase both microbial methanogenesis and methane oxidation rates. The temperature dependence of these processes in the Arctic is not known. However, given the temperature dependence of methanogenesis in temperate sediments,[44,45] it can be shown that an increase in temperature from 1 to 3 °C would result in an increase in methanogenesis in the order of 25–200%. This is particularly important as the supply of organic matter which may fuel methanogenesis in the Arctic Ocean is likely to increase through melting permafrost and fluvial inputs. Thirdly, fluvial inputs from Eurasian and North American rivers (Ob, Yenisei, Lena, Kolyma Mackenzie) may result in direct CH_4 inputs to the shallow continental shelf of the Arctic Ocean. In addition, substantial fluvial supply of organic matter to the Arctic Ocean[46,47] may fuel methanogenesis in estuaries and river plumes. For example, high concentrations of dissolved CH_4 in Arctic Ocean river plumes have been shown for the Kolyma River in the Chukchi Sea[48] and Mackenzie River in the Beaufort Sea.[49] This supply of 'fresh' organic matter may enhance microbial methanogenesis as has been shown for temperate coastal waters.[32,45] Finally, sea-ice functions as a barrier for shelf–basin exchange[50] and sea–air gas exchange, acting as a 'cap' by reducing vertical diffusivity in the water column and turbulent diffusion across the sea–air interface. This increases the water column residence time for biologically labile biogases, such as CH_4, potentially allowing a larger fraction to be microbially oxidised.[51] Currently, sea-ice extent and thickness are in decline in the Arctic Ocean.[52–55] Future projections of further decline in sea-ice cover and increased seawater temperatures in the Arctic Ocean.[56–58] may thus allow rapid exchange of CH_4 with the atmosphere.

With the exception of microbial methane oxidation, these processes result in positive feedback mechanisms which amplify the initial temperature increase through the production and subsequent emission of CH_4 to the atmosphere.

2 Methane Hydrates

2.1 Methane Hydrates and Hydrate Stability

This section provides a brief overview of methane hydrates and their stability. For further detail the reader is referred to the excellent book edited by Michael D. Max.[1] Methane hydrate is a type of clathrate, a metastable solid structure composed of 8 to 46 water molecules (host substance) which form a cage around a single CH_4 molecule (guest substance). CH_4 hydrates may also contain other hydrocarbons, such as ethane (C_2H_6), as the guest substance. The term '*methane clathrate*' is frequently used to describe methane hydrate, though the two terms are not synonymous (hydrate is a particular type of clathrate). Marine CH_4 hydrate is stable under high pressure and low temperature[43] and requires high dissolved CH_4 concentrations in the water or sediment pore-water. The density of CH_4 hydrate is greater than seawater and therefore methane hydrates may form in or on fine sediments with a high organic matter content that can support microbial methanogenesis. As hydrates are only stable under high pressure, they are generally found at depths in excess of one kilo-meter. However, methane hydrates are stable and found at much shallower depth in the Arctic Ocean due to the low ambient seawater temperature. In areas where the criteria for hydrate stability are met, CH_4 hydrates can extend vertically for hundreds of metres in the gas hydrate stability zone (GHSZ). The uppermost limit of this zone is defined by the surrounding temperature and pressure (depth), while the lower boundary of the GHSZ is determined largely by the geothermal gradient, the gradient of the temperature increase with sediment depth. The occurrence or absence of CH_4 hydrate in the GHSZ then depends on the ambient dissolved CH_4 concentration. Figure 1(A) illustrates the GHSZ in a hypothetical Arctic Ocean station with a water column depth of 300 m, seawater temperature of $-1\,°C$ and geothermal gradient of $35\,°C\,km^{-1}$.

The hydrate stability envelope (grey shaded area) is defined by pressure (depth) and temperature.[43] The overlap between ambient temperature and the hydrate stability envelope shows that CH_4 hydrate can extend from just above the sea floor into the sediment until the ambient temperature crosses the hydrate stability curve at 436 m below sea-level, see Figure 1(A). In this example, one would expect methane hydrates to extend from the sea floor

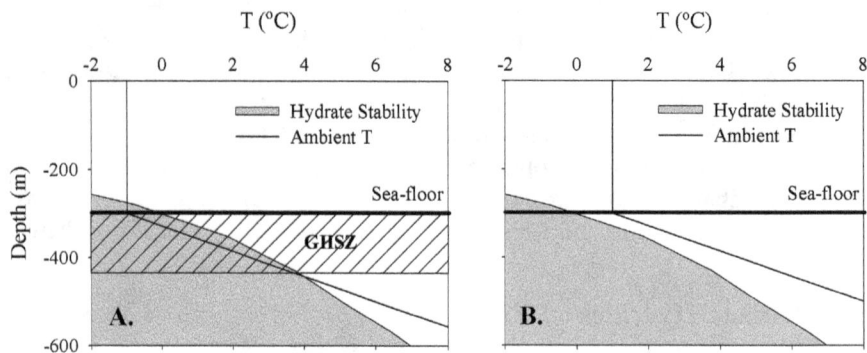

Figure 1   (A) The hydrate stability envelope (grey shaded area) and ambient tem-perature at a hypothetical Arctic Ocean station with a water depth of 300 m, geothermal gradient of $35\,°C\,km^{-1}$ and water column temperature of $-1\,°C$. The gas hydrate stability zone (GHSZ; hatched area) extends from the sea floor (300 m) to the intercept of the hydrate stability curve and ambient temperature (436 m). (B) The same station with a water column temperature of $+1\,°C$, showing that conditions are no longer favourable for the for-mation of a GHSZ.

to a depth of 136 m below the sea floor, given an abundant supply of CH_4. Figure 1(B) shows the same hypothetical Arctic Ocean station, with a water column temperature of 1 °C. This example shows that a 2 °C rise in seawater temperature no longer supports the formation of CH_4 hydrate and thereby illustrates the susceptibility of CH_4 hydrate to dissolution under rising temperatures, *e.g.* under global warming scenarios. Given the size of the CH_4 hydrate reservoir (see Section 2.3) and the global warming potential of CH_4 (Section 1.1), dissolution of CH_4 hydrate has serious consequences for future global warming.

2.2   The 'Clathrate Gun' Hypothesis

The key to understanding the implications of future global warming in the Arctic Ocean with respect to hydrate dissolution can be found in the Earth's geological history. Ice core records have shown that atmospheric CH_4 has increased dramatically during each deglaciation period over the last 650 thousand years.[59,60] Further geochemical data have shown that substantial CH_4 hydrate dissolution occurred during deglaciation events.[61,62] Going further into the geological history of the Earth, geochemical evidence suggests massive CH_4 hydrate dissolution at the Palaeocene–Eocene Thermal Maximum (PETM), *ca.* 55 million years ago.[63,64] It is thought that an initial climatic warming event, or a drop in pressure due to lower sea-levels, may have led to dissolution of hydrates and consequently released large amounts of CH_4 to the atmosphere. In turn, this would have led to further warming in a positive feedback mechanism, where the initial warming perturbation is further enhanced. Due to the irreversible nature of this process once started (like firing a gun), this hypothesis has been called '*The clathrate gun*'[65] or '*Gaia's breath*'[25] in reference to the apparent exhalation of CH_4 by Gaia, the Earth-figure from ancient Greek mythology. A significant observation here is that, according to this hypothesis, methane in the atmosphere is sensitive to climate and *vice versa*. Although the '*clathrate gun*' hypothesis seems plausible, it has been suggested that past, and therefore future, CH_4 hydrate dissolution is more likely to have occurred over thousands of years rather than in a single catastrophic event.[66] Indeed CH_4 hydrate release in response to falling sea-level has also been put forward as a negative feedback mechanism during glaciations, *i.e.* falling sea-level mobilises hydrate and the subsequent release of CH_4 to the atmosphere causes warming and limits the extent of glaciations.[67] Nevertheless, and despite the criticism of the '*clathrate gun*' hypothesis, this remains a 'viable' explanation for the wealth of geochemical data around the PETM.[68]

2.3   Methane Hydrates – Arctic Ocean

Methane hydrates in the Arctic are often found associated with permafrost, particularly on land.[69] Estimating the extent of CH_4 hydrates is extremely difficult, particularly in the Arctic Ocean where sea-ice often prohibits geophysical surveying. Estimates of the magnitude of CH_4 hydrate deposits at the basin-wide and global scales therefore carry large uncertainty and are in the range of 10^{14}–10^{19} g CH_4, with most estimates in the order of 10^{19} g CH_4 (ref. 6,70). By comparison, the total annual sources of CH_4 to the atmosphere at present are estimated to be in the order of 5×10^{14} g $CH_4 a^{-1}$. Soloviev[70] considered the extent of sedimentary basins, sediment thickness, sedimentation rates and suitable geology and suggested that criteria for hydrate accumulation were met over an area of 36×10^6 km^2 globally. The Arctic Ocean accounts for *ca.* 12% of this value[71] despite occupying less than 4% of the global ocean surface area, thereby suggesting that the Arctic Ocean may hold more CH_4 hydrate deposits than other oceans on a unit-area basis. Though hydrate occurrence in the Arctic Ocean has been confirmed, there are only limited estimates of the size of hydrate reservoirs. Thus, the Canadian Arctic is thought to hold 10^{16}–10^{17} g CH_4 (ref. 72), while Alaska is thought to hold 10^{16} g CH_4 (ref. 73).

Comprehensive studies of CH_4 hydrates in the Arctic Ocean have been carried out in the coastal area of the Beaufort Sea, Prudhoe Bay and the Mackenzie River delta. Previous studies in the coastal Beaufort Sea concluded that high CH_4 in the water column likely originated from sedimentary sources, including hydrate dissolution.[49,74] On the basis of its isotopic signature, CH_4 in Prudhoe Bay hydrates is thought to be a mixture of both microbial and thermogenic gas and found overlying known oil fields.[75] The ongoing temperature increase since the last glacial maximum is thought to have led to hydrate destabilisation and CH_4 accumulation in sediments. Where gas accumulation occurs beneath impermeable sediments, the accumulated pressure may in turn create mound-shaped deformities in the sea floor topography which resemble 'pingo-like-features' found on land.[76] Transient gas accumulation and release may account for the change in shape and size of these features which present a known shipping hazard in the area.

2.4   Methane Hydrate Exploitation in the Arctic

Methane hydrates have long been recognised as a potential future energy source.[5,77,78] Efforts to exploit hydrate reservoirs as fossil fuels started in the 1970s and have received growing attention since then, as geopolitical developments have highlighted the need for fuel sufficiency in industrialised economies. However, the extraction of CH_4 gas from CH_4 hydrate reservoirs poses formidable technological challenges. Hydrate exploitation strategies have focused on extracting CH_4 by destabilising CH_4 hydrate either by reducing ambient pressure or increasing temperature in the hydrate reservoir. The possibility of replacing CH_4 with CO_2 as the hydrate guest molecule is of special relevance to carbon capture and storage.[79,80] This is a particularly attractive proposition as the pressure of CO_2 hydrate formation is lower than the pressure of CH_4 hydrate dissociation.[80] It is therefore conceivable that future gas extraction from CH_4 hydrates by depressurisation may be coupled to CO_2

capture in the hydrate field. However, this possibility remains a promise for the future rather than reality at present. A detailed account of the technological innovations required to overcome the challenges of hydrate exploitation is beyond the scope of this chapter. Nevertheless, it is worth giving a brief historical overview of efforts for hydrate exploitation at the onshore Mallik hydrate field in north-west Canada. The Mallik hydrate field site in the Mackenzie River delta was originally investigated by a joint Japanese, Canadian and US consortium in 1998 with the principal aim of developing relevant technologies and research. The project used an 1150 m deep commercial oil exploration well (Mallik 2L-38) from the 1970s (ref. 81). The initial programme was followed by the Mallik 2002 project with broader international participation from Germany and India and involved three new drill sites to 1200 m depth.[82] Several hydrate-bearing layers up to 110 m in thickness, holding up to 80% hydrate by volume, were discovered. The permafrost-associated hydrate reservoir has produced CH_4 both by depressurisation and heating of the hydrate deposits.[83] Detailed geochemical characterisation of the field in this second research phase provided a test-bed for the evaluation of seismic hydrate detection methods.[82,84] More recently, a Japanese–Canadian collaboration successfully recovered $13 \times 10^3 \, m^3$ of gas from the original Mallik 2L-38 well over a six day period.[85] This was the first definitive recovery of gas from gas hydrates (the proposed hydrate origin of gas at the Messoyakha gas field in Siberia is disputed[86] and, in any case, the wells at this location tap into a gas field which may overly hydrate, rather than directly tapping into hydrate deposits). Similar projects in offshore regions remain elusive despite the promise of further finds. However, predicted sea-ice retreat in the near future may open up offshore regions for exploration and exploitation of mineral resources including CH_4 hydrates.

3 Permafrost

3.1 Permafrost Relevance to Methane

Permafrost is frozen soil which remains below 0 °C for at least two years and can be found both on land and under the sea floor. The close association of CH_4 hydrates and permafrost has already been mentioned (Section 2.3). Permafrost may have high organic matter content in the form of peat, root material and other frozen vegetation which was deposited over thousands of years in the Earth's past. Where this is the case, the low temperature in these deposits inhibits microbial mineralisation of organic mater and its carbon content is therefore considered to be 'locked away' from the contemporary C-cycle. Arctic permafrost is thought to hold in the order of 10^{15}–10^{17} g of C.[87,88] The uppermost active layer (AL) of permafrost may undergo seasonal thawing that allows microbial activity to access this large C-store. There is increasing concern that global warming may thaw this vast C-store and mobilise additional greenhouse gases, namely CO_2 and CH_4 (ref. 89).

3.2 Permafrost and Global Warming

Recent evidence suggests that the AL-depth in the Siberian Arctic has increased by up to 0.32 m between 1956 and 1990 (ref. 90). Permafrost thawing may release additional nutrients, thereby enhancing vegetation growth and draw-down of CO_2 from the atmosphere.[91] However, the thawing ground is largely replaced by thermokarst terrain,[92–94] mires which offer suitable conditions for anaerobic mineralisation of organic matter and hence microbial methanogen-esis. Methanogenic archaea have thus been isolated directly from onshore Siberian permafrost horizons[95] and their activity in northern wetlands is thought to account for an efflux of 10 to 44×10^{12} g CH_4 a^{-1} to the atmo-sphere.[96] Christensen *et al.*[94] showed that permafrost thawing in northern Sweden between 1970 and 2000 had resulted in land-cover changes and a net increase of 22–66% in CH_4 efflux to the atmosphere over the same period. In addition to this 'top down' mobilisation of C, permafrost may act as a 'seal' for microbial CH_4 gas in deeper sediment strata.[95] Breaking this 'seal' may therefore release additional accumulated CH_4 to the atmosphere. Melting permafrost is also likely to increase the flux of particulate organic C to Arctic rivers,[97] thereby increasing the supply of organic matter to coastal sediments where methanogenesis may occur. It is thought that northern peatlands, including permafrost regions, have acted as net sinks of CO_2 and net sources of CH_4 during the Holocene, with a negative net radiative forcing impact (cooling) on the atmosphere over the last ten millennia.[98] However, Kvenvolden and Lorenson,[89] estimated that permafrost thawing under global warming scenar-ios for the 21st century may release up to an additional 30×10^{12} g CH_4 a^{-1} to the atmosphere. A similar modelling study has shown that permafrost thawing in Russia may lead to a net increase in CH_4 efflux to the atmosphere in the order of 6 to 8×10^{12} g CH_4 a^{-1} by the middle of the 21st century.[99] These values are comparable to global marine emissions at present. The previous sections have illustrated the close coupling between permafrost, CH_4 hydrates and cli-mate in the wider Arctic environment. The Arctic Ocean is a particular element in this environment which will receive further attention in the following section.

4   Methane in the Arctic Ocean

4.1 Distribution, Sources and Sinks

The location of previous studies in the Arctic Ocean, shown in Figure 2, highlights the relative scarceness of dissolved CH_4 data from the Arctic Ocean.[38,48,49,51,74,100–103]

Furthermore, CH_4 measurements from the central Arctic Ocean basin are conspicuously absent, while published datasets are confined to the marginal seas on the periphery of the Arctic Ocean proper (Figure 2). Previous studies in these areas have demonstrated high CH_4 concentrations and super-saturation with respect to atmospheric equilibrium in surface waters, though under-saturation has also been reported. Thereby, super-saturation up to 2500% has

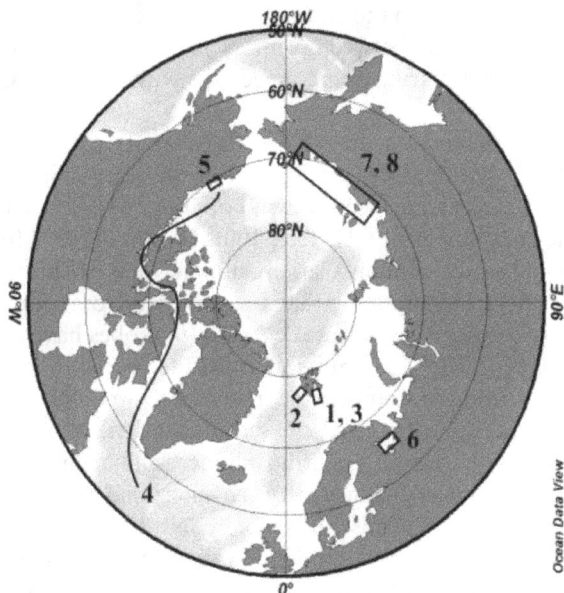

Figure 2 Approximate location of previous studies in the Arctic Ocean reporting dissolved CH_4 concentration in seawater: 1, ref. 100; 2, ref. 101; 3, ref. 102; 4, ref. 51; 5, ref. 49; 6, ref. 38; 7, ref. 103; 8, ref. 48. (This figure was created with Ocean Data View software.)[106]

been found in surface seawater of the East Siberian Sea.[48] Increasing CH_4 concentrations with depth in the Beaufort Sea,[49] East Siberian Sea[48] and off-shore Spitzbergen[101] point to bottom water sources, namely microbial methanogenesis and hydrate dissolution. The proximity of high CH_4 hotspots to river plumes (Mackenzie, Colville, Kolyma and Indigirka rivers) further suggests microbial methanogenesis in coastal sediments where remineralisation of fluvial organic matter inputs takes place or direct CH_4 inputs from permafrost degradation on land.[48–49,51,103] However, near sea floor CH_4 hotspots, away from river plumes, have also been found in the East Siberian Sea[103] and Barents Sea[38] and attributed to hydrate erosion and dissolution at the sea floor. To this effect, bottom water currents, originating from the generation of dense water during sea-ice formation in the surface, may induce turbulence near the sea floor and release CH_4 from sediments into the water column.[102] In addition to sedimentary and fluvial sources, there is evidence that CH_4 is produced in the water column, presumably in anaerobic micro-environments that favour microbial methanogenesis.[51,100] Damm *et al.*[100] found a strong correlation between surface chlorophyll concentration and dissolved CH_4 in coastal waters off Svalbard in the Barents Sea, suggesting a strong link between summertime water column productivity and methanogenesis. Based on a negative correlation between decreasing dimethylsulfoniopropionate (DMSP) and increasing CH_4, they suggested that DMSP produced by algae may act as a direct substrate for microbial methanogenesis in the water column. These findings highlight the close coupling between primary production and methanogenesis, as well as the terrestrial-marine interconnections with regard to CH_4 sources in the Arctic Ocean.

Microbial oxidation and sea–air gas exchange are the major sinks for CH_4 in the Arctic Ocean. Microbial CH_4 oxidation in the water column and in sediments of the Arctic Ocean has been shown both directly[51,104] and inferred from the ^{13}C isotopic enrichment of CH_4 (ref. 100). Damm et al.[100] found apparent ^{13}C enrichment of water column CH_4 in summertime, with an isotopic fractionation factor of 1.017 suggesting that CH_4 emitted from sediments in winter was subsequently oxidised in the water column (microbes have a preference for the lighter ^{12}C-CH_4). Methane emission from the sea to the atmosphere has also been inferred from the summertime increase in atmospheric CH_4 observed in the Arctic, [103] and modelled from the sea–air concentration gradient (excess CH_4 in seawater with respect to atmospheric equilibrium) using turbulent diffusion models.[49,103] The balance between microbial oxidation and loss of CH_4 to the atmosphere depends on sea-ice conditions (see Section 4.2), the magnitude of sources, water column depth and mode of transport through the water column (diffusion vs. ebullition). CH_4 in gas bubbles has a much shorter residence time in the water column than dissolved CH_4 and a larger fraction may therefore escape to the atmosphere. Lammers et al. [38] estimated that 98% of CH_4 emitted from the sea floor at 300 m depth in an area of the Barents Sea, was biologically consumed, with only a small fraction escaping to the atmosphere. However, a larger fraction maybe emitted to the atmosphere in shallow water, particularly where substantial ebullition of CH_4 from sediments occurs.[103] The magnitude of sources affects the water column dissolved CH_4 concentration and thereby influences both microbial oxidation and sea–air exchange, both of which depend on dissolved CH_4 concentration. Sea-ice plays a particular role in the dynamics of CH_4 in the Arctic Ocean and is discussed in the following section.

4.2   Methane and Sea-Ice

Sea-ice, though a porous medium, significantly reduces wind- and tidally-driven turbulence,[50] thereby preventing sea–air exchange of soluble gases, including CH_4. A number of studies have shown that dissolved CH_4 in Arctic Ocean seawater is higher in the presence than in the absence of sea-ice.[49,51,103] CH_4 accumulation under sea-ice increases its residence time in the water column and may therefore allow a larger fraction to be microbially oxidised. As microbial oxidation is a first-order process with respect to CH_4 concentration, i.e. the oxidation rate increases with increasing concentration of CH_4, microbes effectively compete with sea–air gas exchange for the available CH_4 (ref. 51). In the light of global warming and retreating sea-ice scenarios, the atmospheric sea–air flux of CH_4 from the Arctic Ocean is therefore likely to increase in the 21st century, even if permafrost/hydrate mobilisation and microbial methanogenesis remain unchanged. Sea-ice melting also plays a part in CH_4 dynamics in the Arctic Ocean. Though melting may initially release accumulated CH_4, the low salinity, low dissolved CH_4 (relative to atmospheric equilibrium) melt-water may act as a sink for CH_4 reversing the sign of the sea–air gas exchange flux term.[105] It is hoped that improved understanding of such interactions and their incorporation into coupled biogeochemical–climate models will advance our ability to predict their effect on methane cycling and, by extension, climate.

5 Conclusions

The Arctic Ocean methane cycle is intimately linked with methane hydrate and permafrost dynamics both on land and in the marine environment. Climate plays an important part in their interactions through various feedback mechanisms – climate may influence methane dynamics in the Arctic and *vice versa*. These interactions have been invoked to explain geochemical observations relating to climate change events in the Earth's past (*clathrate gun hypothesis*) which are in turn used as analogues for future climate change predictions. Though the *clathrate gun* hypothesis has its critics, it remains a plausible explanation. Previous work has shown high dissolved CH_4 concentrations in surface waters of the Arctic Ocean. Direct fluvial inputs, hydrate dissolution, remineralisation of permafrost, sediment and water column methanogenesis all contribute to CH_4 super-saturation with respect to atmospheric equilibrium and drive a net efflux of CH_4 to the atmosphere. Microbial oxidation on the other hand moderates this flux term, particularly in the presence of sea-ice which increases the residence time of CH_4 in Arctic Ocean surface waters. Our current understanding suggests that the net flux of CH_4 from the Arctic Ocean and wider Arctic environment to the atmosphere will increase over the near future, with consequent adverse effect on global warming.

References

1. M. D. Max, *Natural Gas Hydrate in Oceanic and Permafrost Environments,* Coastal Systems and Continental Margins 5, Kluwer Academic Publishers, Dordrecht, The Netherlands, 2000.
2. K. A. Kvenvolden, *Terra Nova*, 2002, **14**(5), 302.
3. K. A. Kvenvolden, *Org. Geochem.*, 1995, **23**(11–12), 11.
4. K. A. Kvenvolden, G. D. Ginsburg and V. A. Soloviev, *Geo-Mar. Lett.*, 1993, **13**(1), 32.
5. A. Judd and M. Hovland, *Seabed Fluid Flow: The Impact on Geology, Biology and the Marine Environment: The Impact of Geology, Biology and the Marine Environment,* Cambridge University Press, London–New York, 2007.
6. K. C. Hester and P. G. Brewer, *Annu. Rev. Mar. Sci.*, 2009, **1**, 303.

7. P. J. Crutzen, *Nature*, 1991, **350**(6317), 380.
8. S. Solomon, D. Qin, M. Manning, R. B. Alley, T. Berntsen, N. L. Bindoff, Z. Chen, A. Chidthaisong, J. M. Gregory, G. C. Hegerl, M. Heimann, B. Hewitson, B. J. Hoskins, F. Joos, J. Jouzel, V. Kattsov, U. Lohmann, T. Matsuno, M. Molina, N. Nicholls, J. Overpeck, G. Raga, V. Ramaswamy, J. Ren, M. Rusticucci, R. Somerville, T. F. Stocker, P. Whetton, R. A. Wood and D. Wratt, in *Technical Summary*, ed. S. Solomon, D. Qin, M. Manning, Z. Chen, M. Marquis, K. B. Averyt, M. Tignor and H. L. Miller, Cambridge University Press, Cambridge, United Kingdom and New York, NY, USA, 2007.
9. H. W. Bange, U. H. Bartell, S. Rapsomanikis and M. O. Andreae, *Glob. Biogeochem. Cycle*, 1994, **8**(4), 465.
10. K. L. Denman, G. Brasseur, A. Chidthaisong, P. Ciais, P. M. Cox, R. E. Dickinson, D. Hauglustaine, C. Heinze, E. Holland, D. Jacob, U. Lohmann, S. Ramachandran, P. L. d. S. Dias, S. C. Wofsy and X. Zhang, in *Couplings Between Changes in the Climate System and Biogeochemistry*, ed. S. Solomon, D. Qin, M. Manning, Z. Chen, M. Marquis, K. B. Averyt, M.Tignor and H. L. Miller, Cambridge University Press, Cambridge, United Kingdom and New York, NY, USA, 2007.
11. P. Bousquet, P. Ciais, J. B. Miller, E. J. Dlugokencky, D. A. Hauglustaine, C. Prigent, G. R. Van der Werf, P. Peylin, E. G. Brunke, C. Carouge, R. L. Langenfelds, J. Lathiere, F. Papa, M. Ramonet, M. Schmidt, L. P. Steele, S. C. Tyler and J. White, *Nature*, 2006, **443**(7110), 439.
12. T. S. Bates, K. C. Kelly, J. E. Johnson and R. H. Gammon, *J. Geophys. Res. Atmos.*, 1996, **101**(D3), 6953.
13. R. C. Upstill-Goddard, J. Barnes, T. Frost, S. Punshon and N. J. P. Owens, *Glob. Biogeochem. Cycle*, 2000, **14**(4), 1205.
14. M. A. DeAngelis and C. Lee, *Limnol. Oceanogr.*, 1994, **39**(6), 1298.
15. C. Kelley, *Biogeochemistry*, 2003, **65**(1), 105.
16. D. G. Marty, *Limnol. Oceanogr.*, 1993, **38**(2), 452.
17. J. J. Middelburg, J. Nieuwenhuize, N. Iversen, N. Hogh, H. De Wilde, W. Helder, R. Seifert and O. Christof, *Biogeochemistry*, 2002, **59**(1–2), 95.
18. M. R. Winfrey, D. G. Marty, A. J. M. Bianchi and D. M. Ward, *Geomicrobiol. J.*, 1981, **2**(4), 341.
19. N. J. P. Owens, C. S. Law, R. F. C. Mantoura, P. H. Burkill and C. A. Llewellyn, *Nature*, 1991, **354**(6351), 293.
20. M. T. Madigan and J. M. Martino, *Brock Biology of Microorganisms*, Pearson Education Inc., Upper Saddle River, New Jersey, 2006.
21. A. Judd, G. Davies, J. Wilson, R. Holmes, G. Baron and I. Bryden, *Mar. Geol.*, 1997, **137**(1–2), 165.
22. K. A. Kvenvolden, T. D. Lorenson and W. S. Reeburgh, *EOS Trans. Am. Geophys. Union*, 2001, **82**(40), 457.
23. L. Naudts, J. Greinert, Y. Artemov, P. Staelens, J. Poort, P. Van Rensbergen and M. De Batist, *Mar. Geol.*, 2006, **227**(3–4), 177.
24. J. D. Kessler, W. S. Reeburgh, J. Southon, R. Seifert, W. Michaelis and S. C. Tyler, *Earth Planet. Sci. Lett.*, 2006, **243**(3–4), 366.

25. K. A. Kvenvolden and B. W. Rogers, *Mar. Pet. Geol.*, 2005, **22**(4), 579.
26. A. G. Judd, *Environ. Geol.*, 2004, **46**(8), 988.
27. A. Boetius and E. Suess, *Chem. Geol.*, 2004, **205**(3–4), 291.
28. A. Boetius, K. Ravenschlag, C. J. Schubert, D. Rickert, F. Widdel, A. Gieseke, R. Amann, B. B. Jorgensen, U. Witte and O. Pfannkuche, *Nature*, 2000, **407**(6804), 623.
29. K. U. Hinrichs, J. M. Hayes, S. P. Sylva, P. G. Brewer and E. F. DeLong, *Nature*, 1999, **398**(6730), 802.
30. G. Abril and N. Iversen, *Mar. Ecol. Prog. Ser.*, 2002, **230**, 171.
31. M. A. DeAngelis and M. I. Scranton, *Glob. Biogeochem. Cycle*, 1993, **7**(3), 509.
32. V. Kitidis, L. Tizzard, G. Uher, A. Judd, R. C. Upstill-Goddard, I. M. Head, N. D. Gray, G. Taylor, R. Durán, R. Diez, J. Iglesias and S. García-Gil, *J. Mar. Sys.*, 2007, **66**, 258.
33. H. Niemann, J. Duarte, C. Hensen, E. Omoregie, V. H. Magalhaes, M. Elvert, L. M. Pinheiro, A. Kopf and A. Boetius, *Geochim. Cosmochim. Acta*, 2006, **70**(21), 5336.
34. H. Niemann, M. Elvert, M. Hovland, B. Orcutt, A. Judd, I. Suck, J. Gutt, S. Joye, E. Damm, K. Finster and A. Boetius, *Biogeosciences*, 2005, **2**, 335.
35. H. Niemann, T. Losekann, D. de Beer, M. Elvert, T. Nadalig, K. Knittel, R. Amann, E. J. Sauter, M. Schluter, M. Klages, J. P. Foucher and A. Boetius, *Nature*, 2006, **443**(7113), 854.
36. C. J. Schubert, M. J. L. Coolen, L. N. Neretin, A. Schippers, B. Abbas, E. Durisch-Kaiser, B. Wehrli, E. C. Hopmans, J. S. S. Damste, S. Wakeham and M. M. M. Kuypers, *Environ. Microbiol.*, 2006, **8**(10), 1844.
37. T. Treude, A. Boetius, K. Knittel, K. Wallmann and B. B. Jorgensen, *Mar. Ecol. Prog. Ser.*, 2003, **264**, 1.
38. S. Lammers, E. Suess and M. Hovland, *Geol. Rundsch.*, 1995, **84**(1), 59.
39. M. Schmid, M. De Batist, N. G. Granin, V. A. Kapitanov, D. F. McGinnis, I. B. Mizandrontsev, A. I. Obzhirov and A. Wuust, *Limnol. Oceanogr.*, 2007, **52**(5), 1824.
40. M. M. Holland and C. M. Bitz, *Climate Dyn.*, 2003, **21**(3–4), 221.
41. D. A. Wiesenburg and N. L. Guinasso, *J. Chem. Eng. Data*, 1979, **24**(4), 356.
42. K. A. Kvenvolden, *Rev. Geophys.*, 1993, **31**(2), 173.
43. G. R. Dickens and M. S. QuinbyHunt, *Geophys. Res. Lett.*, 1994, **21**(19), 2115.
44. G. B. Avery, R. D. Shannon, J. R. White, C. S. Martens and M. J. Alperin, *Biogeochemistry*, 2003, **62**, 19.
45. H. W. Bange, S. Dahlke, R. Ramesh, L.-A. Meyer-Reil, S. Rapsomanikis and M. O. Andreae, *Estuarine Coastal Shelf Sci.*, 1998, **47**, 807.
46. R. Benner, B. Benitez-Nelson, K. Kaiser and R. M. W. Amon, *Geophys. Res. Lett. 2004*, **31***(5)*, L05305.
47. S. Opsahl, R. Benner and R. M. W. Amon, *Limnol. Oceanogr.*, 1999, **44**(8), 2017.

48. N. Shakhova, I. Semiletov and G. Panteleev, *Geophys. Res. Lett.*, 2005, **32**(9), L09601.
49. K. A. Kvenvolden, M. D. Lilley, T. D. Lorenson, P. W. Barnes and E. McLaughlin, *Geophys. Res. Lett.*, 1993, **20**(22), 2459.
50. E. Carmack and D. C. Chapman, *Geophys. Res. Lett.*, 2003, **30**(14), 1778.
51. V. Kitidis, R. C. Upstill-Goddard and L. G. Anderson, *Mar. Chem.*, submitted.
52. J. Stroeve, M. M. Holland, W. Meier, T. Scambos and M. Serreze, *Geophys. Res. Lett.*, 2007, **34**(9), L09501.
53. J. C. Comiso, J. Y. Yang, S. Honjo and R. A. Krishfield, *J. Geophys. Res. -Oceans*, 2003, **108**(C12), 3384.
54. J. C. Comiso, *Geophys. Res. Lett.*, 2002, **29**(20), 1956.
55. P. Wadhams, *Philos. Trans. R. Soc. London, Ser. A*, 1995, **352**(1699), 301.
56. H. Y. Teng, W. M. Washington, G. A. Meehl, L. E. Buja and G. W. Strand, *Climate Dyn.*, 2006, **26**(6), 601.
57. J. M. Gregory, P. A. Stott, D. J. Cresswell, N. A. Rayner, C. Gordon and D. M. H. Sexton, *Geophys. Res. Lett.*, 2002, **29**(24), 2175.
58. M. M. Holland, C. M. Bitz and B. Tremblay, *Geophys. Res. Lett.*, 2006, **33**(23), L23503.
59. J. R. Petit, J. Jouzel, D. Raynaud, N. I. Barkov, J. M. Barnola, I. Basile, M. Bender, J. Chappellaz, M. Davis, G. Delayque, M. Delmotte, V. M. Kotlyakov, M. Legrand, V. Y. Lipenkov, C. Lorius, L. Pepin, C. Ritz, E. Saltzman and M. Stievenard, *Nature*, 1999, **399**, 429.
60. R. Spahni, J. Chappellaz, T. F. Stocker, L. Loulergue, G Hausammann, K. Kawamura, J. Fluckiger, J. Schwander, D. Raynaud, V. Masson-Delmotte and J. Jouzel, *Science*, 2005, **310**(5752), 1317.
61. T. M. Hill, J. P. Kennett, D. L. Valentine, Z. Yang, C. M. Reddy, R. K. Nelson, R. J. Behl, C. Robert and L. Beaufort, *Proc. Natl. Acad. Sci. U. S. A.*, 2006, **103**(37), 13570.
62. J. P. Kennett, K. G. Cannariato, I. L. Hendy and R. J. Behl, *Science*, 2000, **288**(5463), 128.
63. G. R. Dickens, M. M. Castillo and J. C. G. Walker, *Geology*, 1997, **25**(3), 259.
64. G. R. Dickens, J. R. Oneil, D. K. Rea and R. M. Owen, *Paleoceanography*, 1995, **10**(6), 965.
65. J. P. Kennett, K. G. Cannariato, I. L. Hendy and R. J. Behl, *Methane Hydrates in Quaternary Climate Change: The Clathrate Gun Hypothesis*, American Geophysical Union, Washington, D.C., 2003.
66. D. Archer, *Biogeosciences*, 2007, **4**(4), 521.
67. C. K. Paull, W. Ussler and W. P. Dillon, *Geophys. Res. Lett.*, 1991, **18**(3), 432.
68. G. R. Dickens, *Fire in the Ice*, 2008, Summer 9.
69. T. S. Collett and S. R. Dallimore, in *Permafrost-Associated Gas Hydrate*, ed. M. D. Max, Kluwer Academic Publishers, Dordrecht, The Netherlands, 2000.

70. K. A. Kvenvolden, in *Natural Gas Hydrate: Introduction and History of Discovery,* ed. M. D. Max, Kluwer Academic Publishers, Dordrecht, The Netherlands, 2000.
71. V. A. Soloviev, *Geol. Geofiz.*, 2002, **43**(7), 648.
72. J. A. Majorowicz and K. G. Osadetz, *Am. Assoc. Petroleum Geol. Bull.*, 2001, **85**(7), 1211.
73. T. S. Collett, *Bull. Can. Petroleum Geol.*, 1997, **45**(3), 317.
74. R. W. Macdonald, *Environ. Sci. Technol.*, 1976, **10**, 1241.
75. T. S. Collett, *Am. Assoc. Petroleum Geol. Bull.*, 1993, **77**(5), 793.
76. C. K. Paull, W. Ussler, S. R. Dallimore, S. M. Blasco, T. D. Lorenson, H. Melling, B. E. Medioli, F. M. Nixon and F. A. McLaughlin, *Geophys. Res. Lett.*, 2007, **34**(1), L01603.
77. T. S. Collett, in *Natural Gas Hydrate as a Potential Energy Resource*, ed. M. D. Max, Kluwer Academic Publishers, Dordrecht, The Netherlands, 2000.
78. M. D. Max, in *Hydrate Resource, Methane Fuel, and a Gas-Based Economy?*, ed. M. D. Max, Kluwer Academic Publishers, Dordrecht, The Netherlands, 2000.
79. J. H. Yoon, T. Kawamura, Y. Yamamoto and T. Komai, *J. Phys. Chem. A*, 2004, **108**(23), 5057.
80. T. Komai, Y. Yamamo and K. Ohga, *Ann. N. Y. Acad. Sci.*, 2000, **912**(Gas Hydrates: Challenges for the Future), 272.
81. T. S. Collett, R. E. Lewis, S. R. Dallimore, M. W. Lee, T. H. Mroz and T. Uchida, *Bull. Geol. Surv. Canada*, 1999, **544**, 295.
82. S. R. Dallimore and T. S. Collett, *Scientific Results from the Mallik 2002 Gas Hydrate Production Research Well Program, Mackenzie Delta, Northwest Territories, Canada*, Bulletin 585, Geological Survey of Canada, 2005.
83. G. J. Moridis, T. S. Collett, S. R. Dallimore, T. Satoh, S. Hancock and B. Weatherill, *J. Petroleum Sci. Eng.*, 2004, **43**(3–4), 219.
84. S. Chand, T. A. Minshull, D. Gei and J. M. Carcione, *Geophys. J. Int.*, 2004, **159**(2), 573.
85. K. Yamamoto, *Fire in the Ice*, 2008, Summer, 1.
86. T. S. Collett and G. D. Ginsburg, *Proceedings of the 7th (1997) International Offshore and Polar Engineering Conference*, 1997, **I**, 96.
87. E. Stokstad, *Science*, 2004, **304**(5677), 1618.
88. S. A. Zimov, E. A. G. Schuur and F. S. Chapin, *Science*, 2006, **312**(5780), 1612.
89. K. A. Kvenvolden and T. D. Lorenson, *Chemosphere*, 1993, **26**(1–4), 609.
90. T. J. Zhang, O. W. Frauenfeld, M. C. Serreze, A. Etringer, C. Oelke, J. McCreight, R. G. Barry, D. Gilichinsky, D. Q. Yang, H. C. Ye, F. Ling and S. Chudinova, *J. Geophys. Res. Atmos.*, 2005, **110**(D16), D16101.
91. C. Waelbroeck, P. Monfray, W. C. Oechel, S. Hastings and G. Vourlitis, *Geophys. Res. Lett.*, 1997, **24**(3), 229.
92. F. E. Nelson, *Science*, 2003, **299**, 1673.

93. S. Payette, A. Delwaide, M. Caccianiga and M. Beauchemin, *Geophys. Res. Lett.*, 2004, **31**(18), L18208.

94. T. R. Christensen, T. R. Johansson, H. J. Akerman, M. Mastepanov, N. Malmer, T. Friborg, P. Crill and B. H. Svensson, *Geophys. Res. Lett.*, 2004, **31**(4), L04501.

95. E. Rivkina, V. Shcherbakova, K. Laurinavichius, L. Petrovskaya, K. Krivushin, G. Kraev, S. Pecheritsina and D. Gilichinsky, *FEMS Microbiol. Ecol.*, 2007, **61**(1), 1.

96. K. M. Walter, S. A. Zimov, J. P. Chanton, D. Verbyla and F. S. Chapin, *Nature*, 2006, **443**(7107), 71.

97. L. Guo, C. L. Ping and R. W. Macdonald, *Geophys. Res. Lett.*, 2007, **34**(13), L13603.

98. S. Frolking and N. T. Roulet, *Global Change Biol.*, 2007, **13**(5), 1079.

99. O. A. Anisimov, *Environ. Res. Lett.*, 2007, **2**(4), 045016.

100. E. Damm, R. P. Kiene, J. Schwarz, E. Falck and G. Dieckmann, *Mar. Chem.*, 2008, **109**(1–2), 45.

101. E. Damm, A. Mackensen, G. Budeus, E. Faber and C. Hanfland, *Continental Shelf Res.*, 2005, **25**(12–13), 1453.

102. E. Damm, U. Schauer, B. Rudels and C. Haas, *Continental Shelf Res.*, 2007, **27**(12), 1692.

103. N. Shakhova and I. Semiletov, *J. Mar. Sys.*, 2007, **66**(1–4), 227.

104. T. D. Lorenson and K. A. Kvenvolden, *Methane in Coastal Seawater, Sea Ice and Bottom Sediments, Beaufort Sea, Alaska,* US Geological Survey, Menlo Park, CA, 1995.

105. K. U. Heeschen, R. S. Keir, G. Rehder, O. Klatt and E. Suess, *Glob. Biogeochem. Cycle*, 2004, **18**(2), GB2012.

106. R. Schlitzer, *Ocean Data View.* http://odv.awi.de 2008.

Subject Index

expansion, 214
soils, 190
World
Business Council on Sustainable
    Development, 133, 136
Energy Outlook, 41
population, 4

Zero
Emission
  Kiln, 141
  power plants, 27
  Technology Platform, 54
Gen, 80–81

www.ingramcontent.com/pod-product-compliance
Lightning Source LLC
Chambersburg PA
CBHW050127240326
41458CB00124B/1632